Risk Analysis in Engineering

Techniques, Tools, and Trends

Risk Analysis
in Engineering
Techniques, Tools, and Trends

Mohammad Modarres

Taylor & Francis
Taylor & Francis Group
Boca Raton   London   New York

A CRC title, part of the Taylor & Francis imprint, a member of the
Taylor & Francis Group, the academic division of T&F Informa plc.

Published in 2006 by
CRC Press
Taylor & Francis Group
6000 Broken Sound Parkway NW, Suite 300
Boca Raton, FL 33487-2742

No claim to original U.S. Government works
Printed in the United States of America on acid-free paper
10 9 8 7 6 5 4 3 2 1

International Standard Book Number-10: 1-57444-794-7 (Hardcover)
International Standard Book Number-13: 978-1-57444-794-1 (Hardcover)

Library of Congress Cataloging-in-Publication Data

Catalog record is available from the Library of Congress

Taylor & Francis Group
is the Academic Division of Informa plc.

Visit the Taylor & Francis Web site at
http://www.taylorandfrancis.com

and the CRC Press Web site at
http://www.crcpress.com

Preface

Engineering systems are becoming more complex, interconnected, nonlinear, geographically widespread, and expensive. Consequences of failures of such systems are exceedingly damaging and destructive to humans, environment, and economy. Failures of engineering systems are unavoidable. They are due to internal causes or have external origins. Of special concerns are externally caused system failures due to malicious terror acts, which disrupt or fail key infrastructure systems, leading to large and widespread damage.

For minimizing occurrences and controlling consequences of engineering system failures, limitations are often imposed on their uses in the form of safety, health, environmental, and security regulations. As the developed nations become more prosperous, the demands for safety, health, security, and environmental protection equally rise, leading to stricter requirements and demand for increased performance of engineering systems. To address this trend, formal engineering risk analysis emerged in the late 1960's and made substantial progress into a well-established discipline. Two leading industries that made substantial contributions to this growth are the nuclear and aerospace industries. However, more recent applications of such techniques cover a broad spectrum of industries, including ground and air transportation, chemical process and oil, food processing, defense, financial, and energy production and distribution. The concerns in these industries vary widely and include physical security, information security, asset protection, human safety and health, environmental protection, and component and structural health management.

Ideally, the risk which contemplates possible system failures and consequences in the form of adverse events leading to capital losses, fatalities, and environmental contamination can be estimated directly from historical occurrences of such adverse events. This is possible only in engineering systems for which such data and information are readily available. Relying on historical adverse events is, for example, the principal risk assessment approach in the insurance industries. Other examples include assessment of land transportation risks, in which performance and accident data are available. However, there are many engineering systems for which adequate amounts of historical failure data and events needed to assess their risks may not be available. Therefore, it is often necessary to estimate the risks of complex engineering systems from models of such systems — especially for those systems involving low-frequency failure events, leading to adverse consequences, such as nuclear power plants. Risk models used in engineering systems are often probabilistic in nature, because one needs to estimate the associated low frequency of failures and adverse events.

In the mid-1970's the nuclear industry led by Professor Norman C. Rasmussen of MIT introduced the Probabilistic Risk Assessment (PRA) methodology to assess risks associated with the operation of the nuclear power plants in the United States. This model-based methodology has gained considerable attention and has gone through substantial improvements since its inception. The PRA methodology, while originally focused on estimating the frequency of some adverse outcomes (consequences), has changed to become a life-cycle tool for supporting design, manufacturing, construction, operation, maintenance, security, and decommissioning of complex engineering systems. Recently, PRA methodology is used for devising risk-based and risk-informed regulatory and policy requirements that are commensurate with the risks of the engineering systems.

Three primary components of risk analysis namely risk assessment, risk management, and risk communication, involve separate, but important interrelated activities. While the primary purpose of PRA is to assess the risks of engineering systems, it also provides systematic and transparent risk models and assessment tools for better risk management and risk communication.

This book has been developed as a guide for practitioners of risk analysis and as a textbook for graduate-level and senior-level undergraduate courses in risk analysis and presents the engineering approach to probabilistic risk analysis. Particularly, the book emphasizes methods for comprehensive PRA studies, including formal decision techniques needed for risk management. A critical part of any PRA is the treatment of uncertainties in the models and parameters used in the analysis. Since the PRA is a best-estimate approach, the uncertainties associated with the results and risk management strategies must be characterized. For this reason, the book has a special emphasis on uncertainty characterization.

The book assumes no prior knowledge of risk analysis and provides the necessary mathematical and engineering foundations. However, some basic knowledge of undergraduate senior-level probability and statistics and basic engineering is helpful. While the book is not written with a particular application in mind, because of it pioneering roles in developing the PRA techniques, the book relies heavily on examples from the nuclear industry. To make it a complete guide, this book uses related materials from my previous book, *Reliability Engineering and Risk Analysis: A Practical Guide* (Marcel Dekker, 1999).

Chapter 1 describes the basic definitions and the notions of risk, safety, and performance. Chapter 2 presents the elements of risk analysis and their applications in engineering. Chapter 3 focuses on methods for performing PRAs. This is a key chapter because it describes the structure of a PRA and how to build the necessary logic models, and how to solve such models to find the risk values. Chapter 4 describes how to assess and measure performance of the building blocks of the PRAs, such as reliability of hardware subsystems, structures, components, human actions, and software. It also shows how performance and risk interrelate. Since the performance measurements and the PRA models involve uncertainties, Chapter 5 describes methods of characterizing such uncertainties, and methods for propagating them through the PRA model to estimate uncertainties of the results. Chapters 4 and 5 cover critical topics of PRAs. Chapter 6 reviews a topic that makes the transition from risk assessment to risk management. This chapter describes ways to identify and rank important and sensitive contributors to the estimated risk using the PRA and performance assessment models. Contributors to risk are the most important results of any risk assessment. They focus the analyst's attention to those elements of the engineering systems that highly affect risks and uncertainties, and prioritize candidates for risk management actions. Chapters 7 and 8 are primarily related to probabilistic risk management techniques, with Chapter 7 describing risk acceptance criteria used as reference levels to accept, compare, and reduce estimated risks. Chapter 8 summarizing the formal methods for making decisions related to risk management options and strategies. The book ends with a brief review of the main aspects, issues, and methods of risk communication presented in Chapter 9. Finally, Appendix A provides a comprehensive review of the necessary probability and statistics techniques for model-based risk analyses, and Appendix B provides necessary statistical tables.

This book came from notes, homework problems, and examinations of a graduate course entitled "Risk Assessment for Engineers" offered by me at the University of Maryland, College Park for the past 20 years. I would like to thank all the students who participated in this course, many of whom made valuable inputs to the contents of the book. I would like to thank Drs. Mark Kaminskiy and Vasily Krivtsov who were my coauthors with this book's predecessor, *Reliability Engineering and Risk Analysis: A Practical Guide*. Many graduate students and research staff working with me helped in the development of the materials and examples of this book. Particularly, I am indebted to Jose L. Hurtado, Reza Azarkhail, Sam Chamberlain, M. Pour Goul Mohamad, Kristine Fretz, Carlos Baroetta, and Genebelin Valbuena for their valuable contributions and examples. Finally, the technical graphics and formatting of the book would have not been possible without tireless efforts of Miss Willie M. Webb.

Mohammad Modarres
College Park, MD, USA

Author

Mohammad Modarres is professor of nuclear engineering and reliability engineering and director of the Center for Technology Risk Studies at the University of Maryland, College Park. He pioneered and developed the internationally leading graduate program in Reliability Engineering at the University of Maryland. His research areas are probabilistic risk assessment, uncertainty analysis, and physics of failure modeling. In the past 23 years he has been with the University of Maryland, he has served as a consultant to several governmental agencies, private organizations, and national laboratories in areas related to risk analysis, especially applications to complex systems and processes such as the nuclear power plants. Professor Modarres has over 200 papers in archival journals and proceedings of conferences and three books in various areas of risk and reliability engineering. He is a University of Maryland Distinguished Scholar-Teacher. He received his Ph.D. in Nuclear Engineering from MIT and his M.S in Mechanical Engineering also from MIT.

Table of Contents

1 Introduction and Basic Definitions

1.1 INTRODUCTION

Risk is a measure of the potential loss occurring due to natural or human activities. Potential losses are the adverse consequences of such activities in form of loss of human life, adverse health effects, loss of property, and damage to the natural environment. Risk analysis is the process of characterizing, managing, and informing others about existence, nature, magnitude, prevalence, contributing factors, and uncertainties of the potential losses. In engineering systems,[1] the loss may be external to the system, caused by the system to one or more recipients (e.g., humans, organization, economic assets, and environment). Also the loss may be internal to the system, causing damage only to the system itself. For example, in a nuclear power plant, the loss can be damage to the plant due to partial melting of the reactor core or damage due to release of radioactivity into the environment by the power plant. The former case is a risk internal to the system, and the latter case represents a loss caused by the system (the nuclear plant) to the environment. From an engineering point of view, the risk or potential loss is associated with exposure of the recipients to hazards, and can be expressed as a combination of the *probability* or *frequency* of the hazard and its consequences. Consequences to be considered include injury or loss of life, reconstruction costs, loss of economic activity, environmental losses, etc. In engineering systems, risk analysis is performed to measure the amount of potential loss and more importantly the elements of the system that contribute most to such losses. This can be performed either *explicitly* or *implicitly*. When explicitly addressed, targets should be set in terms of the *acceptable risk levels*. However, usually engineers do not make the decision about risk acceptance of systems. Decisions are made by risk managers, policy makers, and politicians who are influenced by the prevailing economic environment, press, public opinion, interest groups, and so on. This aspect also underlines the importance of risk communication between the various parties and stakeholders involved.

To better understand risk analysis in the context of *complex engineering systems*, it is appropriate to begin with the nature of system complexity.

1.2 IMPORTANCE OF RISK ANALYSIS

1.2.1 COMPLEXITY AND CHARACTERISTICS OF ENGINEERING SYSTEMS AND THEIR MODELS

Adaptability of systems to cope with internal and external environments leads to evolution and emergence of systems that are progressively complex, autonomous, and intelligent.

---

[1] An engineering system is defined as an entity composed of hardware, software, and human organization.

Abundant examples point to this increasing complexity in engineering systems such as transportation and energy systems. In fact, many researchers characterize natural evolution as a kind of progression toward complexity. Although the growth of complexity during evolution is not universal, this appears to be the case in our complex interconnected world. For example, considering the tightly interconnected public-engineered infrastructures such as communications, energy, transportation, and water resources, one may conclude that simplification in one area may still lead to increasing complexity of the whole infrastructure. For example, simplification of communication (faster and easier) methods leads to enhanced access to information by terrorists leading to less secure energy or water sources, and subsequent requirements for additional layers of protection, prevention, and mitigation. Concurrently, as the public gets better educated, more knowledgeable, and wealthier, it demands for far higher standards of safety, health, and security for the ever-increasing complex systems. It is imperative that in a democratic society, the government and the private industry understand these public demands and make their policies consistent with them. In recent years, risk analysis has been considered a powerful approach to address these public concerns and to develop sound policy and design strategy.

For analysis of any complex engineering system, a model would be needed. Such a model must be consistent with the primary characteristics of complex engineering systems. The characteristics of complex engineering systems are: *evolving, integrated, dynamic, large,* and *intelligent*. The first three are the primary characteristics and the last two are adjunct. The evolving characteristic requires the risk model that has provisions for internal and external feedbacks, and allows for opportunistic and incremental improvements. Further, the *uncertainty*[2] and *ambiguity*[3] associated with the characteristics and properties of the various elements and their relationships in complex systems should also be incorporated. Note that in complex engineering systems, we must also characterize *meta-uncertainty* or the uncertainty to estimate the uncertainty itself.

The model should allow for representation of system element's couplings and integration. Such a representation should accommodate the ever-increasing complexity along with the system structure, functions, and goals. The model should allow for specification and updating of relationships between system elements (connections). The model should also recognize uncertain, nonlinear, and counterintuitive relationships.

Further, the continuous feedback along time (dynamic characteristic) in the system must also be captured by such a model. The model should allow for incorporation and integration of diverse but related subsystems (substructures, subfunctions, and intermediate goals), without imposing constraints to the size. In effect, such a model would be large.

Finally, the model should exhibit abilities to capture system's properties such as self-organization and learning. As such, since the system benefits from its feedback capabilities and learns from its environments to continually remain fit, so should the model of the system.

---

[2] Uncertainty is a state of mind about a proposition (e.g., existence of characteristics of a state, event, process, or phenomenon) that disappears when the condition of truth for that proposition exists (i.e., observed). A mathematical representation of uncertainty comprises three things: (1) axioms, specifying properties of uncertainty; (2) interpretations, connecting axioms with observables; (3) measurement procedure, describing methods for interpreting axioms. Probability is one of several ways to express uncertainty. For more information, see Bedford and Cooke [1].

[3] Ambiguity exists linguistically due to our inability to express a meaningfully declarative sentence. Hence, we may have ambiguity with the truth condition of a proposition. For example, Noam Chomsky gives an example of a syntactically correct, but ambiguous sentence: "Colorless green ideas sleep furiously." That is, the meaning of the individual words does not suggest that the truth condition exists. Therefore, ambiguity must be removed before one can properly discuss uncertainty. See Bedford and Cooke [1].

1.2.2 DEMAND FOR RISK ANALYSIS

Evidence shows that while the world around us is becoming more complex, we also live longer, healthier, and wealthier lives than at any time in the past. Some economists and risk analysts such as Morgan [2] argue that we worry more about risk today exactly because we have more to lose and we have more disposable income to spend on risk reduction. Such factors undeniably exert pressure on the manufacturers, policy makers, and regulators to provide and assure systems, technologies, products, and strategies that are safe, healthy, and environmentally friendly.

Demonstration and design of safety in complex systems has shaped the approaches to formal risk analysis. While formal methods of risk analysis are relatively new, the concept of measuring or guessing the risk has been around for centuries (see Covello and Mumpower [3]).

1.2.3 EMERGENCE OF FORMAL RISK ANALYSIS FOR MANAGING AND REGULATING RISKS

As noted earlier, engineering systems are becoming more complex and demand for risk analysis is greater than ever. A mechanism to control and avert risk has been to regulate manufacturing, operation, and construction of complex systems. Unfortunately, many decisions and regulations have not relied on formal risk analysis. As a result, the regulations are also complex. The complexity has been compounded by a combination of several factors, the most notable of which are vague laws, politics, and the influence of interest groups. Despite of their good intentions, some governmental regulations control marginal risks at an exorbitant cost. Federal, state, and local risk regulations are aimed primarily at reducing fatalities from cancers and accidents. While about 500,000 cancer deaths and 100,000 accidental deaths are observed each year in the U.S., it turns out that only a small fraction of them are preventable by regulations. For example, less than 10% of cancer deaths could be averted by imposing regulations which completely eliminate certain risks. While it is possible to avert some of these deaths, this may come at a prohibitively high cost. Regulations should be imposed when their societal cost is consistent with their risk reduction potential. More discussions on this topic will be provided in Chapter 8.

The conventional view of safety risk regulation is that the existence of risks is undesirable and, with appropriate technological interventions, we can eliminate those risks. However, this perspective does not recognize the risk reduction costs involved; the fact that a no-risk society would be so costly and infeasible (see VanDoren [4]). It is clear that regulations do not come for free — the society ultimately pays for them in form of higher product prices, higher taxes, loss of global competitiveness, and lower income. The net effect of some costly regulations is a reduction in the gross national product. As such, control of regulations should be an important goal of a fiscally responsible society. Risk analysis methods provide the kind of results for attaining this objective. Recently, many legislators including U.S. Congress have advocated greater reliance on the use of risk information regulation by the executive agencies.

Risk analysis and especially probabilistic risk assessment (PRA) can play pivotal roles in making design, manufacturing, operation, policy, and regulatory decisions. Progress in the field of risk analysis and especially in PRA has been enormous. It hardly existed three decades ago, but now it is ready to make important contributions to the characterization and control of risk (e.g., in risk management and regulations). A great advantage of risk analysis is that it brings calculations out into the open, encourages informed dialog, and can greatly improve public confidence in the process.

1.2.4 CATEGORIES OF RISK

Risk can be categorized in several ways. Different disciplines often categorize risk differently, using terms such as hazards or risk exposures. The categorization can be done on the basis of the causes of risk or the nature of loss (consequences), or both. Fundamentally, as noted in Section 1.1, risk can be defined as the potential for loss. Such a loss can be ultimately measured in economic terms and thus view risk as a potential economic loss. However, a more appropriate categorization is based on five broad categories that account for potential losses. These risk categories are: health, safety, security, financial, and environmental:

1. *Health risk analysis* involves estimating potential diseases and losses of life affecting humans, animals, and plants.
2. *Safety risk analysis* involves estimating potential harms caused by accidents occurring due to natural events (climatic conditions, earthquakes, brush fires, etc.) or human-made products, technologies, and systems (i.e., aircraft crashes, chemical plant explosions, nuclear plant accidents, technology obsolescence, or failure).
3. *Security risk analysis* involves estimating access and harm caused due to war, terrorism, riot, crime (vandalism, theft, etc.), and misappropriation of information (national security information, intellectual property, etc.).
4. *Financial risk analysis* involves estimating potential individual, institutional and societal monetary losses such as currency fluctuations, interest rates, share market, project losses, bankruptcy, market loss, misappropriation of funds, and property damage.
5. *Environmental risk analysis* involves estimating losses due to noise, contamination, and pollution in ecosystem (water, land, air, and atmosphere) and in space (space debris).

There are also interrelations between these categories. For example, environmental risks may lead to financial risks.

1.2.5 TRENDS IN THE USE OF RISK ANALYSIS METHODS IN ENGINEERING

A traditional approach to risk analysis has been to design or to regulate engineering systems conservatively to avoid risk (i.e., through over-design). These include for example the philosophy of defense-in-depth in the nuclear industry, which includes multiple safety barriers, large safety margins, quality control, and frequent inspections. Experience and research have shown that this philosophy, while reasonably assures safety, often leads to expensive systems, products, and technologies that the society and market would not be able to afford. Further, studies have also shown that while some designs and regulations based on the conservative approaches appear to reduce risk of complex engineering systems and products, this may come at an exorbitant cost and still does not guarantee safety. Recognizing these problems, industries and regulatory agencies have been steadily relying on formal risk analysis techniques to evaluate contributors to risk, and to improve safety of engineering systems more formally. For example, the U.S. Nuclear Regulatory Commission has been a pioneer in using risk-informed techniques in devising or augmenting its regulations derived from conservative defense-in-depth methods with risk analysis results. The nuclear industry and more recently transportation (land and air), space, and food safety industries promote a greater use of risk analysis in their operations, and policy decision making.

Risk analysis can be used in all stages of design, development, construction, and operation of engineering systems. For example, the following applications are possible:

Conceptual design
Compare alternative design options
Design
Provide barriers to prevent, minimize or eliminate harm
Minimize life-cycle cost

Apportion risk limits and performance goals
Development
Identify systems or subsystems with most contribution to safety and risk
Test safety and risk significant elements of the design
Quality assurance
Warranty development
Regulation
Regulate consistent with the significance of the elements of the system that contribute
   most to risk
Set monitoring and performance criteria
Perform inspections
Operation
Optimize cost of maintenance and other operational activities
Define surveillance requirements and schedules
Replacement policies and decisions
Aging estimation and management
Developing security measures
Decommissioning
Assess safety of possible decommissioning alternatives
Select most appropriate disposal method
Assess long-term liability issues

1.3 ELEMENTS AND TYPES OF RISK ANALYSIS

National Research Council [5] defines risk analysis as having three core elements of risk
assessment, risk management, and risk communication. Interactions and overlap between the
three elements are depicted in Figure 1.1.

   The first core element of risk analysis is risk assessment — the process through which the
probability or frequency of a loss by or to an engineering system is estimated and the
magnitude of the loss (consequence) is also measured or estimated. Risk management is
the process through which the potential (likelihood or frequency) of magnitude and contributors to risk are estimated, evaluated, minimized, and controlled. Risk communication is the
process through which information about the nature of risk (expected loss) and consequences,

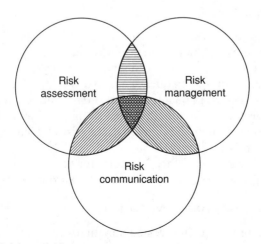

FIGURE 1.1 Elements of risk analysis.

risk assessment approach and risk management options are exchanged, shared, and discussed between the decision makers and other stakeholders.

Risk analysis is about estimating the potential and magnitude of any loss and ways to control it from or to a system. If there are adequate historical data on such losses, then the risk can be directly estimated from the statistics of the actual loss. This approach is often used for cases where data on such losses are readily available such as car accidents, cancer risks, and frequency of certain natural events such as storms and floods. The other option is for cases where there is not enough data on the actual losses. In this case, the loss is "modeled" in the risk analysis. Therefore the potential loss (i.e., the risk) is estimated. In most cases, data on losses are small or even unavailable, especially for complex engineering systems. Therefore, the analyst should model and predict the risk. The risk analysis attempts to measure the magnitude of a loss (consequences) associated with complex systems, including evaluation, risk reduction, and control policies. Generally, three types of risk analysis are: quantitative, qualitative, and a mix of the two. All these methods are widely used, each with different purposes, strengths, and weaknesses.

1.3.1 QUANTITATIVE RISK ANALYSIS

The quantitative risk analysis attempts to estimate the risk in form of the probability (or frequency) of a loss and evaluates such probabilities to make decisions and communicate the results. In this context, the "uncertainty" associated with the estimation of the frequency (or probability) of the occurrence of the undesirable events and the magnitude of losses (consequences) are characterized by using the probability concepts. When evidences and data are scarce, uncertainties associated with the quantitative results play a decisive role in the use of the results (i.e., meta-uncertainties or uncertainties about the expected loss).

Quantitative risk analysis is clearly the preferred approach when adequate field data, test data, and other evidences exist to estimate the probability (or frequency) and magnitude of the losses. The use of quantitative risk analysis has been steadily rising in the recent years, primarily due to availability of quantitative techniques and tools, and our ability to make quantitative estimation of adverse events and scenarios in complex systems from limited data. However, the use of quantitative risk analysis has been restricted to large scope risk analyses, because quantitative risk analysis is complicated, time-consuming, and expensive.

In this book, the main emphasis is on quantitative risk analysis and associated methods and tools. In particular, the book focuses on PRA.

1.3.2 QUALITATIVE RISK ANALYSIS

This type of risk analysis is perhaps the most widely used one, just because it is simple and quick to perform. In this type, the potential loss is qualitatively estimated using linguistic scales such as low, medium, and high. In this type of analysis, a matrix is formed which characterizes risk in form of the frequency (or likelihood) of the loss versus potential magnitudes (amount) of the loss in qualitative scales. The matrix is then used to make policy and risk management decisions. Because this type of analysis does not need to rely on actual data and probabilistic treatment of such data, the analysis is far simpler and easier to use and understand, but is extremely subjective.

Qualitative risk analysis is the method of choice for very simple systems such as a single product safety, simple physical security, and straightforward processes.

1.3.3 MIXED QUALITATIVE–QUANTITATIVE ANALYSIS

Risk analysis may use a mix of qualitative and quantitative analyses. This mix can happen in two ways: the frequency or potential for loss is measured qualitatively, but the magnitude of the loss (consequence) is measured quantitatively or *vice versa*. Further, it is possible that both the frequency and magnitude of the loss are measured quantitatively, but the policy setting and decision making part of the analysis relying on qualitative methods, for example, using qualitative policy measures for quantitative ranges of loss. Also, quantitative risk values may be augmented by other quantitative or qualitative risk information to arrive at a decision. This latter case is the method of choice for the U.S. Nuclear Regulatory Commission's new regulatory paradigm called "risk-informed" regulation, whereby risk information from formal PRAs are combined with other quantitative and qualitative results obtained from deterministic analyses and engineering judgments to set regulatory decisions and policies.

In this book, we will elaborate on specifics and examples of quantitative, qualitative, and mixed risk analyses.

1.4 RISK ASSESSMENT

Risk assessment is a formal and systematic analysis to identify or quantify frequencies or probabilities and magnitude of losses to recipients due to exposure to hazards (physical, chemical, or microbial agents) from failures involving natural events and failures of hardware, software, and human systems.

Generally speaking, a risk assessment amounts to addressing three very basic questions posed by Kaplan and Garrick [6].

1. What can go wrong?
2. How likely is it?
3. What are the losses (consequences)?

The answer to the first question leads to identification of the set of undesirable (e.g., accident) scenarios. The second question requires estimating the probabilities (or frequencies) of these scenarios, while the third estimates the magnitude of potential losses. This triplet definition emphasizes development of accident scenarios as an integral part of the definition and assessment of risk. Risk scenarios are in fact one of the most important products of risk assessment.

Development of risk scenarios starts with a set of "initiating events" (IEs) that perturb the system (i.e., events that change normal operating envelope or configuration of the system). For each IE, the analysis proceeds by determining the additional events (e.g., in form of hardware, software, or human errors) that may lead to undesirable consequences. Then, the end effects of these scenarios are determined (e.g., the nature and magnitude of any loss). The probability or frequency of each scenario is also determined using quantitative or qualitative methods. The expected consequence (loss value) is then estimated. Finally, the multitude of such scenarios are put together to create a complete picture risk profile of the system.

The risk assessment process is, therefore, primarily one of scenario development, computing the expected consequences from each possible scenario. Because the risk assessment process focuses on scenarios that lead to hazardous events, the general methodology becomes one that allows the identification of all possible scenarios, calculation of their individual probabilities, and a consistent description of the consequences that result from each. In engineering systems, scenario development requires a set of descriptions on how a barrier confining a hazard is threatened (i.e., an initiating event occurs), how the barrier fails, and the effects on recipients (health, safety, environment, etc.) when it is exposed to the uncontained hazard. This means that one needs to formally accomplish the following steps (far more detail discussion on these steps will be provided in Chapters 2–5).

1.4.1 IDENTIFICATION OF HAZARDS

Hazard is a condition or physical situation with a potential for an undesirable consequence (loss). In risk assessment, a survey of the processes under analysis should be performed to identify the hazards of concern. These hazards can be categorized as follows:

- Chemical (e.g., toxins, corrosive agents, smoke)
- Biological (e.g., viruses, microbial agents, biocontaminants)
- Thermal (e.g., explosions, fire)
- Mechanical (e.g., impact from a moving object, explosions)
- Electrical (e.g., electromagnetic fields, electric shock)
- Ionizing radiation (e.g., x-rays, gamma rays)
- Nonionizing radiation (e.g., microwave radiation, cosmic rays)
- Information (e.g., propaganda, computer virus)

Presumably, each of these hazards will be part of the system of interest and normal system barriers will be used as means of their containment (e.g., using firewalls to prevent access to information). This means that, provided there is no disturbance in the system, the barrier that contains the hazard remains unchallenged. However, in a risk scenario one postulates events and actions that remove or degrade such barriers, and estimates the final consequence from these challenges. Therefore, development of a scenario involves identification of hazards, barriers, potential challenges to such barriers, and amount of hazards exposed.

1.4.2 IDENTIFICATION OF SYSTEM BARRIERS

In an engineering system, each of these applicable hazards must be examined to determine all active and passive barriers that contain, prevent, or mitigate undesirable exposures to such hazards. These barriers may physically surround and isolate the hazard (e.g., passive structures — walls, pipes, valves, fuel clad structures); they can be based on a specified distance from a hazard source to minimize exposure to the hazard; they may provide direct shielding of the recipient from the hazard (e.g., protective clothing, bunkers), or they may mitigate the condition to minimize exposure to the hazard (e.g., a cooling unit, a sprinkler system, an emergency evacuation system).

1.4.3 IDENTIFICATION OF CHALLENGES TO BARRIERS

Identification of each of the individual barriers is followed by a concise definition of the requirements for maintaining functionality of each. This can be done by developing an analytical model of the barrier or just a qualitative description of such requirements. One can also simply identify what is needed to maintain the integrity of each barrier. These are due to the internally or externally induced degradation of the strength or endurance of the barrier that can no longer withstand applied loads and stresses:

- Barrier strength or endurance degrades because of
    reduced thickness (due to deformation, erosion, corrosion, wear, etc.),
    changes in material properties (e.g., fracture toughness, yield strength)
- Stress or damage on the barrier increases by
    internal agents such as forces or pressure
    penetration or distortion by external objects or forces

Above examples of causes of system degradation are often the results of one or more of the following conditions:

- Malfunction of process equipment (e.g., the emergency cooling system in a nuclear power plant)
- Problems with human–machine interface
- Poor design and maintenance
- Adverse natural phenomena
- Adverse human-made environments

1.4.4 ESTIMATION OF FREQUENCY OR PROBABILITY OF A HAZARD EXPOSURE

The next step in the risk assessment procedure is to identify those scenarios in which all the barriers may be breached and the hazard may reach the recipients (human, environment, productions line, organization, etc.), followed by the best possible estimation of the probability or frequency for each scenario. For convenience, those scenarios that pose similar levels of hazard with similar hazard dispersal and exposure behaviors may be grouped together, and combine their respective probabilities or frequencies.

1.4.5 CONSEQUENCES EVALUATION

The losses produced by exposure to the hazard may encompass, for example harm to people, damage to assets, or contamination of land. These losses are evaluated from knowledge of the behavior of the particular hazards when recipients are exposed to them and the amount of such exposure for each scenario.

From the generic nature of risk analysis, there appears to be a common approach for understanding the ways in which hazard exposure occurs. This understanding is critical to the development of scenarios that can then be solved. Quantitative and qualitative solutions can provide estimates of barrier adequacy and methods of effective enhancement. This formalization provides a basis for the practice of PRA. This technique, pioneered by the nuclear industry, is the basis of a large number of formal engineering risk assessments today. This approach will be described in detail in Chapter 2.

1.5 RISK MANAGEMENT

As risk assessment focuses on identifying, quantifying, and characterizing uncertainties with losses, risk management essentially turns into an effort to manage such uncertainties. Risk management is a practice involving coordinated activities to prevent, control, and minimize losses incurred due to a risk exposure, weighing alternatives, and selecting appropriate actions by taking into account risk values, economic and technology constraints, legal and political issues. Risk management utilizes a number of formal techniques, methods and tools including trade-off analysis, cost–benefit analysis, risk effectiveness, multiattribute decision analysis and predictive failure analysis (e.g., condition monitoring). The primary focus in risk management throughout the life cycle of a complex system involves proactive decision making to:

- Continually assess the risk (what could go wrong?)
- Decide which risks are significant to deal with
- Employ strategies to avert, control, or minimize risks
- Continually assess effectiveness of the strategies and revise them, if needed

Risk management is the most important and diverse part of risk analysis. It involves many disciplines from subject matter experts to risk and decision analysts. A good risk management effort, even in the presence of potentially risk-significant systems and technologies, can be very effective to avert, control, or minimize expected losses. As the name is also evident, this is a continual effort starting with results of risk assessment. As complex system's internal and external environment changes, so does its risk assessment. As such, risk management starts with an accurate estimation of the temporal (instantaneous or averaged) risk. As the configuration of the systems and other internal and external factors also change, it is normal that significant contributors to risk also change. Risk management involves identifying the prime contributors to risk. Routinely, complex systems show that the 80:20 rules or the "Pareto's Principle" applies. That is, more than 80% of the risk is contributed by less than 20% of risk scenarios or elements of the complex system. Risk management involves identification of these 20% or so risk contributors, to avert control and minimize them. That is, to achieve the highest risk reduction with the limited resources available. For selecting major risk contributors, formal techniques such as importance measures may be employed. Chapter 6 discusses importance measures and methods for identifying and ranking risk contributors.

When the main risk contributors are known, the process of identifying and analyzing strategies to avert, control, and minimize risk starts. The first task is identifying whether the risk contributors are so significant that drive the total risk or cause risk acceptance limit violated. Risk acceptance limits, especially when dealing with human health and environmental risks are greatly contentious issues and difficult to establish. In this book, we will elaborate on this topic in more detail in Chapter 7. Once the nature and impact of the significant risk contributors are known, alternative strategies for their management should be proposed. Usually a simple analysis in form of expert judgment, brain storming, or relying on written guidelines is used to propose these strategies. The next task is to assess and select the most promising and effective strategy to avert risk (such as relying on insurance, redesigning, etc.), to control risk (evacuation, escape strategies, protective clothing, etc.), and to minimize risk (add redundancies, diversities, exposure reduction barriers into the system design, operation, maintenance). The process of finding the best strategy involves subjective risk-informed judgments supplemented by formal techniques such as cost–benefit, risk effectiveness, multi-objective decision analysis, and a host of other formal optimization techniques. These methods will be discussed in Chapter 8. Strategies are often in form of assigning responsibility, determining appropriate risk approach (more research, accept, mitigate, or monitor); if risk will be mitigated, one must define mitigation level (e.g., action item list or more detailed task plan) and goal.

When the best strategy is selected, it should be monitored over time to measure its impact and make adjustments and revisions when deemed necessary. This element of risk management makes it a continuous process and not just the once effort. Monitoring and tracking of risk include acquiring updating, compiling, analyzing, and organizing risk data; report results; and verify and validate mitigation actions.

Morgan [2] defines four procedures to implement risk management options. These are tort and other common laws (laws related to negligence, liability, nuisance, trespass, etc.), insurance (private or governmental to pass risk to others), voluntary standards (such as National Fire Protection Association), and mandatory government standards and regulations. The latter option is a relatively new phenomenon, but perhaps the fastest growing one. It is noteworthy to point that the first technological risk regulated by the U.S. government was that of steamboat boiler explosions, which took the Congress 50 years (after several hundred accidents and 300 deaths) to enact the regulations.

It is clear that risk assessment and management are closely intertwined and synergize each other. As depicted in Figure 1.2, the risk assessment is used to understand contributors to risk and measure changes in such contributors over the time.

Risk management uses the contributors to risk values to rationally and formally strategize and utilize scarce resources to avert, control, minimize, and track significant contributors.

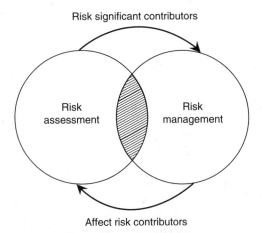

Risk significant contributors

Risk
assessment

Risk
management

Affect risk contributors

FIGURE 1.2 Relationship between risk assessment and risk management.

Once a new strategy is adopted, the risk would change and may lead to new and different risk contributors (with less or of the same significance). This can be validated by further risk assessment efforts, and thus the synergistic cycle continues.

1.6 RISK COMMUNICATION

Risk communication is the activity of transferring, exchanging or sharing data, information and knowledge about risk, risk assessment results, and risk management approach between the decision makers, analysts and the rest of stakeholders. The information can relate to the existence, form, likelihood, frequency, severity, acceptability, controllability, or other aspects of risk. This topic will be discussed in more detail in Chapter 9. Depending on what is to be communicated and to whom, risk communications contain information on the following.

1.6.1 THE NATURE OF THE RISK

This involves communicating important characteristics, nature, form, and importance of the hazard of concern, severity of the risk, and the urgency of the risk situation. Further, the quantitative and dynamic characteristics of risk should be specified. These include the trend in risk (increasing, decreasing, or remain constant), the probability or frequency of risk exposure, the amount of exposure that constitutes a significant risk, and the nature and size of the population at risk. Finally, specific risk recipients who have the greatest risk should be recognized.

1.6.2 THE NATURE OF THE BENEFITS

Often a key question in risk management is whether or not the *benefits* outweigh the *risk*. If we only associate risk with losses (not benefits) then we are risk averse, i.e., we only control and reduce our risks. Note that actions taken to reduce a risk can sometimes be considered benefits in the sense that possible losses are reduced. No matter how risk averse we are, we need to consider direct or indirect benefits of risk. These include the nature of the benefits such as the actual or expected benefits associated with the risk, and direct and indirect beneficiaries of risk. Further, the point of balance between risks and benefits, magnitude and importance of the benefit, and the total amount of benefit to all affected must be clarified in a quantitative and qualitative manner.

1.6.3   UNCERTAINTIES IN RISK ASSESSMENT

Any predictive assessment is based on a model of the world which is inherently uncertain. Further, the process of building such models (for example in estimating the parameters of the model) from raw data and experience also involves uncertainty. Risk assessment is not exception too. The question is not often whether or not uncertainty is involved in an assessment, management strategy, or analysis, rather it is how much uncertainty is involved. In identifying the uncertainties in risk assessment, it is imperative to discuss the methods, models, and expertise used to assess the risk. This should also include a characterization of importance of each of the uncertainties, the weaknesses of, or inaccuracies in the available data. Another common property of any analysis is a host of assumptions. The basis and the sensitivity of the risk and risk management strategies (including decisions or regulations) to changes in assumptions should also be discussed. Chapter 5 discusses the methods for estimating such uncertainties.

1.6.4   RISK MANAGEMENT OPTIONS

A key element of any effective communication is elaboration on the nature of the options considered to arrive at a strategy, policy, or regulation and what actions are or will be taken to manage the risk. What actions, if any, are available to reduce the risk exposure to the stakeholders? Further, the communication must elaborate on the justification for choosing a specific risk management option and effectiveness of the option. As stated above, risk management options can also be viewed in terms of their benefits, such that the significance and beneficiaries of the option should be identified. It is also vital to communicate the cost of managing the risk, and who pays (directly or indirectly) for it. Finally, the risks that remain after a risk management option is implemented should be clearly identified.

REFERENCES

1. Bedford, T. and Cooke, R., *Probabilistic Risk Analysis: Foundations and Methods*, Cambridge University Press, London, 2001.
2. Morgan, M.G., Probing the question of technology-induced risk, *IEEE Spectrum*, 18(11), 58, 1981.
3. Covello, V. and Mumpower, J., Risk analysis and risk management: a historical perspective, *Risk Analysis*, 5(2), 103, 1985.
4. VanDoren, P., Cato Handbook for Congress: Policy Recommendations for 108th Congress, Cato Institute Report, 2003.
5. National Research Council, *Risk Assessment in the Federal Government: Managing the Process*, National Academy Press, Washington, DC, 1983. This guideline has recently been augmented by a new National Research Council study, *Science and Judgment in Risk Assessment*, National Academy Press, Washington DC, 1994.
6. Kaplan, S. and Garrick, B.J., On the quantitative definition of risk, *Risk Analysis*, 1(1), 11, 1981.

2 Elements of Risk Assessment

2.1 TYPES OF RISK ASSESSMENT

In this chapter, we discuss the main elements of engineering risk assessment. The focus will be on the general techniques for risk assessment along with simple examples of using these techniques for qualitative and quantitative risk assessments. Risk assessment techniques have been used by both government and industry to estimate the safety, reliability, and effectiveness of various products, processes, and facilities. A risk assessment may focus on the health effects that occur when toxic chemicals and biological contaminants are released into the environment or consumed by people. This type of risk assessment is referred to as a health risk assessment. A risk assessment may focus on the adverse environmental, health, and economic effects that can occur when an "engineered" system fails, due to natural or human-initiated events followed by failure of protective or mitigative barriers, culminating in human, environmental, and economical consequences (losses). This type of risk assessment is usually referred to as engineering risk assessment. While we talk about both types of risk assessments, the primary focus of this book and this chapter is on engineering risk assessment with emphasis on safety risk assessment techniques. In an engineering risk assessment, the analyst considers both the frequency of an event, initiating a scenario of subsequent failure events and the probabilities of such failures within the engineering system. But in a health risk assessment, the analyst assesses consequences from situations involving chronic releases of certain amount of chemical and biological toxicants to the environment with no consideration of the frequency or probability of such releases.

The ways for measuring consequences (amount of losses) are also different in health and engineering risk assessments. Health risk assessment focuses on specific toxicants and contaminants and develops a model (deterministic or probabilistic) of the associated exposure amount (dose) and resulting health effects (response), or the so-called dose–response models. The consequences are usually in the form of cancer fatality. In engineering risk assessment, the consequence varies. Common consequences include worker health and safety, economic losses to property (environment, facility, equipment), immediate or short-term loss of life, and long-term loss of life from cancer (e.g., due to exposure to radiation, chemical toxins, or biological agents).

The most useful applications of engineering risk assessment are not the estimation of so-called "bottom-line" risk numbers such as fatalities and other losses, rather they are the insights gained from a systematic consideration of what can go wrong with a system and what parts of the system are the most significant or pivotal contributors to such losses.

2.2 RISK, HAZARD, PERFORMANCE, AND ENGINEERING RISK ASSESSMENT

2.2.1 RISK AND HAZARD

The terms "hazard" and "risk" often by mistake are used interchangeably, but in formal risk analysis they have exact definitions. The term "hazard" refers to the potential for producing an undesired consequence (loss) without regard to the frequency or probability of the loss. For example, a nuclear power hazard is release of large amounts of radioactivity to the environment due to an accident that could lead to a number of possible undesired consequences such as land contamination. Therefore, large radioactive release can be considered a *hazard* associated with a nuclear power plant. The term "risk" conveys not only the occurrence of an undesired consequence, but also how likely (or probable) such a consequence will occur. Thus two plants may hold the same hazards but can pose vastly different risks, depending upon the likelihood of such hazards getting released. This, for example, depends upon the effectiveness of the barriers (such as safety systems) that are in place to avert such a release.

2.2.2 ENGINEERING RISK ASSESSMENT

To estimate the value of risk, a risk assessment is performed. Engineering risk assessment (from now on we refer to engineering risk assessment as simply risk assessment) consists of answering the following questions posed by Kaplan and Garrick [1] and discussed in Section 1.3:

1. What can go wrong that could lead to a hazard exposure outcome?
2. How likely is this to happen?
3. If it happens, what consequences are expected?

To answer question 1, a list of outcomes (or more precisely scenarios of events leading to the outcome) should be defined. The frequency, probability, or likelihood[1] of these scenarios should be estimated (i.e., answer to question 2), and the consequence (amount of losses) due to each scenario should be calculated (i.e., answer to question 3). According to this definition, risk can be defined quantitatively as the following set of triplets:

$$R = \langle S_i P_i C_i \rangle, \quad i = 1, 2, \ldots, n \tag{2.1}$$

where S_i is a scenario of events that lead to hazard exposure, P_i is the likelihood of scenario i, and C_i is the consequence due to occurrence of events in scenario i (i.e., a measure of the amount of loss).

Since (2.1) involves estimation of the probability or frequency of occurrence of events (e.g., failure of barriers for preventing hazard exposure such as safeguard systems), risk analyst engineer should have a good knowledge of probability concepts. This topic has been discussed in Appendix A.

---

[1] The *probability* that an event which has already occurred would yield a specific outcome is called *likelihood*. The concept differs from that of a probability in that a probability refers to the occurrence of future events, while likelihood refers to past events with known outcomes. Eric W. Weinstein, "Likelihood," from *MathWorld* — A Wolfram Web Resource. http://mathworld.wolfram.com/Likelihood.html

A simple mathematical representation of risk (expected loss) consistent with (2.1) is the form commonly found in the literature as

$$\text{Risk}\left(\frac{\text{Consequence}}{\text{Unit of time or space}}\right) = \text{Frequency}\left(\frac{\text{Event}}{\text{Unit of time or space}}\right)$$

$$\times \text{Magnitude}\left(\frac{\text{Consequence}}{\text{Event}}\right) \qquad (2.2)$$

Thus if we wished to calculate the annual fatality risk due to automobile accidents in the U.S., we calculate this as

$$\underbrace{\left(15 \times 10^6 \frac{\text{accidents}}{\text{year}}\right)}_{\text{frequency}} \underbrace{\left(\frac{1 \text{ fatality}}{300 \text{ accidents}}\right)}_{\text{consequence}} = \underbrace{50,000 \frac{\text{fatalities}}{\text{year}}}_{\text{expected loss}}$$

Clearly in this case, the answer could have been obtained directly by looking up the number of automobile accident fatalities in the records. However, if we wanted to know the risk from nuclear accidents, LNG explosions, etc., there is no statistical base, so the formula becomes useful. Of course, one needs a formal method of estimating the frequency of events leading to hazard exposure (accident) and the magnitude of these consequences. This is what a PRA attempts to do. Chapter 3 will discuss PRA methods in more detail.

Example 2.1

Consider health hazards of *Salmonella enteritidis* (ES), which is the second most common cause of food poisoning after Campylobacter. It has been found primarily in unpasteurized milk, eggs, egg products, meat, and poultry. ES can survive if food is not cooked properly. In the U.S., there are 19 ES-related human illnesses per 10^6 shell eggs consumed, 710 deaths per million ES-related illnesses, 47 billion shell eggs consumed per year, the average cost of the illness is US$ 400 per case. Calculate economic risk (expected losses) caused by ES illnesses and mortality risks due to consumption of shell eggs in the U.S.

Solution

Frequency of illness: 893,000 illnesses per year
Magnitude of death: 710 deaths per million illnesses (0.07% per illness)
Mortality risk value (expected losses): $893,000 \times (710/10^6) = 634$ deaths/year
Cost per illness (regardless of outcome): US$ 400
Economic risk value (expected losses): $893,000 \times \$400 = $ US$ 357,200,000 per year.

Example 2.2

According to the U.S. Department of Transportation, in 2003, there were 6.3 million automobile accidents: 1 in 3 of such accidents resulted in injury, and 1 in 165 resulting in death. Assuming average loss of US$ 450,000 per death and US$ 25,000 of property damage for accidents involving fatality; average loss of US$ 15,000 per injury and property loss of US$ 10,000 for accidents involving injury; average property loss of US$ 3,000 for all other accidents, calculate the total monetary risk (expected losses) of automobile accidents per driver in the U.S. Assume that the U.S. population exposed to automobile accidents is 250 million.

Solution

Risk Contributor	Fatality	Injury	Other	Total
Probability per person per accident	$6.3 \times 10^6/250 \times 10^6 = 0.025$	0.025	0.025	
Probability of events given accident	1/165	1/3	109/165	
Probability of consequence per person	$1/165 \times 0.025 = 1.53 \times 10^{-4}$	8.4×10^{-3}	1.66×10^2	
Magnitude of consequence (US$ at risk)	$450,000 + 25,000$	$15,000 + 10,000$	3000	
Risk (expected loss)	US$ 72.54	US$ 210.00	US$ 49.94	US$ 332.49 per person– year

An insurance company may use this risk value to set insurance premium for drivers.

Note that the risk unit is the key to characterizing the true severity of a hazard. For example in Figure 2.1, one can see that while the annual fatality risk of automobile has remained constant over the years (i.e., risk over time), in Figure 2.2, the risk per 100 million miles traveled (i.e., risk over space) has actually been reduced. Therefore, in characterizing risks of hazards, the unit of risk value is critical in highlighting the risk and contributors to risk. Also, the way risk results are shown is important in risk communications. This topic will be further discussed in Chapter 9.

In general, (2.2) can be represented in a more general form of

$$R = \sum_i f_i c_i \tag{2.3}$$

where f_i is the frequency of the scenario i and c_i is the consequence of the scenario.

One useful way to represent the final risk values is by using the so-called Farmer's curves [2], proposed as a method for siting nuclear power plants. In this approach, the consequence is plotted against the complementary cumulative distribution of the event (scenario) frequency. Although this definition is slightly different than that given by Farmer, these graphs are often very useful and common in representing the risk profile of engineering systems. As an example, a Farmer's curve for risk scenarios shown in Table 2.1 is plotted in Figure 2.3.

Although the risk value calculated by (2.3) is widely used, it has some serious shortcomings if used as a measure of risk for comparing various options. To illustrate this problem, consider the following example. Suppose the event of exposure of a hazard has a frequency of occurrence of 10^{-6} per year with an average consequence of 10^6 deaths per event. Further,

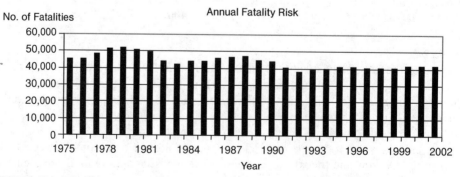

FIGURE 2.1 Total fatalities per year due to automobile accidents (1975–2003).

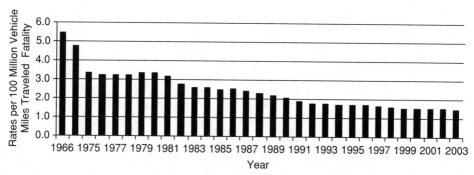

FIGURE 2.2 Motor vehicle fatality rates per 100 million vehicle miles traveled (1966–2003).

suppose that a second hazard has a frequency of exposure of 0.1 per year, with an average of 10 deaths per accident. From (2.3), the risk values (expected loss) of these two events are $10^{-6} \times 10^6 = 1$ death/year and $0.1 \times 10 = 1$ death/year. That is, the two events are "equally risky." However, most would prefer to take the latter risk than the former (as risk perception is widely different than the actual risk for these two risk situations). This is due to the fact that society or individuals are not risk-neutral, rather they are either mostly "risk-averse" to large consequence events (few may be "risk-neutral" or even "risk-seekers"). To reflect this type of risk aversion, the mathematical relationship expressed by (2.3) would have to be changed to something other than a simple product to account for the degree of risk aversion, if the results are to be used to justify acceptability or relative significance of various hazard exposures. A number of other functional relations between frequency and consequence can be found in the literature. However, none has yet gained any widespread acceptance.

It has become a common practice in PRA to calculate the risk as defined above, but also to calculate a distribution function that expresses the frequency of risk scenario versus magnitude of its consequence. Then the decision-making process must decide how to account for risk aversion (for more discussions see Chapter 8).

A key in determining the frequency (f_i) in (2.3) is to understand the notion of probability. In the technical community, there exist two quite different definitions of probability. The classical statisticians think of probability as a property of a process or event that can be determined from an infinite population of data. For example, the probability of "heads" (P_H) in a coin toss is defined as

$$P_H = \frac{N_H}{N} \quad \text{as } N \to \infty$$

TABLE 2.1
Risk Scenarios

Scenario	Frequency	Consequence	Complementary Cumulative Frequency
s_1	f_1	C_1	$F_1' = f_N + f_{N+1} + \cdots + f_1$
s_2	f_2	C_2	$\vdots$
s_3	f_3	C_3	$\vdots$
$\vdots$	$\vdots$	$\vdots$	$\vdots$
s_{N-1}	f_{N-1}	C_{N-1}	$F_{N-1}' = f_N + f_N + 1$
s_N	f_N	C_N	$F_N' = f_N$

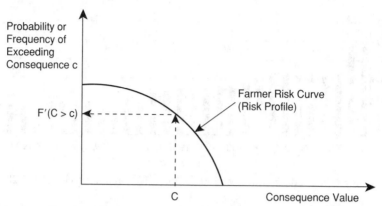

FIGURE 2.3 An example of a risk profile.

where N_H is the number of heads and N is the number of tosses. Statisticians have devised techniques for making estimates of P_H, and of the confidence in this estimate using less than an infinite population of data. Those who accept this definition are often called *frequentists* and believe that P_H is a precise value and that information needed to make estimates of it can come only from observation of the process or event. Therefore if in 100 tosses, 48 heads have been observed, the frequentist offers a point estimate probability of 0.48 that in a single toss a head will be observed.

The other school of thought, advocated by those called *subjectivists*, holds that P_H has a value at any time that represents the total available knowledge about the process or event at that particular time. Thus, if the coin were examined carefully before the first toss and found to have a head and a tail and be well balanced, and an examination of the tossing process showed it to be fair, the subjectivist would state his belief of the value of $P_H = 0.5$. Note that the value of $P_H = 0.5$ is based on the prior knowledge of the coin and the flipping process, but not solely based on any particular observations.

Now suppose the coin is flipped ten times and we observe seven heads and three tails. At this point, the frequentists' best estimate of P_H is 0.7. However, the subjectivist will add this new observation to his prior knowledge in a logical and consistent manner. A well-known relationship called Bayes' theorem (discussed in Chapter 3) describes the mechanics of combining beliefs based on prior knowledge and actual observations. Using Bayes' theorem, the subjectivist finds his prior estimate of P_H is only slightly modified upward when this new information is considered. If it is in fact a fair coin toss, then after a large number of tosses both groups will eventually obtain a value of P_H that approaches 0.5.

The use of PRA in engineering systems with large undesirable consequences of low probability must employ the logic of the subjectivist (or Bayesian) approach since rarely will enough data exist to use the frequentists' definition. Although some controversy still exists between these two schools of thought, the Bayesian approach is gaining wider acceptance. For a good discussion of this point see Glickman and Gough [3].

2.2.3 PERFORMANCE AND PERFORMANCE ASSESSMENT

Performance of a system is referred to the ability of the system to realize its intended functions at all times. The "ability" of the system can be characterized either qualitatively or quantitatively. In this book, we are only concerned with quantitative performance.

Quantitatively performance is not a deterministic notion. That is, there is a probability that the intended function can be realized at a time of need or during a period of interest. For example, consider the bulkhead of an aircraft structural frame whose function is to hold the wings of the aircraft at all times. The ability to realize the function of holding the aircraft wings under a given condition at a specific time, is determined by its capacity (in this case strength) to withstand (applied stresses) or endure (amount of damage) all the time, and during all the aircraft missions. Since the capacity and challenges are random (due to randomness in both the applied load and material strength and toughness), performance is therefore intrinsically a probabilistic attribute. That is, the system is able to realize its intended function only when its capacity exceeds the challenges. For example, in the case of the aircraft bulkhead, the probability that it can endure the fatigue caused by random applied loads can be considered as measure of performance.

There are subsidiary attributes of performance that can be used to better characterize system performance. While not unique, three subsidiary attributes of performance are among the important ones: capability, efficiency, and availability (which in turn has two elements of reliability and maintenance).

Capability is a measure of the likelihood or probability that the system attains its functions (not fails) under all internal and external conditions during missions. The fact that internal and external conditions and missions could impact the number and magnitude of the challenges and capacity of the system, the probability that the system attains it functions (works) changes by time or changes by the level of challenges (such as stresses applied) and capacity (such as endurance for fatigues). Due to randomness of challenges and system capacity and changes over the time, the reliability of the system in effect considers not only the degree to which capacity exceeds challenges, but also changes (primarily deterioration and degradation) in both capacity and challenges over time and under different conditions.

Failures[2] are the results of the existence of *challenges* and conditions occurring in a particular scenario. The system has an inherent *capacity* to withstand or endure such challenges. Capacity may be reduced by specific internal or external conditions over time or cycles of application. When challenges surpass the capacity of the system, failure may occur. Specific models use different definitions and metrics for capacity and challenge. "Adverse conditions" generated artificially or naturally, internally or externally, may increase or induce challenges to the system, or reduce the capacity of the system to withstand challenges.

Figure 2.4 and Figure 2.5 depict elements of a framework to construct failure models. Several simple failure models have been discussed by Dasgupta and Pecht [4] based on this framework.

Efficiency is a measure of the effectiveness of realizing the intended function as measured by a comparison of input and output of the system (as in converting energy and consuming time, effort, and money to attain system's functions). Therefore a system that is too heavy, has low yield, is costly and cumbersome to operate, consumes too much energy, or pollutes the environment.

Availability shows the state of the system measured probabilistically, for example, in terms of the fraction of time that the system may be down due to incipient failures (e.g., repairing cracks initiated due to fatigue in the bulkhead of an aircraft) and unable to realize its intended functions.

---

[2] For PRA purposes, *failures* are regarded as basic (and undesired) events which render a barrier (component, subsystem, or system) incapable of performing its intended function and require numerical estimates of probability of such events. *Faults*, on the other hand, are higher order events representing the occurrence or existence of an undesired state of a barrier (e.g., component or set of components) which is analyzed further, and ultimately resolved.

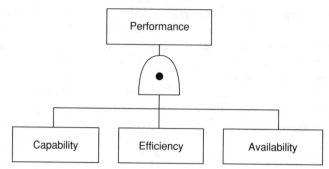

FIGURE 2.4 Elements of performance.

2.2.4　Elements of a Formal Risk Assessment

The two types of risk assessments are qualitative and quantitative risk assessments. In qualitative risk assessment, the potential and the consequence of losses are measured through qualitative measures such as high, medium, and low, or through numerical grades (such as fuzzy values). Quantitative risk assessment uses probability to measure the potential for loss and the actual amount of loss to represent the consequence. The process of risk assessment is, however, the same for both cases. As illustrated in Figure 2.6, risk assessment involves understanding the sources of hazards, barriers that prevent or mitigate exposure of such hazards, and recipients and consequences of exposure to such hazards.

2.2.4.1　Steps in Risk Assessment

The following major steps are common to both qualitative and quantitative risk assessments:

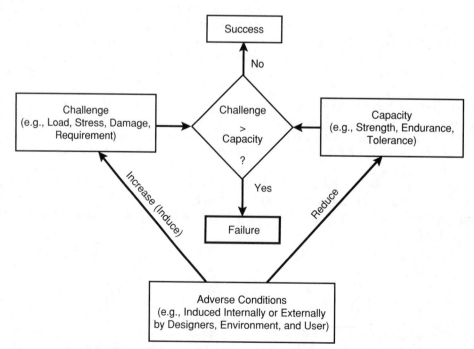

FIGURE 2.5 A failure framework.

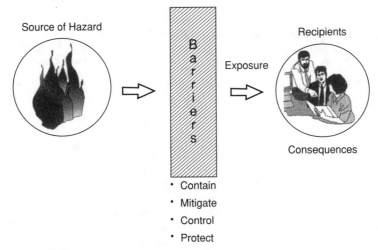

FIGURE 2.6 General illustration of risk assessment concept.

Hazard Identification
In this step, the sources, amount, and intensity of hazards that could cause losses to humans, environment, and property are identified. Generally, there are two classes of hazards: naturally occurring and artificially made. Table 2.2 summarizes classification and examples of these hazards. The hazards identified in Table 2.2 are not comprehensive as there are many types of actual hazards.

Barrier Identification
Barriers can be viewed as obstacles that perform the function of containing, removing, neutralizing, preventing, mitigating, controlling, or warning release of hazards. There are two kinds of barriers: active and passive. Active barriers are referred to those physical equipments, systems, and devises as well as human and organizational systems that are activated to realize one or more of the barrier functions above. Passive barriers continually provide the necessary functions using for example natural phenomena such as natural circulation to provide the barrier functions.

Examples of active barriers include human operator, cooling systems, control systems, advanced warning systems, information and intelligence systems, radioactive scrubbing system, ventilation system, fire squads, escape vehicles, human guards, and emergency response systems.

Examples of passive barriers include pipes, containment structures and vessels, cooling tanks (which flows by gravity), casing, protective clothing, bunkers, and natural air circulating system.

Often one or more of these barriers should work effectively to provide the necessary barrier function. If the barrier's performance is adequate, then the hazard would not be exposed or minimally exposed. If not, the exposure could lead to consequences and losses. As such, an important element of a risk assessment is the assessment and adequacy of performance of the barriers.

Barriers Performance Assessment
Barriers are not perfect and thus they may fail or prove incapable of providing their intended functions. These could be due to failure or degradation in active or passive elements of the barrier. For example, a container of natural gas or radioactivity can fail if one or more of the following conditions occur:

TABLE 2.2
Classification and Examples of Natural and Artificial Hazards

Natural Hazards	Artificial Hazards
Weather and ecosystem	*Chemicals*
Flood	1. Flammable
Storm	Hydrogen
Tornado	Ammonia
Hurricane	Propane
Drought	Carbon monoxide
Landslide	2. Reactive
Wildfire	Fluorine (gas)
Earth	Hydrogen peroxide (liquid)
Earthquake	Nitrites (solid)
Volcanic eruption	3. Corrosive
Tsunami	Strong acids
Landslide	Nonmetal chlorides
Radon	Halogens
Space	*Carcinogenic or toxic*
Cosmic rays	Asbestos
Meteorite	Tobacco smoke
Asteroids and comets	Estrogens
Biological	Tamoxifen
Infectious diseases (due to naturally	Hydrogen cyanide
growing bacteria, viruses, and fungi)	Carbon monoxide
Animal diseases	*Thermal*
Plant pests	Extreme temperatures in solids, liquids, or gases
Allergens	(due to radiation, combustion, condensation,
	fire, explosion, vaporization, etc.)
	Mechanical
	High velocity moving or rotating solids, liquids, and gases
	Vibrating objects
	Electromagnetic
	High voltage
	High magnetic field
	Ionizing radiation
	Radioactive solids, gases, and liquids
	Nonionizing radiation
	Microwave
	Radio frequency
	Ultraviolet
	Infrared
	Laser
	Biological
	Biotech, human grown bacteria, viruses, fungi,
	mycobacteria, wood dust, parasites

- Applied stress caused by the natural gas exceeds its strength
- Accumulated damage to the barrier (e.g., crack growth due to fatigue) exceeds endurance (e.g., fracture toughness of the tank)
- Barrier strength (or endurance) degrades because of some underlying chemical or mechanical mechanisms:
  (a) reduced thickness (for example due to geometrical change caused by mechanisms such as fatigue, erosion, or corrosion)
  (b) change in material properties (e.g., reduction toughness due to radiation damage mechanism)
- Malfunction of process equipment (e.g., the emergency cooling system of a nuclear plant fails because its pumps did not start when needed)
- Human errors due to poor man–machine interface
- Human errors due to poor organizational communications
- Poor maintenance which does not restore the machinery properly
- Adverse natural phenomena
- Adverse operating environment.

A key aspect of any risk assessment is to assess and characterize, qualitatively or quantitatively, the performance (for example, reliability and thus probability or possibility of failure of a barrier) for each scenario involving failure of one or more barriers. The calculation should always include the uncertainties associated with the estimated values. Many methods are employed in risk assessment for calculating the performance of barriers. In this book, we will elaborate on these methods in Chapter 4.

Exposure Assessment
If the barriers to hazard exposure are compromised, then some or all hazards will be released and potentially expose recipients. The step of exposure assessment tries to assess the amount and characteristics (toxicity, concentration, temperature, etc.) resulting from the release of the hazards. In a qualitative assessment, this step is often done either by making simple, order of magnitude type calculations, or entirely relying on expert judgment. In quantitative assessment, models of barrier failure are usually developed and the amount of exposure is estimated. Always the quantitative estimation involves characterization of uncertainties associated with the risk values.

Risk Characterization
The last step is to measure the frequency and magnitude of the consequences from the exposure to hazards. The first step is to identify the exposure paths, and the recipients of the exposure. This is followed by determining the correlation between amount of exposure received by the recipients and the consequences incurred because of the exposure. For example, if we are interested in human cancer risks due to the release of radiation into the environment, caused by failure of multiple barriers in a nuclear power plant, the receipt in this case is human and the consequence is radiation-induced cancer. Here, one can use the correlations of whole body gamma radiation hazard exposure to cancer (or dose–response) which states that 10,000 person-rem causes one statistical cancer case. This means that if 100 people each receive 100 rems (rem is a large scale for measuring ionizing radiation) or 200 people each receiving 50 rems, you would expect one statistical case of cancer. This is a probabilistic (quantitative) correlation. Other correlations (qualitative) or threshold type (nonprobabilistic) limits may also be used. For example, 1000 rems or more exposed to a specific organ of the body may be considered as permanent loss of that organ, but not below this threshold.

After determining the consequence, the risk value is calculated using (2.1). The frequency involves multiplication of qualitative or quantitative estimates of the likelihood of the exposure for each scenario of hazard exposure by the amount of consequence given the exposure to the recipients.

In a report by the National Research Council [5], also known as the "Red Book," the specific steps in a "health" risk assessment is formally defined as:

Hazard identification: Determining the identities and quantities of environmental contaminants present that can pose a hazard to human health.

Hazard characterization: Evaluating the relationship between exposure concentrations and the incidence of adverse effects in humans. A qualitative or quantitative evaluation of the nature of the adverse effects associated with biological, chemical, and physical agents that may be present.

Exposure assessment: Determining the conditions under which people could be exposed to contaminants and the doses that could occur as a result of exposure. Including the qualitative and quantitative evaluation of the degree of intake would likely occur.

Risk characterization: Describing the nature of adverse effects that can be attributed to contaminants, estimate their likelihood in exposed populations, and evaluate the strength of the evidence and the uncertainty associated with them. This includes the integration of hazard identification, hazard characterization, and exposure assessment into an estimation of the adverse effects likely to occur in a given population, including uncertainties.

2.2.5 QUALITATIVE RISK ASSESSMENT

Sometimes, it is easier to perform a qualitative risk analysis because it does not require gathering precise data. In this approach, rank-ordered approximations are sufficient and are often quickly estimated. Rank-ordered approximations of probability and consequence (defined qualitatively) can yield useful approximations of risk. For instance, one approximation scale defines probability from high to low. Qualitative probability categories and their accompanying definitions are as follows:

1. Frequent — Likely to occur often during the life of an individual item or system or very often in operation of a large number of similar items.
2. Probable — Likely to occur several times in the life of an individual item or system or often in operation of a large number of similar items.
3. Occasional — Likely to occur sometime in the life of an individual item or system, or will occur several times in the life of a large number of similar components.
4. Remote — Unlikely, but possible, to occur sometime in the life of an individual item or system, or can reasonably be expected to occur in the life of a large number of similar components.
5. Improbable — Very unlikely to occur in the life of an individual item or system that it may be assumed not to be experienced, or it may be possible, but unlikely, to occur in the life of a large number of similar components.
6. Incredible — Considered as an "Act of God" or physical events that are not expected to occur in the life of many similar large facilities or systems.

Similarly, consequence can be defined in descending order of magnitude. Table 2.3 describes one example of the risk values associated with each frequency–consequence category.

TABLE 2.3
Qualitative Risk Assessment Matrix

Frequency of Occurrence	Frequency (per Year)	Severity of Consequence			
		Catastrophic	Critical	Marginal	Negligible
Frequent	>1	H	H	H	I
Probable	$1-10^{-1}$	H	H	I	L
Occasional	$10^{-1}-10^{-2}$	H	H	L	L
Remote	$10^{-2}-10^{-4}$	H	H	L	L
Improbable	$10^{-4}-10^{-6}$	H	I	L	T
Incredible	$<10^{-6}$	I	I	T	T

Note: The category definitions and values used in this matrix are illustrative only.
H, high risk; I, intermediate risk; L, low risk; T, trivial risk.

For this example, the severity of the consequence categories are defined as:

1. Catastrophic — involving many deaths, loss of system or plant, such as significant loss of production, significant public interest and regulatory intervention occur.
2. Critical — involving a few severe injuries, major system damage or other event which causes some loss of production, affects more than one department, or region or could have resulted in catastrophic consequences under different circumstances.
3. Marginal — minor injury, minor system damage, or other event generally confined to one department or region.
4. Negligible — an event that can almost be ignored.

In most instances, individuals can usually reach agreement (within a one- or two-category span) about which category best describes a situation. "Multiplying" the probability by the consequences yields risk levels. The following example shows a qualitative risk assessment.

Example 2.3
Chamberlain and Modarres [6] describe a typical compressed natural gas (CNG) bus system shown in Figure 2.7, comprised of the following major subsystems and activities:

- Natural gas supply
- The compression and storage station
- Dispensing facility
- CNG bus
- Operator interactions with equipment and maintenance activities.

Perform a qualitative risk assessment for fatalities resulting from fire and explosion events.

Solution
Step 1   Hazards
The hazard is the natural gas (primarily methane gas) fire and explosions leading to passenger and nonpassenger fatalities.
Step 2   Barriers
Barriers are CNG storage tanks, pressure control systems, operators, warning and gas detection devices, and preventive maintenance activities.
Step 3   Barrier Performance
Several possible failures of barriers leading to fire are possible. Barriers performance in critical risk scenarios:

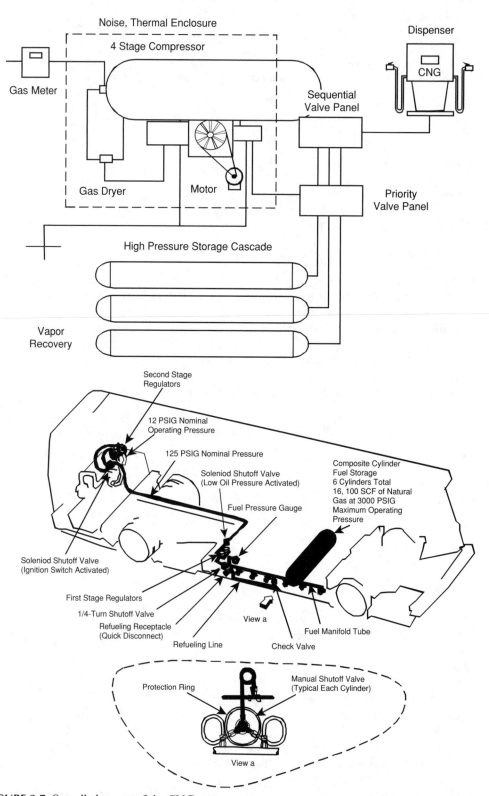

FIGURE 2.7 Overall elements of the CNG system.

1. CNG tank or control system catastrophic failures (internally caused failures), leading to instantaneous release of CNG in the presence of an ignition source.
2. CNG tank or control system degraded failures (e.g., internally caused leak), resulting in gradual release of CNG in the presence of an ignition source.
3. CNG tank, control system, or human errors, leading to release of CNG and ignition due to electrostatic discharge sparks.
4. Accidental impact of CNG tank and other hardware with external bodies (e.g., due to collisions with other vehicles), resulting in gas release in the presence of an ignition source.
5. Operator/driver error resulting in the release of CNG in the presence of an ignition source.

Specific scenarios involving failure of critical barriers have been listed in Table 2.4.

TABLE 2.4
Barrier Failures and Their Effects

Failure Category of the Barrier	Failures Internal to the Barrier	Failures External to the Barrier
Mechanically induced failure	Pressure gauge failing cause missile damage	Explosion of CNG fuel tank due to failure of the compressor to shut off
	Fire resulting from gas release through pressure relive system because of pressure/temperature cycling	Missile damage from flying parts generated during CNG tank disassembly
	CNG tank falling hazards	High speed flying object from fueling hose failure
	CNG fuel tank explosion or other fires due to vehicle impact in traffic accident	Fire due to gas dispenser damage from vehicle collision
	Vehicle fire caused by fuel system leaks due to poor design or improper installation	Fire from damaged gas supply pipelines
	Vehicle fire due to undetected fuel system leaks from failed components	Fire resulting from drive away during fueling
	Vehicle explosion resulting from fuel system leaks	Missile damage from catastrophic compressor failure causing damages and injuries/fatalities
	Building explosion from vehicle, dispenser, or storage vessel leaks	
	Explosion due to mechanical damage to vehicle storage cylinder	
Chemically induced failures	Explosion due to catastrophic tank rupture resulted from external corrosion of CNG fuel tank	Fire from corrosion of gas supply pipeline resulting in leaks
	Explosion of vehicle tank due to internal corrosion	Fire and explosion from corrosion of storage vessels
Electrically induced failures	Vehicle fire due to electrical failures not related to CNG system	Electric shock from power supply to natural gas compressor stations
		Fire from electrostatic charge occurred during tank venting

TABLE 2.5
Matrix of Gas Release and Dispersion Scenarios

CNG Release Mode	Ignition Mode	Expected Consequence
Instantaneous	Immediate	Fireball
	Delayed	Vapor cloud explosion or flash fire
Gradual	Immediate	Jet flame
	Delayed	Vapor cloud explosion or flash fire

Step 4   Exposure
Failures described in Table 2.4 can lead to one of the four possible fire characteristics shown in Table 2.5.

For each of the four fire scenarios in Table 2.5, the approximate size and temperature of the resulting fire leads to the assessment of exposure types and the likelihood of fatalities due to fire. Severity of each exposure type can be categorized into the ones summarized in Table 2.6. Therefore, each of the scenarios in Table 2.4 can lead to one of the exposures with a severity listed in Table 2.6.

Step 5   Risk Characterization
Considering the scenarios involving important barrier failures listed in Table 2.4, it is important to know the relative frequency of occurrence of each failure scenario. Table 2.7 shows the guideline used to assign a relative qualitative frequency to each of the scenarios listed in Table 2.4.

After assigning a severity and frequency to each of the barrier failure scenarios in Table 2.4, a "Risk Matrix" shown in Table 2.8 can be found. In this matrix, each element shows the number of scenarios having the indicated severity and frequency. Clearly, the scenarios that fall in the upper left quadrant of the matrix are "risk significant" and require protection and mitigation risk management strategies, and the ones in the lower right quadrant are unimportant to risk. The rest may need additional evaluations to determine whether or not they call for additional risk management strategies.

2.2.6   Quantitative Risk Assessment

This method follows the same steps as qualitative risk assessment (QRA), except in Steps 3–5, the performance of the hazard barriers, amount of exposures and consequences of exposures must be quantified. This book will elaborate on quantitative methods in Chapters 3–5, and

TABLE 2.6
Severity Description from Exposure of the Fire

Severity Category	Severity Category Description
Catastrophic	CNG release involving catastrophic fire or explosion
Critical	Unconfined CNG release with critical fire or explosive potential
Marginal	Small CNG release with marginal ignition potential or fire effects
Minor	Failure with minor fire potential and only loss of system operation

TABLE 2.7
Relative Frequency Categories for Fire Scenarios

Frequency Category	Frequency Category Description
(A) Frequent	Likely to occur within 1 year or less
(B) Probable	Likely to occur within 10 years or less
(C) Unlikely	Probable within the expected life of 20 years for a bus or station
(D) Remote	Possible but not likely during the expected life of 20 years

in this chapter only a simple qualitative–quantitative method has been illustrated through Example 2.4 that follows.

Example 2.4
Repeat Example 2.3 with a quantitative risk assessment.

Solution
Consider failure of the CNG tank as a possible initial cause of a failure leading to a gas release and fire scenario. Figure 2.8 depicts the scenarios, frequencies, and consequences. Also, marginal risk contribution due to each scenario is calculated along with the risk contributors due to all scenarios. Clearly the risk is calculated as

Risk = Frequency of a barrier failure
     × Probability of gas release given barrier failure
     × Probability of expansion and ignition given gas release
     × Probability of a particular fire dispersion type given ignition of the gas
     × Probability of a particular fire type given a specific dispersion
     × Probability that fire occurs in a specific location
     × Consequence

The consequence is also a conditional type which depends on the characteristics of the scenario. Data given in Table 2.9 are from generic sources of failures.

Clearly, the probability that other equipment or operator error initiates a gas release scenario must also be considered. When all other possible errors are considered, the total results have been summarized in Table 2.9.

TABLE 2.8
Risk Matrix Showing the Number of Scenarios Falling into the Various Risk Categories

	Catastrophic	Critical	Marginal	Minor
Likely	0	0	0	4
Probable	1	8	6	15
Unlikely	3	7	12	19
Remote	4	3	2	3

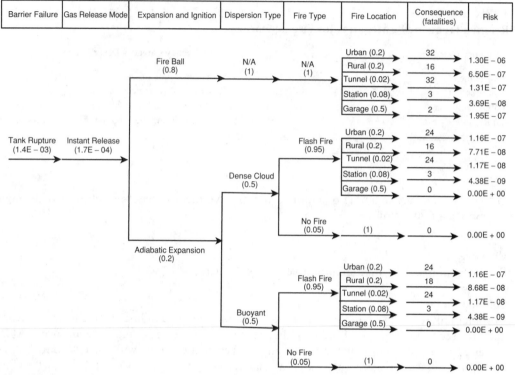

FIGURE 2.8 Scenarios involving a CNG tank failure.

Exercises

2.1 The following table shows the data calculated in a risk study by Keeney, Kulkarni and Nair [7] for assessing the risk of an LNG terminal.

Group	Expected Fatalities per Year	Number of People Sharing the Risk	Risk per Person per Year	Probability of Exceeding (x)	Number of Fatalities (x)
Permanent population in Port O'Connor	2.0×10^{-8}	800	2.5×10^{-11}	3.3×10^{-5}	80
Permanent population in Indianola	1.30×10^{-7}	80	1.7×10^{-9}	3.3×10^{-5}	800
Transient daytime visitors	2.50×10^{-6}	2500	9.9×10^{-10}	3.3×10^{-5}	2500
Individuals in boats	1.35×10^{-5}	3000	4.5×10^{-9}	3.0×10^{-5}	3000
All individuals exposed to risk	1.70×10^{-5}	9000	1.9×10^{-9}	1.7×10^{-5}	9000

(a) Based on the data in this table, plot the risk profile in terms (annual frequency of exceeding the given number of fatalities versus number of fatalities). That is the so-called Farmer's curve.

(b) What is the frequency of exceeding 100 fatalities?

2.2 A new manufacturing process is proposed for installation at a factory. It is believed that this process releases a known chemical carcinogen gas during its operation. If the expected toxicity of the gas is 1.44×10^{-4} g carcinogen per cubic meter of air and a worker's breathing rate is 2.32×10^{-4} m^3/s, determine and plot the *individual* cancer risk (cancer likelihood here) as a function of time (e.g., years). Use the following data:

TABLE 2.9
Risk Matrix for CNG Fueled Buses for all Scenarios*

CNG Bus Fire Scenarios Involving Failure of the Following Class of Barriers	Frequency of Occurrence/Bus/Year	Risk (Fatalities/Bus/Year)	Risk (Fatalities/100 Million Miles of Travel)
Bus hardware (such as the gas tank)	1.4×10^{-3}	2.7×10^{-6}	2.8×10^{-2}
Refueling station hardware	3.7×10^{-3}	7.5×10^{-6}	7.8×10^{-2}
Electrostatic discharge of CNG	1.4×10^{-5}	3.7×10^{-6}	3.9×10^{-2}
Impact failures due to collisions	3.6×10^{-2}	4.6×10^{-6}	4.8×10^{-2}
Non-CNG hardware	3.6×10^{-4}	3.1×10^{-6}	3.2×10^{-2}
Operator error	4.0×10^{-2}	3.5×10^{-7}	3.5×10^{-3}
Total fire fatality risk	—	2.2×10^{-5}	2.3×10^{-1}

*Assuming ~11,000 miles of travel / bus / year.

(a) There are 2000 h of work per year.
(b) Exposure–consequence relationship is linear and can be obtained from

$$P_c = 5 \times 10^{-4} \text{ ea}$$

where P_c is probability of consequence (cancer here) and ea is accumulated grams of the carcinogen in the lung.
(c) The inhaled carcinogen of 10% will absorb and remain in the lung and the rest will immediately exit the body.

2.3 Which of the following two technologies is riskier?

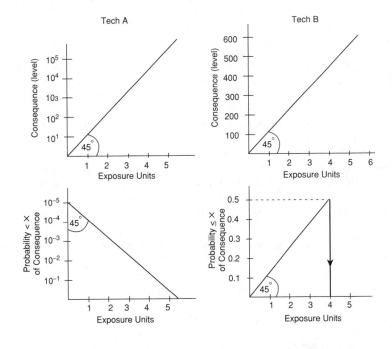

2.4 The frequency of a scenario of events leading to accidental exposure of a toxic chemical is 3×10^{-4} per year. Such an accident can expose people to toxic air containing 10 $\mu g/m^3$ of a carcinogen.

(a) What is the individual annual risk of such an accident to an exposed person?

Use the following information:

Human breathing rate	5 m³/h
Body absorption fraction of the toxic material by the body	0.1
Health effects	10 cancer death/gram absorbed
Average exposed time in a year	10 min
Population exposed	1500

(b) What are the "odds" of cancer death in this case?
(c) How long an exposure could lead to an individual annual risk of 10^{-6}?

2.5 The following information about risk of school bus accidents are known. There are 448,000 school buses in the U.S., annually 130 accidental deaths occur. Approximately 3% of all fatalities are fire fatalities of which 8% are occupants of the bus, the rest are pedestrians and occupants of other cars involved in the accidents with the buses. Each bus travels an average of 9500 miles/year. Determine:

(a) Frequency of fire-related fatalities for both occupants and nonoccupants of the school buses.
(b) Fire-related fatality for occupants and nonoccupants per unit of distance traveled (e.g., per 100 million miles traveled).
(c) Mean length of operation time per bus so that the total fire risk reaches 10^{-6} per person.

REFERENCES

1. Kaplan, S. and Garrick, B.J., On the quantitative definition of risk, *Risk Analysis*, 1(1), 11, 1981.
2. Farmer, F.R., Reactor safety and siting: a proposed risk criterion, *Nuclear Safety*, 8, 539, 1967.
3. Glickman, T.S. and Gough, M., Eds., *Resources for the Future*, Reading in Risk, Washington DC, 1990.
4. Dasgupta, A. and Pecht, M., Material failure mechanisms and damage models, *IEEE Transactions on Reliability*, 40(5), 531, 1991.
5. National Research Council, *Risk Assessment in the Federal Government: Managing the Process*, National Academy Press, Washington DC, 1983. This guideline has recently been augmented by a new National Research Council study, *Science and Judgment in Risk Assessment*, National Academy Press, Washington DC, 1994.
6. Chamberlain, S. and Modarres, M., Risk analysis of compressed natural gas school buses, *Risk Analysis*, 25(2), 2005.
7. Keeney, R., Kulkarni, R., and Nair, K., Assessing the risk of an LNG Terminal, *Technology Review*, 81(1), 1978.

3 Probabilistic Risk Assessment

3.1 INTRODUCTION

Probabilistic risk assessment (PRA) is a systematic procedure for investigating how complex systems are built and operated. The PRAs model how human, software, and hardware elements of the system interact with each other. Also, they assess the most significant contributors to the risks of the system. The PRA procedure involves quantitative application of the triplets discussed by (2.1), in that probabilities (or frequencies) of scenarios of events leading to exposure of hazards are estimated, and the corresponding magnitude of health, safety, environmental, and economic consequences for each scenario are predicted. The risk value (i.e., expected loss) of each scenario is often measured as the product of the scenario frequency and its consequences. The main result of the PRA is not the actual value of the risk computed (the so-called bottom-line number); rather it is the determination of the system elements that substantially contribute to the risks of that system, uncertainties associated with such estimates, and effectiveness of various risk reduction strategies available. That is, the primary value of a PRA is to highlight the system design and operational deficiencies and optimize resources that can be invested on improving the design and operation of the system.

In this chapter, a review of the steps involved and the description of methods used in each step have been presented. In Section 3.2, discusses 11 steps in performing a PRA, Section 3.4 will explain the techniques and examples for three key steps: scenario development, logic modeling, and quantification and integration. Methods and techniques used in other steps will be discussed in Chapters 4–6.

3.2 STEPS IN CONDUCTING A PRA

The following subsections provide a discussion of essential components of PRA as we walk our way through the steps that must be performed. We will also describe the methods that are useful for PRA analysis. The *NASA PRA Guide* [1] describes the components of the PRA a modified version of which is shown in Figure 3.1. Each component of PRA will be discussed in detail more in the following:

3.2.1 OBJECTIVES AND METHODOLOGY DEFINITION

Preparing for a PRA begins with a review of the objectives of the analysis. Among the many objectives possible, the most common ones include design improvement, risk acceptability, decision support, regulatory and oversight support, and operations and life management. Once the objective is clarified, an inventory of possible techniques for the desired analyses should be developed. The available techniques range from required computer codes to system experts and analytical experts.

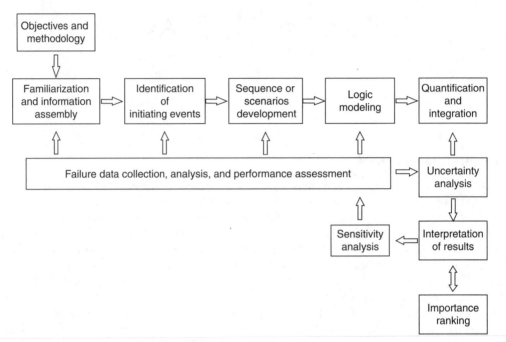

FIGURE 3.1 Components of the overall PRA process [1].

This, in essence, provides a road map for the analysis. The methods described in the following chapters (especially for analyzing the system logic, data, human interactions, software errors, risk ranking, and uncertainty considerations) should be selected and studied.

The resources required for each analytical method should be evaluated, and the most effective option selected. The basis for the selection should be documented, and the selection process reviewed to ensure that the objectives of the analysis will be adequately met. In this book, there will be no further elaboration on this step.

3.2.2 FAMILIARIZATION AND INFORMATION ASSEMBLY

A general knowledge of the physical layout of the overall system (e.g., facility, design, process, aircraft, or spacecraft), administrative controls, maintenance and test procedures, as well as barriers and subsystems, whose job is to protect, prevent, or mitigate hazard exposure conditions, is necessary to begin the PRA. All subsystems, structures, locations, and activities expected to play a role in the initiation, propagation, or arrest of a hazard exposure condition must be understood in sufficient detail to construct the models necessary to capture all possible scenarios. A detailed inspection of the overall system must be performed in the areas expected to be of interest and importance to the analysis.

The following items should be performed in this step:

1. Major critical barriers, structures, emergency safety systems, and human interventions should be identified.
2. Physical interactions among all major subsystems (or parts of the system) should be identified and explicitly described. The result should be summarized in a dependency matrix.

3. Past major failures and abnormal events that have been observed in the system or facility should be noted and studied. Such information would help ensure inclusion of important applicable scenarios.
4. Consistent documentation is key to ensuring the quality of the PRA. Therefore, a good filing system must be created at the outset, and maintained throughout the study.

With the help of the designers, operators, and owners, the analysts should determine the ground rules for the analysis, the scope of the analysis, and the configuration and phases of the operation of the overall system to be analyzed. One should also determine the faults and conditions to be included or excluded, the operating modes of concern, the freeze date of the system or facility, and the hardware configuration on the system freeze date. The freeze date is an arbitrary date after which no additional changes in the overall system design and configuration will be modeled. Therefore, the results of the PRA are only applicable to the overall system at the freeze date. In this book, there will be no further elaboration on this step.

3.2.3   IDENTIFICATION OF INITIATING EVENTS

This task involves identifying those events (abnormal events or conditions) that could, if not correctly and timely responded to, result in hazard exposure. The first step involves identifying sources of hazard and barriers around these hazards. The next step involves identifying events that can lead to a direct threat to the integrity of the barriers.

A system may have one or more operational modes which produce its output. In each operational mode, specific functions are performed. Each function is directly realized by one or more systems by making certain actions and behaviors. These systems, in turn, are composed of more basic units (e.g., subsystems, components, hardware) that accomplish the objective of the system. As long as a system is operating within its design parameter tolerances, there is little chance of challenging the system boundaries in such a way that hazards will escape those boundaries. These operational modes are called normal operation modes.

During normal operation mode, loss of certain functions or systems will cause the process to enter an off-normal (transient) state transition. Once in this transition, there are two possibilities. First, the state of the system could be such that no other function is required to maintain the process or overall system in a safe condition (safe refers to a mode where the chance of exposing hazards beyond the system boundaries is negligible). The second possibility is a state wherein other functions (and thus systems) are required to prevent exposing hazards beyond the system boundaries. For this second possibility, the loss of the function or the system is considered as an initiating event. Since such an event is related to the normally operating equipment, it is called an operational initiating event.

Operational initiating events can also apply to various modes of the system (if it exists). The terminology remains the same since, for each mode, certain equipment, people, or software must be functioning. For example, an operational initiating event found during the PRA of a test nuclear reactor was low primary coolant system flow. Flow is required to transfer heat produced in the reactor to heat exchangers and ultimately to the cooling towers and the outside environment. If this coolant flow function is reduced to the point at which an insufficient amount of heat is transferred, core damage could result (thus the possibility of exposing radioactive materials — the main source of hazard in this case). Therefore, another system must operate to remove the heat produced by the reactor (i.e., a protective barrier). By definition, then, low primary coolant system flow is an operational initiating event.

One method for determining the operational initiating events begins with first drawing a functional block diagram of the system. From the functional block diagram, a hierarchical

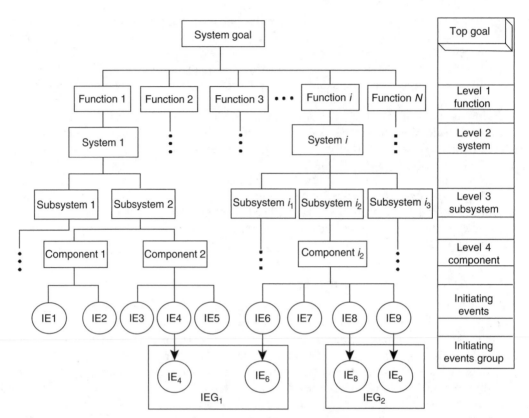

FIGURE 3.2 Function system subsystem initiating event relationship.

relationship is produced, with the process objective being successful completion of the desired system. Each function can then be decomposed into its subsystems and components, and can be combined in a logical manner to represent operations needed for the success of that function (Figure 3.2 adopted from Ref. [1], illustrates this hierarchical decomposition).

Potential initiating events are events that result in failures of particular functions, subsystems, or components, the occurrence of which causes the overall system to fail. These potential initiating events are "grouped" such that members of a group require similar subsystem responses to cope with the initiating event. These groupings are the operational initiator categories. An alternative, but very similar functional decomposition approach for identifying the initiating events has been shown in Figure 3.3. This figure shows a partial functional decomposition model for identifying initiating events in a nuclear plant.

An alternative to the use of functional hierarchy for identifying initiating events is the use of failure mode and event analysis (FMEA) (see Stamatis [2]). The difference between these two methods is noticeable; namely, the functional hierarchies are deductively and systematically constructed, whereas FMEA is an inductive and experiential technique. The use of FMEA for identifying initiating events consists of identifying failure events (modes of failures of equipment, software, and human) whose effect is a threat to the integrity and availability of the hazard barriers of the system. In both of the above methods, one can always supplement the set of initiating events with generic initiating events (if they are known). For example, see Sattison et al. [3] for these initiating events for nuclear reactors, and *NASA Guide* [1] for space vehicles.

To simplify the process, after identifying all initiating events, it is necessary to combine those initiating events that pose the same threat to hazard barriers and require the same

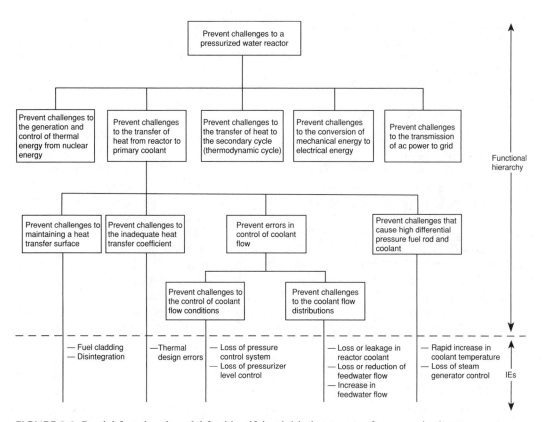

FIGURE 3.3 Partial functional model for identifying initiating events of a pressurized water reactor.

mitigating functions of the process to prevent hazard exposure. The following inductive procedures should be followed when grouping initiating events:

1. Combine the initiating events that directly break all hazard barriers.
2. Combine the initiating events that break the same hazard barriers (not necessarily all the barriers).
3. Combine the initiating events that require the same group of mitigating human or automatic control actions following their occurrence.
4. Combine the initiating events that simultaneously disable the normal operation as well as some of the available mitigating human, software, or automatic actions.

Events that cause off-normal operation of the overall system and require other systems to operate so as to maintain hazards within their desired boundaries, but are not directly related to a hazard mitigation, protection, or prevention function, are nonoperational initiating events. Nonoperational initiating events are identified with the same methods used to identify operational events. One class of such events of interest are those that are primarily external to the overall system or facility. These so-called "external events" will be discussed later in more detail in this section. In this book, there will be no further elaboration about this step in PRA.

The following procedures should be followed in this step of the PRA:

1. Select a method for identifying specific operational and nonoperational initiating events. Two representative methods are functional hierarchy and FMEA. If a generic list of initiating events is available, it can be used as a supplement.

2. Using the method selected, identify a set of initiating events.
3. Group the initiating events having the same effect on the system, for example, those requiring the same mitigating functions to prevent hazard exposure are grouped together.

3.2.4 SEQUENCE OR SCENARIO DEVELOPMENT

The goal of scenario development is to derive a complete set of scenarios that encompasses all of the potential exposure propagation paths that can lead to loss of containment or confinement of the hazards, following the occurrence of an initiating event. To describe the cause and effect relationship between initiating events and subsequent event progression, it is necessary to identify those functions (e.g., safety functions) that must be maintained, activated or terminated to prevent loss of hazard barriers. The scenarios that describe the functional response of the overall system or process to the initiating events are frequently displayed by event trees.

Event trees order and depict (in an approximately chronological manner) the success or failure of key mitigating actions (e.g., human actions or mitigative hardware actions) that are required to act in response to an initiating event. In PRA, two types of event trees can be developed: functional and systemic. The functional event tree uses operation or loss of mitigating functions as its events. The main purpose of the functional tree is to better understand the scenario of events at an abstract level, following the occurrence of an initiating event. The functional tree also guides the PRA analyst in the development of a more detailed systemic event tree. The systemic event tree reflects the scenarios of specific events (specific human actions, protective or mitigative subsystem, operations or failures) that lead to a hazard exposure. That is, the functional event tree can be further decomposed to show failure of specific hardware, software, or human actions that perform the functions described in the functional event tree. Therefore, a systemic event tree fully delineates the overall system response to an initiating event and serves as the main tool for further analyses in the PRA. In Section 3.4 of this chapter, specific tools and techniques used for this purpose will be discussed in more detail.

There are two kinds of external events. First kind refers to events that originate from the facility or the overall system (but outside of the physical boundary of the facility), which are called *internal events*, external to the process of the system. Events that adversely affect the facility or overall system and occur external to its physical boundaries, but can still be considered as part of the system, are defined as internal events, external to the system. Typical internal events external to the system are internal conditions such as fire from fuel stored within a facility or floods occurring due to rupture of tank which is part of the overall system. The effects of these events should be modeled with event trees to show all possible scenarios.

The second kind of external events are those that originate outside of the overall system, and are called *external events*. Examples of external events are fires and floods that originate from outside of the system. Examples include seismic events, extreme heat, extreme drought, transportation events, volcanic events, high wind events, terrorism, and sabotage. Again, this classification can be used in developing and grouping the event tree scenarios.

The following procedures should be followed in this step of the PRA:

1. Identify the mitigating functions for each initiating event (or group of events).
2. Identify the corresponding human actions, systems, or hardware operations associated with each function, along with their necessary conditions for success.
3. Develop a functional event tree for each initiating event (or group of events).
4. Develop a systemic event tree for each initiating event, delineating the success conditions, initiating event progression phenomena, and end effect of each scenario.

3.2.5 LOGIC MODELING

Event trees commonly involve branch points which shows if a given subsystem (or event) either works (or happens) or does not work (or does not happen). Sometimes, failure of these subsystems (or events) is rare and there may not be an adequate record of observed failure events to provide a historical basis for estimating frequency of their failure. In such cases, other logic-based analysis methods such as fault trees or master logic diagrams (MLDs) may be used, depending on the accuracy desired. The most common method used in PRA to calculate the probability of subsystem failure is fault tree analysis. This analysis involves developing a logic model in which the subsystem is broken down into its basic components or segments for which adequate data exist. In Section 3.4, we shall discuss how a fault tree can be developed to represent the event headings of an event tree.

Different event tree modeling approaches imply variations in the complexity of the logic models that may be required. If only main functions or systems are included as event tree headings, the fault trees become more complex and must accommodate all dependencies among the main and support functions (or subsystems) within the fault tree. If support functions (or systems) are explicitly included as event tree headings, more complex event trees and less complex fault trees will result. Section 3.4 presents methods and techniques used for logic modeling along with examples.

The following procedures should be followed as a part of developing the fault tree:

1. Develop a fault tree for each event in the event tree heading for which actual historical failure data do not exist.
2. Explicitly model dependencies of a subsystem on other subsystems and intercomponent dependencies (e.g., common cause failures (CCFs) that are described in Chapter 4).
3. Include all potentially reasonable and probabilistically quantifiable causes of failure, such as hardware, software, test and maintenance, and human errors, in the fault tree.

The following steps should be followed in the dependent failure analysis:

1. Identify the hardware, software, and human elements that are similar and could cause dependent or CCFs. For example, similar pumps, motor-operated valves, air-operated valves, human actions, software routine, diesel generators, and batteries are major components in process plants, and are considered important sources of CCFs.
2. Items that are potentially susceptible to CCF should be explicitly incorporated into the corresponding fault trees and event trees of the PRA where applicable.
3. Functional dependencies should be identified and explicitly modeled in the fault trees and event trees.

3.2.6 FAILURE DATA COLLECTION, ANALYSIS, AND PERFORMANCE ASSESSMENT

A critical building block in assessing the reliability and availability of complex systems is the data on the performance of its barriers to contain hazards. In particular, the best resources for predicting future availability are past field experiences and tests. Hardware, software, and human reliability data are inputs to assess performance of hazard barriers, and the validity of the results depends highly on the quality of the input information. It must be recognized, however, that historical data have predictive value only to the extent that the conditions under which the data were generated remain applicable. Collection of the various failure data consists fundamentally of the following steps: collecting and assessing generic data, statistically evaluating facility- or overall system-specific data, and developing failure probability distributions using test or facility- and system-specific data. The three types of events identified during the risk scenario definition and system modeling must be quantified for the event trees and fault trees to estimate the frequency of occurrence of sequences: initiating events, component failures, and human error.

The quantification of initiating events and hazard barriers and components failure probabilities involve two separate activities. First, the probabilistic failure model for each barrier or component failure event must be established; then the parameters of the model must be estimated. Typically the necessary data include time of failures, repair times, test frequencies, test downtimes, CCF events. Further uncertainties associated with such data must also be characterized. Chapter 4 discusses available methods for analyzing data to obtain the probability of failure of barriers or the probability or frequency of occurrence of equipment failure. Chapter 5 presents characterization of uncertainties associated with performance data, models, and probabilistic assessments. Establishment of the database to be used will generally involve collection of some facility- or system-specific data combined with the use of generic performance data when specific data are absent or sparse.

To attain the very low levels of risk, the systems and hardware that comprise the barriers to hazard exposure must have very high levels of performance. This high performance is typically achieved through the use of well-designed systems with adequate margin of safety considering uncertainties, redundancy, and diversity in hardware, which provides multiple success paths. The problem then becomes one of ensuring the independence of the paths since there is always some degree of coupling between agents of failures such as those activated by failure mechanisms, either through the operating environment (events external to the system) or through functional and spatial dependencies. In Chapter 4, we will elaborate on the nature and analysis of these dependencies. Treatment of dependencies should be carefully included in both event tree and fault tree development in the PRA. As the reliability of individual subsystems increases due to redundancy, the contribution from dependent failures becomes more important; in certain cases, dependent failures may dominate the value of overall reliability. Including the effects of dependent failures in the reliability models used in the PRA is a difficult process and requires development of some sophisticated, fully integrated models be developed and used to account for unique failure combinations that lead to failure of subsystems and ultimately exposure of hazards. The treatment of dependent failures is not a single step performed during the PRA; it must be considered throughout the analysis (e.g., in event trees, fault trees, and human reliability analyses).

The following procedures should be followed as part of the data analysis task:

1. Determine generic values of material strength or endurance, load or damage agents, failure times, failure occurrence rate, and failures on demand for each item (hardware, human action, or software) identified in the PRA models. This can be obtained either from facility- or system-specific experiences, from generic sources of data or both (see Chapter 4 for more detail on this subject).
2. Gather data on hazard barrier tests, repair, and maintenance data primarily from experience, if available. Otherwise use generic performance data.
3. Assess the frequency of initiating events and other probability of failure events from experience, expert judgment, or generic sources (see Chapter 4).
4. Determine the dependent or CCF probability for similar items, primarily from generic values. However, when significant specific data are available, they should be primarily used (see Chapter 4).

3.2.7 QUANTIFICATION AND INTEGRATION

Fault trees and event trees are integrated and their events are quantified to determine the frequencies of scenarios and associated uncertainties in the calculation of the final risk values.

The approach for integration is discussed in Section 3.4. The approach depends somewhat on the manner in which system dependencies have been handled. We will describe the more complex situation in which the fault trees are dependent, i.e., there are physical dependencies (e.g., through support units of the main hazard barriers such as those providing motive, proper working environment, and control functions).

Normally, the quantification will use a Boolean reduction process to arrive at a Boolean representation for each scenario. Starting with fault tree models for the various systems or event headings in the event trees, and using probabilistic estimates for each of the events modeled in the event trees and fault trees, the probability of each event tree heading (often representing failure of a hazard barrier) is calculated (if the heading is independent of other headings). The fault trees for the main subsystems, support units (e.g., lubricating and cooling units, power units) are merged where needed, and the equivalent Boolean expression representing each event in the event tree model is calculated. The Boolean expressions are reduced to arrive at the smallest combination of basic failures events (the so-called *minimal cut sets*) that lead to exposure of the hazards. These minimal cut sets for each of the main subsystems (barriers) — that are often identified as event headings on the event trees — are also obtained. The minimal cut sets for the event tree event headings are then appropriately combined to determine the cut sets for the event tree scenarios. This process is further described in Section 3.4.

If possible, all minimal cut sets must be generated and retained during this process; unfortunately in complex systems and facilities this leads to an unmanageably large collection of terms and a combinatorial outburst. Therefore, the collection of cut sets are often truncated (i.e., probabilistically small and insignificant cut sets are discarded based on the number of terms in a cut set or on the probability of the cut set). This is usually a practical necessity because of the overwhelming number of cut sets that can result from the combination of a large number of failures, even though the probability of any of these combinations may be vanishingly small. The truncation process does not disturb the effort to determine the dominant scenarios since we are discarding scenarios that are extremely unlikely.

A valid concern is sometimes voiced that even though the individual cut sets discarded may be at least several orders of magnitude, less probable (or frequent) than the average of those retained, the large number of them discarded may sum to a significant part of the risk. The actual risk might thus be larger than that indicated by the PRA results. This can be discussed as part of the modeling and computational uncertainty characterization. Detailed examination of a few PRA studies of very complex systems or facilities, for example nuclear power plants shows that cut set truncation will not introduce any significant error in the total risk assessment results. The process of quantification is generally straightforward, and the methods used are described in Section 3.4. More objective truncation methods are discussed by Modarres and Dezfuli [4].

Other methods for evaluating scenarios also exist that directly estimate the frequency of scenario without specifying cut sets. This is often done in highly dynamic systems whose configuration changes as a function of time leading to dynamic event tree and fault trees. For more discussion on these systems, see Chang et al. [5], *NASA Procedures PRA Guide* [1], and Dugan et al. [6].

The following procedures should be followed as part of the quantification and integration step in the PRA:

1. Merge corresponding fault trees associated with each failure or success event modeled in the event tree scenarios (i.e., combine them in a Boolean form). Develop a reduced Boolean function for each scenario (i.e., truncated minimal cut sets).

2. Calculate the total frequency of each sequence, using the frequency of initiating events, the probability of barrier failure including contributions from test and maintenance frequency (outage), CCF probability, and human error probability.
3. Use the minimal cut sets of each sequence for the quantification process. If needed, simplify the process by truncating based on the cut sets or probability.
4. Calculate the total frequency of each scenario.
5. Calculate the total frequency of all scenarios of all event trees.

3.2.8 UNCERTAINTY ANALYSIS

Uncertainties are part of any assessment, modeling, and estimation. In engineering calculations, we routinely ignored the estimation of uncertainties associated with failure models and parameters, because either the uncertainties are very small, but more often because analyses are done conservatively (e.g., by using high safety factor, design margin). Since PRAs are primarily used for decision making and management of risk, it is critical to incorporate uncertainties in all facets of the PRA.

Also, risk management decisions that consider PRA results must consider estimated uncertainties. In PRAs, uncertainties are primarily shown in the form of probability distributions. For example, the probability of failure of a subsystem (e.g., a hazard barrier) may be represented by a probability distribution showing the range and likelihood of risk values.

Because the uncertainty analysis is an important part of the PRA, the subject is discussed in far more detail in Chapter 5. The process involves characterization of the uncertainties associated with frequency of initiating events, probability of failure of subsystems (or barriers), probability of all event tree headings, strength or endurance of barriers, applied load or incurred damage by the barriers, amount of hazard exposures, consequences of exposures to hazards, and sustained total amount of losses. Other sources of uncertainties are in the models used. For example, the fault tree and event tree models, stress–strength and damage–endurance models used to estimate failure or capability of some barriers, probabilistic failure models of hardware, software, and humans, correlation between amount of hazard exposure and the consequence, exposure models and pathways, and models to treat inter- and intrabarrier failure dependencies. Another important source of uncertainty is the incompleteness of the risk models and other failure models used in the PRAs. For example, the level of detail used in decomposing subsystems using fault tree models, scope of the PRA, and lack of consideration of certain scenarios in the event tree just because they do not know or have experienced before.

Once uncertainties associated with hazard barriers have been estimated and assigned to models and parameters, they must be "propagated" through the PRA model to find the uncertainties associated with the results of the PRA; primarily with the bottomline risk calculations, and with the list of risk-significant elements of the system. Propagation is done using one of the several techniques, but the most popular method used is Monte Carlo simulation. The results are then shown and plotted in form of probability distributions.

Steps in uncertainty analysis include:

1. Identify models and parameters that are uncertain and the method of uncertainty estimation to be used for each.
2. Describe the scope of the PRA and significance and contribution of elements that are not modeled or considered.
3. Estimate and assign probability distributions depicting model and parameter uncertainties in the PRA.

4. Propagate uncertainties associated with the hazard barrier models and parameters to find the uncertainty associated with the risk value.
5. Present the uncertainties associated with risks and contributors to risk in an easy way to understand and visually straightforward to grasp.

3.2.9 SENSITIVITY ANALYSIS

Sensitivity analysis is the method of determining the significance of choice of a model or its parameters, assumptions for including or not including a barrier, phenomena or hazard, performance of specific barriers, intensity of hazards, and significance of any highly uncertain input parameter or variable to the final risk value calculated. The process of sensitivity analysis is straightforward. The effects of the input variables and assumptions in the PRA are measured by modifying them by several folds, factors or even one or more order of magnitudes one at a time, and measure relative changes observed in the PRA's risk results. Those models, variables, and assumptions whose change leads to the highest change in the final risk values are determined as "sensitive." In such a case, revised assumptions, models, additional failure data, and more mechanisms of failure may be needed to reduce the uncertainties associated with sensitive elements of the PRA.

Sensitivity analysis helps to focus resources and attentions to those elements of the PRA that need better attention and characterization. A good sensitivity analysis strengthens the quality and validity of the PRA results. Usually elements of the PRA that could exhibit multiple impacts on the final results, such as certain phenomena (e.g., pitting corrosion, fatigue cracking, and CCF) and uncertain assumptions are usually good candidates for sensitivity analysis. The methods and important aspects of sensitivity analysis have been discussed in more detail in Chapter 6. The steps involved in the sensitivity analysis are:

1. Identify the elements of the PRA (including assumptions, failure probabilities, models, and parameters) that analysts believe might be sensitive to the final risk results.
2. Change the contribution or value of each sensitive item in either direction by several factors in the range of 2–100. Note that certain changes in the assumptions may require multiple changes of the input variables. For example, a change in failure rate of similar equipments requires changing of the failure rates of all these equipments in the PRA model.
3. Calculate the impact of the changes in Step 2 one-at-a-time and list the elements that are most sensitive.
4. Based on the results in Step 3, propose additional data, any changes in the assumptions, use of alternative models, modification of the scope of the PRA analysis.

3.2.10 RISK RANKING AND IMPORTANCE ANALYSIS

Ranking the elements of the system with respect to their risk or safety significance is one of the most important outcomes of a PRA. Ranking is simply arranging the elements of the system based on their increasing or decreasing contribution to the final risk values. Importance measures estimate hazard barrier, subsystems, or more basic elements of them usually based on their contribution to the total risk of the system. The ranking process, while very important, should be done with much care. In particular, during the interpretation of the results, since formal importance measures are context dependent and their meaning varies depending on the intended application of the risk results, the choice of the ranking method is important.

There are several unique importance measures in PRAs. For example, Fussell–Vesely, risk reduction worth (RRW), and risk achievement worth (RAW) are identified as appropriate measures for use in PRAs, and all are representative of the level of contribution of various elements of the system as modeled in the PRA and enter in the calculation of the total risk of the system. For example, Birnbaum importance measure represents changes in total risk of the system as a function of changes in the basic event probability of one component at a time. If simultaneous changes in the basic event probabilities are being considered, a more complex representation would be needed. The subject of importance measures as used in the PRAs is discussed in more detail in Chapter 6.

Importance measures can be classified based on their mathematical definitions. Some measures have fractional type definitions and show changes (in number of folds or factors) in system total risk under certain conditions with respect to the normal operating or use condition (e.g., as is the case in RRW and RAW measures). Some measures calculate the changes in system total risk as failure probability of hazard barriers and other elements of the system or conditions under which the system operates change. This difference can be normalized with respect to the total risk of the system or even expressed in a percentage change form. There are other types of measures, which account for the rate of changes in the system risk with respect to changes in the failure probability of the elements of the system. These measures can be interpreted mathematically as partial derivative of the risk as a function of failure probability of its elements (barriers, components, human actions, phenomena, etc.). For example, the Birnbaum measure falls under this category.

Another important set of importance measures focus on ranking the elements of the system or facility with the most contribution to the total uncertainty in the risk results obtained from PRAs. This process is called "uncertainty ranking" and is different from component, subsystem, and barrier ranking. In this importance ranking, the analyst is only interested to know which of the system elements drive the final risk uncertainties, so that resources can be focused on reducing important uncertainties.

There is another classification of importance measures in which they can be divided into two major categories: absolute and relative. Absolute measures are representative of fixed importance of one element of the system, independent of the importance of other elements, while relative importance expresses significance of one element with respect to weight of importance of other elements. Absolute importance can be used to estimate the impact of component performance on the system regardless of how important other elements are, while relative importance estimates the significance of the risk-impact of the component in comparison to the effect or contribution of others.

Absolute measures are useful when we speculate on improving systems, since they directly show the impact on the total risk of the system. Relative measures are preferred when resources or actions to improve or prevent failures are taken in a global and distributed manner.

Applications of importance measures may be categorized into the following areas:

1. (Re)Design: To support decisions of the system design or redesign by adding or removing elements (barriers, subsystems, human interactions, etc.).
2. Test and maintenance: To address questions related to the plant performance by changing the test and maintenance strategy for a given design.
3. Configuration and control: To measure the significance or the effect of failure of a component on risk or safety or temporarily taking a component out of service.
4. Reduce uncertainties in the input variables of the PRAs.

The following are the major steps of importance ranking:

1. Determine the purpose of the ranking and select appropriate ranking importance measure that has consistent interpretation for the use of the ranked results.
2. Perform risk ranking and uncertainty ranking, as needed.
3. Identify the most critical and important elements of the system with respect to the total risk values and total uncertainty associated with the calculated risk values.

3.2.11 INTERPRETATION OF RESULTS

When the risk values are calculated, they must be interpreted to determine whether any revisions are necessary to refine the results and the conclusions. There are two main elements involved in the interpretation process. The first is to understand whether or not the final values and details of the scenarios are logically and quantitatively meaningful. This step verifies the adequacy of the PRA model and the scope of analysis. The second is to characterize the role of each element of the system in the final results. This step highlights additional analyses data and information gathering that would be considered necessary.

The interpretation process heavily relies on the details of the analysis to see whether the scenarios are logically meaningful (for example, by examining the minimal cut sets of the scenarios), whether certain assumptions are significant and greatly control the risk results (using the sensitivity analysis results), and whether the absolute risk values are consistent with any historical data or expert opinion available. Based on the results of the interpretation, the details of the PRA logic, its assumptions, and scope may be modified to update the results into more realistic and dependable values.

The ranking and sensitivity analysis results may also be used to identify areas where gathering more information and performing better analysis (for example, by using more accurate models) is warranted. The primary aim of the process is to reduce uncertainties in the risk results.

The interpretation step is a continuous process with receiving information from the quantification, sensitivity, uncertainty, and importance analysis activities of the PRA. The process continues until the final results can be best interpreted and used in the subsequent risk management steps.

The basic steps of the PRA results interpretation are:

1. Determine accuracy of the logic models and scenario structures, assumptions, and scope of the PRA.
2. Identify system elements for which better information would be needed to reduce uncertainties in failure probabilities and models used to calculate performance.
3. Revise the PRA and reinterpret the results until attaining stable and accurate results.

3.3 STRENGTH OF PRA

The most important strengths of the PRA, as the formal engineering approach to risk assessment are:

- PRA provides an integrated and systematic examination of a broad set of design and operational features of a complex system.
- PRA incorporates the influence of system interactions and human–system interfaces.
- PRA provides a model for incorporating operating experience with the complex system and updating risk estimates.
- PRA provides a process for the explicit consideration of uncertainties.

- PRA permits the analysis of competing risks (e.g., of one system versus another or of possible modifications to an existing system).
- PRA permits the analysis of (assumptions, data) issues via sensitivity studies.
- PRA provides a measure of the absolute or relative importance of systems, components to the calculated risk value.
- PRA provides a quantitative measure of overall level of health and safety for the engineered system.

Major errors may result from weak models, or associated data of potentially important factors in the risk of the system including cases where

- Initiating events with very low frequencies of occurrence
- Human performance models and interactions with the system are highly uncertain
- Failures occurring from a CCF such as an extreme operating environment are difficult to identify and model.

Figure 3.4 shows a graphical depiction of the relationship between the PRA steps.

3.4 SCENARIO AND LOGIC MODELING, DEVELOPMENT, AND QUANTIFICATION

In this section, more detailed aspects of scenario and logic model development and evaluation such as fault tree, event tree, and MLD analysis have been discussed.

3.4.1 FAULT TREE METHOD

The operation of a system (or a hazard barrier) can be considered from two opposite viewpoints: the various ways in which a system fails (fault tree analysis), or the various ways in which a system succeeds (success tree analysis). Most of the construction and analysis methods used in the literature are, in principle, the same for both fault trees and success trees. PRAs primarily use fault trees for modeling subsystem and barrier failure probabilities. Therefore, we focus on the fault tree method and highlight how the results of fault tree analysis can be modified and used for success tree analysis results.

The fault tree development is a deductive process by means of which an undesirable event, called the top event, is postulated, and the possible ways for this event to occur are systematically deduced. For example, a typical top event could be "failure of control circuit A to send a signal when it should." The deduction process is performed so that the fault tree considers all component failures that contribute to the occurrence of the top event. It is also possible to include individual failure modes of each element of the system (or subsystem such as a given barrier) as well as human and software errors (and the interactions between the two) during the system operation. The fault tree itself is a graphical representation of the various combinations of failures that lead to the occurrence of the top event.

A fault tree does not necessarily contain all possible failure modes of the components (or more basic units) of the system. Only those failure modes which contribute to the existence of the top event are modeled. For example, consider a failed-safe control circuit. If loss of the dc power to the circuit causes the circuit to open a contact, which in turn sends a signal to another system for operation, a top event of "control circuit fails to generate a safety signal" would not include the "failure of dc power source" as one of its events, even though the dc power source (e.g., batteries) is part of the control circuit. This is because the top event would not occur due to the loss of the dc power source.

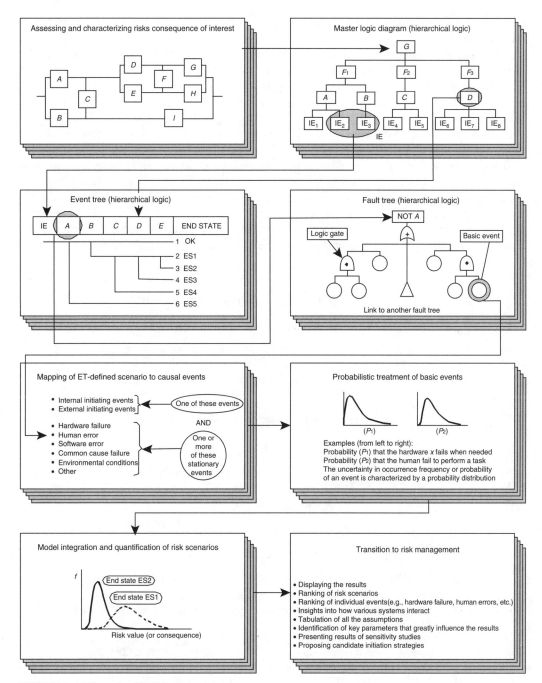

FIGURE 3.4 The overall PRA process.

The postulated fault events that appear on the fault tree structure may not be exhaustive. Only those events considered important can be included. However, it should be noted that the decision for inclusion of failure events is not arbitrary; it is influenced by the fault tree construction procedure, system design and operation, operating history, available failure data, and experience of the analysts. At each intermediate point, the postulated events represent the immediate, necessary, and sufficient causes for the occurrence of the intermediate (or top) events.

The fault tree itself is a logical model, and thus represents the qualitative characterization of the system logic. There are, however, many quantitative algorithms to evaluate fault trees. For example, the concept of cut sets can be applied to fault trees by using the Boolean algebra method. Cut sets are combination of system elements modeled in the fault tree that cause system (e.g., the hazard barrier) failure. By using $\Pr(C_1 \cup C_2 \cup \cdots \cup C_m)$, where C_1, C_2, ..., C_m are the minimal cut sets of the fault tree, the probability of occurrence of the top event can be determined.

First let us understand the symbols used to build the logic trees, including fault trees. In essence, there are three types of symbols: events, gates, and transfers. Basic events, undeveloped events, condition events, and external events are sometimes referred to as primary events. When postulating events in the fault tree, it is important to include not only the undesired component states (e.g., applicable failure modes), but also time when they occur. Figure 3.5 shows the typical symbols used in the fault tree.

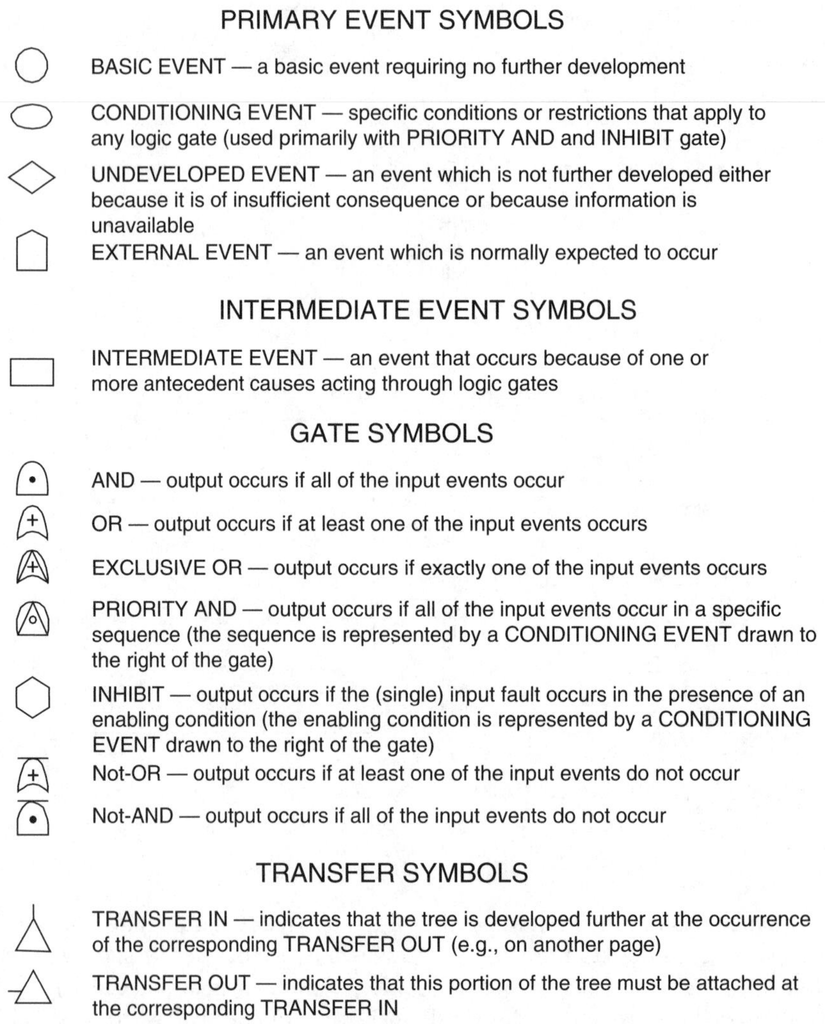

PRIMARY EVENT SYMBOLS

BASIC EVENT — a basic event requiring no further development

CONDITIONING EVENT — specific conditions or restrictions that apply to any logic gate (used primarily with PRIORITY AND and INHIBIT gate)

UNDEVELOPED EVENT — an event which is not further developed either because it is of insufficient consequence or because information is unavailable

EXTERNAL EVENT — an event which is normally expected to occur

INTERMEDIATE EVENT SYMBOLS

INTERMEDIATE EVENT — an event that occurs because of one or more antecedent causes acting through logic gates

GATE SYMBOLS

AND — output occurs if all of the input events occur

OR — output occurs if at least one of the input events occurs

EXCLUSIVE OR — output occurs if exactly one of the input events occurs

PRIORITY AND — output occurs if all of the input events occur in a specific sequence (the sequence is represented by a CONDITIONING EVENT drawn to the right of the gate)

INHIBIT — output occurs if the (single) input fault occurs in the presence of an enabling condition (the enabling condition is represented by a CONDITIONING EVENT drawn to the right of the gate)

Not-OR — output occurs if at least one of the input events do not occur

Not-AND — output occurs if all of the input events do not occur

TRANSFER SYMBOLS

TRANSFER IN — indicates that the tree is developed further at the occurrence of the corresponding TRANSFER OUT (e.g., on another page)

TRANSFER OUT — indicates that this portion of the tree must be attached at the corresponding TRANSFER IN

FIGURE 3.5 Primary event, gate, and transfer symbols used in logic trees.

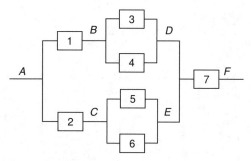

FIGURE 3.6 Block diagram of a simple parallel–series system.

To better understand the fault tree concept, let us consider the block diagram shown in Figure 3.6. Let us also assume that the block diagram models a circuit in which the arrows show the direction of electric current. A top event of "no current at point *F*" is selected, and all events that cause this top event are deductively postulated. Figure 3.7 shows the resulting fault tree.

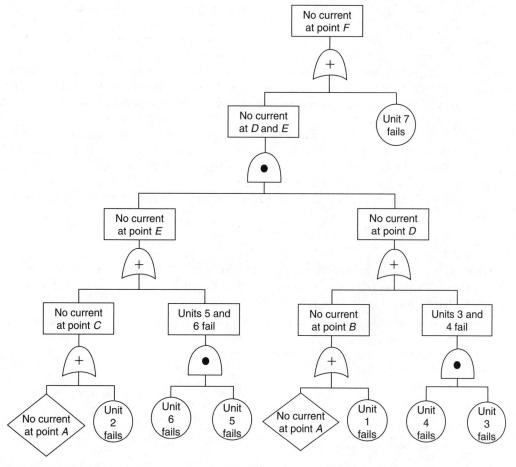

FIGURE 3.7 Fault tree for the complex parallel–series system in Figure 3.6.

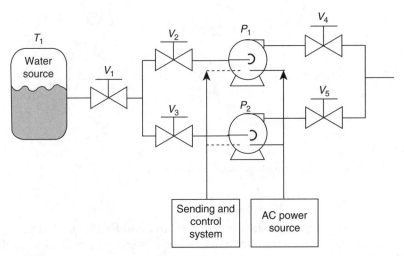

FIGURE 3.8 An example of a pumping system.

As another example, consider the pumping system shown in Figure 3.8. Sufficient water is delivered from the water source T_1 when only one of the two pumps, P_1 or P_2, works. All the valves V_1 through V_5 are normally open. The sensing and control system S senses the demand for the pumping system and automatically starts both P_1 and P_2. If one of the two pumps fails to start or fails during operation, the mission is still considered successful, if the other pump functions properly. The two pumps and the sensing and control system use the same ac power source AC. Assume the water content in T_1 is sufficient and available, there are no human errors, and no failure in the pipe connections is considered important.

It is clear that the system's mission is to deliver sufficient water when needed. Therefore, the top event of the fault tree for this system should be "no water is delivered when needed." Figure 3.9 shows the fault tree for this example. In Figure 3.9, the failures of AC and S are shown with undeveloped events. This is because one can further expand the fault tree if one knows what makes up the failures of AC and S, in which case these events will be intermediate events.

Example 3.1
Consider the simple closed circuit shown in the figure below. When the switch (SW) is turned on, the flow of dc power (B) runs the motor (M). Develop the corresponding fault tree model. Ignore failure of wires and human errors in turning the switch on and off.

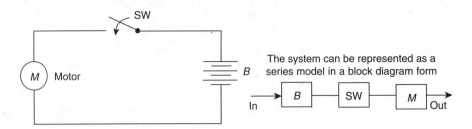

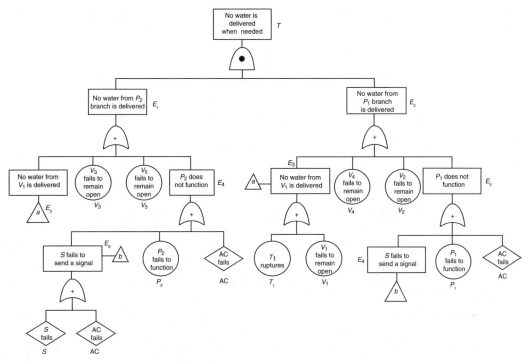

FIGURE 3.9 Fault tree for the pumping system in Figure 3.8.

Solution

Starting from the "output" of the system, deductively follow the source of the failures. The "output" is the mechanical energy received from the motor M. Thus, either the motor itself fails (considered as complex enough to represent unit M failure as being an "undeveloped" event), or no dc current runs to the motor. Since there are multiple causes for the event "no dc current running to the motor," the event should be considered as "intermediate." This intermediate event is further decomposed to only two possibilities of the SW failing (for example, being in open position), or the batteries fail. The fault tree is shown below.

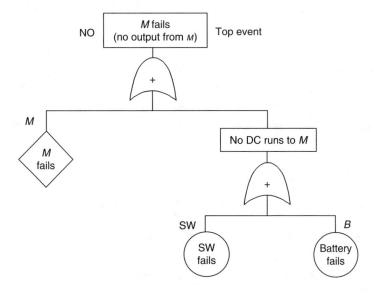

Since $T = M \cup (SW \cup B) = M \cup SW \cup B$, the fault tree can be equally represented in the form below.

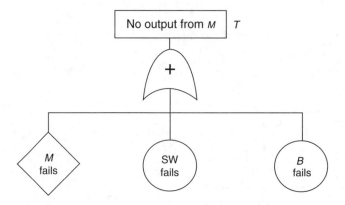

However, because enough information (e.g., failure characteristics and probabilities) about these events is known, we may stop further development at this stage. Although the development of the fault tree in Figure 3.9 is based on a strict deductive procedure (i.e., systematic decomposition of failures starting from "sink" and deductively proceeding toward "source"), one can rearrange it to a simpler equivalent form shown in Figure 3.10. While the development of the fault tree in Figure 3.9 requires only a minimum understanding of the overall functionality and logic of the system, direct development of more compact versions requires a

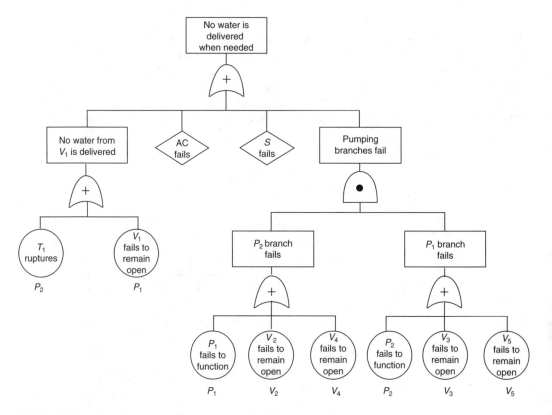

FIGURE 3.10 More compact form of the fault tree in Figure 3.9.

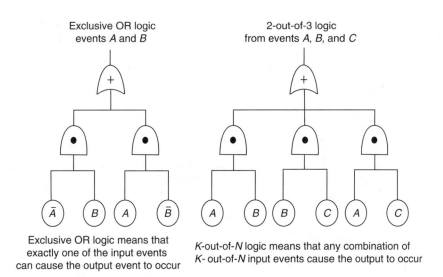

Exclusive OR logic
events A and B

2-out-of-3 logic
from events A, B, and C

Exclusive OR logic means that
exactly one of the input events
can cause the output event to occur

K-out-of-N logic means that any combination of
K- out-of-N input events cause the output to occur

FIGURE 3.11 Exclusive OR and K-out-of-N logics.

much better understanding of the overall system logic. If more complex logical relationships are required, other logical representations can be described by combining the two basic AND and OR gates. For example, the K-out-of-N and exclusive OR logics can be described, as shown in Figure 3.11.

For a more detailed discussion of the construction and evaluation of fault trees, refer to the *NASA Fault Tree Handbook* [7].

3.4.2 EVALUATION OF LOGIC TREES

The evaluation of logic trees (e.g., fault trees, success trees, and MLDs) involves two distinct aspects: logical or qualitative evaluation and probabilistic or quantitative evaluation. Qualitative evaluation involves the determination of the logic tree cut sets, path sets, or logical evaluations to rearrange the tree logic for computational efficiency (similar to the rearrangement presented in Figure 3.10 for a fault tree). Determining the logic tree cut sets or path sets involves some straightforward Boolean manipulation of events. However, there are many types of logical rearrangements and evaluations, such as fault tree modularization that are beyond the scope of this book. The reader is referred to the *NASA Fault Tree Handbook* [7] for a thorough discussion of fault tree analysis. In addition to the traditional Boolean analysis of logic trees, the so-called truth table or combinatorial approach will also be discussed. The latter technique generates mutually exclusive cut or path sets.

3.4.2.1 Boolean Algebra Analysis of Logic Trees

The quantitative evaluation of logic trees involves the determination of the probability of the occurrence of the top event. The qualitative evaluation of logic trees through the generation of cut or path sets is conceptually very simple. The tree OR-gate logic represents the union of the input events. That is, any input events must occur to cause the output event to occur.

For example, an OR gate with input events A and B, and the output event Q can be represented by the equivalent Boolean expression, $Q = A \cup B$. The occurence of either A, B, or both must occur will cause the output event Q to occur. For convenience, instead of the union symbol $\cup$, the equivalent "+" symbol may be used. Thus, $Q = A + B$ is read as "A or B."

Generally, for an OR gate with n inputs, $Q = A_1 + A_2 + \cdots + A_n$. The AND gate can be represented by the intersect logic. Therefore, the Boolean equivalent of an AND gate with two inputs A and B would be $Q = A \cup B$ (or $Q = A \cdot B$).

Figure 3.12 shows the Boolean representation of fault tree gates and associated Venn diagram representation of the gates.

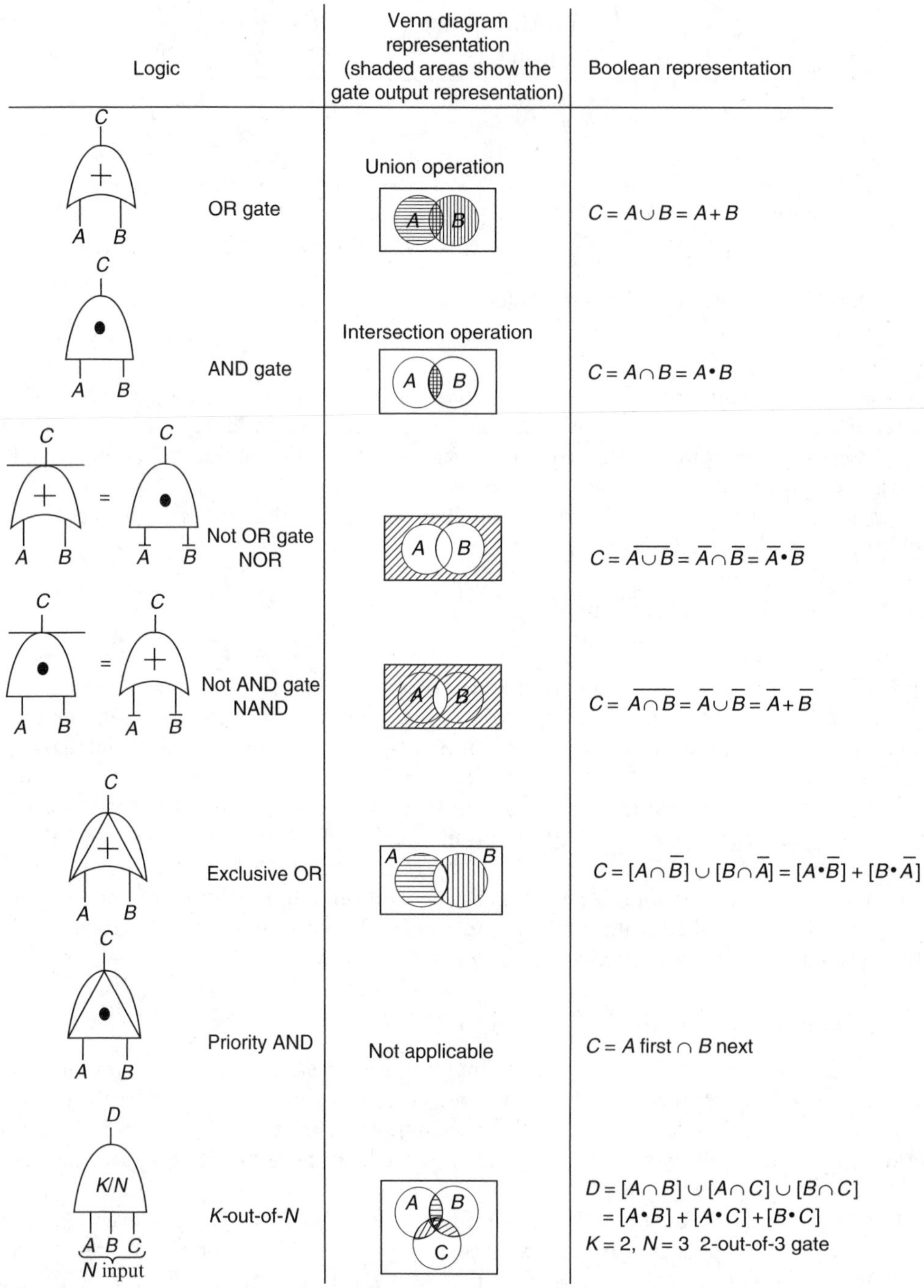

Logic	Venn diagram representation (shaded areas show the gate output representation)	Boolean representation
OR gate	Union operation	$C = A \cup B = A + B$
AND gate	Intersection operation	$C = A \cap B = A \cdot B$
Not OR gate NOR		$C = \overline{A \cup B} = \overline{A} \cap \overline{B} = \overline{A} \cdot \overline{B}$
Not AND gate NAND		$C = \overline{A \cap B} = \overline{A} \cup \overline{B} = \overline{A} + \overline{B}$
Exclusive OR		$C = [A \cap \overline{B}] \cup [B \cap \overline{A}] = [A \cdot \overline{B}] + [B \cdot \overline{A}]$
Priority AND	Not applicable	$C = A \text{ first} \cap B \text{ next}$
K-out-of-N		$D = [A \cap B] \cup [A \cap C] \cup [B \cap C]$ $= [A \cdot B] + [A \cdot C] + [B \cdot C]$ $K = 2, N = 3$ 2-out-of-3 gate

FIGURE 3.12 Boolean representations of important logic gates.

Determination of cut sets using Boolean and other logical expressions is possible through several algorithms. These algorithms include the top–down or bottom–up successive substitution method, the modularization approach, Boolean binary diagrams, truth tables, and Monte Carlo simulation. The *NASA Fault Tree Handbook* [7] describes the underlying principles of most of these qualitative evaluation algorithms. The oldest and most widely used and straightforward algorithm is the successive substitution method. In this approach, the equivalent Boolean representation of each gate in the logic tree is determined such that only primary events remain. Various Boolean algebra rules are applied to reduce the Boolean expression to its most compact form, which represents the minimal path or cut sets of the logic tree. The substitution process can proceed from the top of the tree to the bottom or *vice versa*. Depending on the logic tree and its complexity, the former approach, the latter approach, or a combination of the two, can be used.

As an example, let us consider the fault tree shown in Figure 3.9. Clearly, each node represents a failure. The step-by-step, top–down Boolean substitution (see Appendix A for Boolean operations) of the top event is presented below.

Step 1: $T = E_1 \cdot E_2$.

Step 2: $E_1 = E_3 + V_3 + V_5 + E_4$,
$E_2 = E_3 + V_4 + V_2 + E_5$,
$T = E_3 + V_3 \cdot V_4 + V_3 \cdot V_2 + V_5 \cdot V_4 + V_5 \cdot V_2 + E_4 \cdot V_4 + E_4 \cdot V_2 + E_4$.
$E_5 + V_3 \cdot E_5 + V_5 \cdot E_5$. (*T* has been reduced by using the Boolean identities
$E_3 \cdot E_3 = E_3$, $E_3 + E_3 \cdot X = E_3$, and $E_3 + E_3 = E_3$.)

Step 3: $E_3 = T_1 + V_1$,
$E_4 = E_6 + P_2 + AC$,
$E_5 = E_6 + P_1 + AC$,
$T = T_1 + V_1 + AC + V_3 \cdot V_4 + V_3 \cdot V_2 + V_5 \cdot V_4 + V_5 \cdot V_2 + V_4 \cdot P_2 + P_2$
$\cdot V_2 + E_6 + P_2 \cdot P_1 + V_3 \cdot P_1 + V_5 \cdot P_1$.
(Again, identities such as $AC + AC = AC$ and $E_6 + V_3 \cdot E_6 = E_6$ have been used to reduce *T*.)

Step 4: $E_6 = AC + S$,
$T = S + AC + T_1 + V_1 + V_3 \cdot V_4 + V_3 \cdot V_2 + V_5 \cdot V_4 + V_5 \cdot V_2 + V_4 \cdot P_2 + P_2$
$\cdot V_2 + P_2 \cdot P_1 + V_3 \cdot P_1 + V_5 \cdot P_1$.

The Boolean expression obtained in Step 4 represents four minimal cut sets with one failure event (cut set of size 1) and nine minimal cut sets with two failure events (cut set of size 2). The size 1 cut sets are occurrence of failure events S, AC, T_1, and V_1. The size 2 cut sets are failure events V_3 and V_4; V_3 and V_2; V_5 and V_4; V_5 and V_2; V_4 and P_2; P_2 and V_2; P_2 and P_1; V_3 and P_1; and V_5 and P_1. A simple examination of each cut set shows that its occurrence guarantees the occurrence of the top event (failure of the system). For example, the cut set consisting of simultaneous failure of valve V_2 and pump P_2, causes the two flow branches of the system to be lost, which in turn disables the system. The same substitution approach can be used to determine the path sets. In this case, the events are success events representing adequate realization of described functions.

It is clear from this fault tree example that the evaluation of a large logic tree by hand can be a formidable job. To remedy this problem, a number of software tools have been developed and available commercially for the analysis of logic trees. Quantitative evaluation of the cut sets or path sets is straightforward. For example, for the set of cut sets if $C = \{c_1, \ldots, c_n\}$, then

$$\Pr(C) = \Pr(c_1, \ldots, c_n) \tag{3.1}$$

and if $c_1, \ldots, c_n$ are independent minimal cut sets of the fault tree, then

$$\Pr(C) = \Pr(c_1) \cdots \Pr(c_n)$$

The same concept applies to path sets too.

3.4.2.2 Truth Table or Combinatorial Technique for Evaluation of Logic Trees

Unlike the substitution technique, which is based on Boolean reduction, the binary combinatorial method or the truth table approach does not convert the tree logic into Boolean equations to generate cut or path sets. Rather, this method relies on an algorithm to exhaustively generate all probabilistically significant combinations of both "failure" and "success" basic failure events, and subsequently propagates effect of each combination on the logic tree to determine the state of the top event. As successes and failures are combined, all combinations are mutually exclusive. The quantification of logic trees based on the combinatorial method yields a more exact result.

To illustrate the combinatorial approach, consider the fault tree in Figure 3.13. All possible combinations of success or failure events should be generated. Because there are four events and two states (success or failure) for each event, there are $2^4 = 16$ possible system states (i.e., actual physical states). Some of these states constitute system operation (when top event T does not happen), and some states constitute system failure (when top event T does happen). These 16 states are illustrated in Table 3.1. In this table, the subscript S denotes the nonoccurrence of a failure event (success), and subscript F refers to the failure or occurrence of the event in the fault tree.

Only combinations 14, 15, and 16 lead to the occurrence of the top event T, which results in system failure probability of $\Pr(T) = 0.0018 + 0.0008 + 0.0002 = 0.00028$. This is the exact value (provided that the events are independent).

Clearly, this is consistent with the exact calculation by the Boolean reduction method. Note that sum of the probabilities of all possible combinations (16 of them in this case) are unity because the combinations are all mutually exclusive and cover the entire event space (universal set). Combinations 14, 15, and 16 are mutually exclusive cut sets.

In order to visualize the difference between the results generated from the Boolean reduction and the combinatorial approach, the Venn diagram technique is helpful. Again consider the simple system in Figure 3.13 consisting of four events A, B, C, and D. The Boolean reduction process results in the minimal cut sets corresponding to system failure. These minimal cut sets are $A \cdot B \cdot C$ and $A \cdot B \cdot D$.

The left side of Figure 3.14 represents a Venn diagram for the two cut sets above. Each cut set is represented by one shaded area.

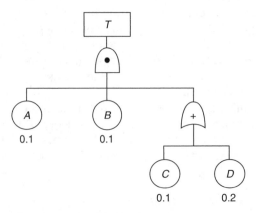

FIGURE 3.13 Example fault tree for truth table development.

TABLE 3.1
Truth Table of Evaluating Fault Tree in Figure 3.13

Combination Number	Combination Definition (System States)	Probability of C_i	System Operation T
1	$A_S B_S C_S D_S$	0.5832	S
2	$A_S B_S C_S D_F$	0.1458	S
3	$A_S B_S C_F D_S$	0.0648	S
4	$A_S B_S C_F D_F$	0.0162	S
5	$A_S B_F C_S D_S$	0.0648	S
6	$A_S B_F C_S D_F$	0.0162	S
7	$A_S B_F C_F D_S$	0.0072	S
8	$A_S B_F C_F D_F$	0.0018	S
9	$A_F B_S C_S D_S$	0.0648	S
10	$A_F B_S C_S D_F$	0.0162	S
11	$A_F B_S C_F D_S$	0.0072	S
12	$A_F B_S C_F D_F$	0.0018	S
13	$A_F B_F C_S D_S$	0.0072	S
14	$A_F B_F C_S D_F$	0.0018	F
15	$A_F B_F C_F D_S$	0.0008	F
16	$A_F B_F C_F D_F$	0.0002	F

$\sum C_i = 1.0000$

S = Success; F = Failure.

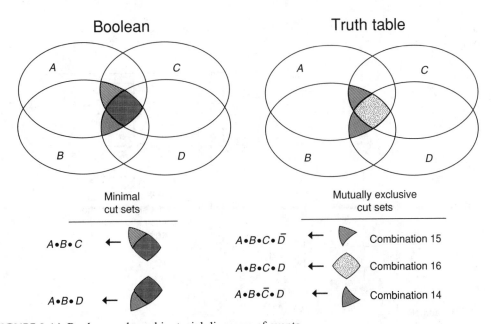

FIGURE 3.14 Boolean and combinatorial diagrams of events.

The two shaded areas are overlapping indicating that the cut sets are not mutually exclusive. Now consider how combinations 14, 15, and 16 are represented in the Venn diagram (right side of the Figure 3.14). Again, each shaded area corresponds to a combination. In this case, there is no overlapping of the shaded areas. That is, the combinatorial approach generates mutually exclusive sets, and those sets that lead to system failure are exclusive sets. Therefore, when the rare event approximation is used, the contributions generated by the combinatorial approach have no overlapping area and produce the exact probability. Since for size problems, usually the rare event approximation is the only practical choice, if the exact probabilities are desired, or failure probabilities are greater than 0.1, then the combinatorial approach is preferred.

A typical logic model may contain hundreds of events. For n events, there are 2^n combinations. Obviously, for a large n (e.g., $n > 20$), the generation of this large number of combinations is impractical; a more efficient method would be needed. An algorithm to generate combinations, in which probabilities exceed some cutoff limit (e.g., 10^{-7}), is proposed by Dezfuli et al. [8]. The algorithm generates combinations that are referred to as probabilistically significant combinations (see Figure 3.15).

In this combinatorial algorithm, the total number of events is first determined. Each event has an associated probability of occurrence. A combination represents the status (i.e., failed or not failed) of every event in the entire logic diagram. The collection of all failed blocks within a combination is referred to as a "failure set" (FS). A FS may have zero elements, meaning there is no failure event in the combination. This set is called the NIL combination. The objective is to generate other probabilistically significant combinations. The following assumptions are made:

1. The failure events are independent.
2. The NIL combination is a significant combination.

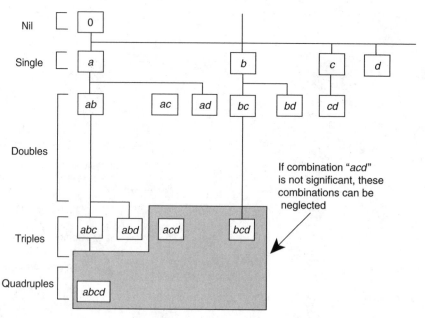

FIGURE 3.15 Computer algorithms for combinatorial approach.

Given a combination C, the assumption of the independence implies that the probability of the combination is

$$P_C = \prod_{i \in \text{FS}} P_i \prod_{i \notin \text{FS}} (1 - P_i) \tag{3.2}$$

where P_i is the probability of an individual failure event. Consider the combination C', which differs from the combination C in such a way that an event j is added to its failure set (i.e., transition of a success event to a failure event). From the above results, it can be concluded that

$$P'_C = P_C \times \frac{P_j}{1 - P_j} \tag{3.3}$$

Note that adding a block j to the failed set increases the probability of a combination if $P_j > 0.5$, and decreases the probability of a combination if $P_j < 0.5$. Consider also the combination C'', which differs from the combination C in that block j is replaced with block k (i.e., the replacement of a block in the failed set with another block). Therefore,

$$P''_C = P_C \times \frac{P_j}{1 - P_j} \times \frac{1 - P_k}{P_k} \tag{3.4}$$

This shows that replacing an event of a failed set in a combination with an event that has a lower failure probability results in a combination of lower probability, and replacing an event with an event that has a higher failure probability results in a combination of higher probability.

As such, the events are sorted in a decreasing order of probability. Each event is identified by its position in this ranking such that $P_i > P_j$ when $i < j$. Each combination is identified by a list of the events it contains in the failed set. To make the correspondence between combinations and lists unique, the list must be in ascending rank order, which corresponds to decreasing probability order.

Now consider a list representing a combination. Define a descendant of the list to be a list with one extra event appended to the failed set. Since the list must be ordered, this extra event must have a higher rank (lower probability) than any events in the original list. If there is no such event, there is no descendant. One should generate all descendants of the input list, and recursively generate all subsequent descendants. Since the algorithm begins with an empty list, it is clear that the algorithm will generate all possible lists. Figure 3.15 illustrates this scheme for the simple case of four events.

To generate only significant lists, we first need to prove that if a list is not significant; its descendants are not significant. The NIL set is significant. According to (3.3), at least one item of the list must have a probability lower than 0.5. Any failure event added to form the descendant would also have a probability lower than 0.5. Therefore, the probability of the descendant would be lower than that of the original set, and therefore, cannot be significant.

The algorithm takes advantage of this property. The descendants are generated in an increasing rank (decreasing probability) order of the added events. Equation (3.4) shows that the probability of the generated combinations is also decreasing. Each list is checked to see whether it is significant. If it is not significant, the routine exits without any recursive operation and without generating any further descendants of the original input list. Figure 3.15 shows the effect, if the state consisting of events a, c, and d is found to be insignificant, all the indicated combinations are immediately excluded from further consideration.

3.4.2.3   Binary Decision Diagrams

The technique described in detail in *NASA Fault Tree Handbook* [7] is one of the most recent techniques for evaluation of logic trees. Advances in binary logic has recently led to the development of a new logic manipulation procedure for fault trees called binary decision diagrams (BDDs) which works directly with the logical expressions. A BDD in reality is a graphical representation of the tree logic. The BDD used for logic tree analysis is referred to as the reduced ordered BDD (i.e., minimal form in which the events appear in the same order at each path). Sinnamon and Andrews [9] provide more information on the BDD approach to fault tree analysis.

The BDD is assembled using the logic tree recursively in a bottom–up fashion. Each basic event in the fault tree has an associated single-node BDD. For example, the BDD for a basic event B is shown in Figure 3.16. Starting at the bottom of the tree, a BDD is constructed for each basic event, and then combined according to the logic of the corresponding gate. The BDD for the OR gate "B or C" is constructed by applying the OR Boolean function to the BDD for event B and the BDD for event C. Since B is first in the relation, it is considered as the "root" node. The union of C BDD is with each "child" node of B.

First consider the terminal node 0 of event B in Figure 3.16. Accordingly, since $0 + X = X$ and $1 + X = 1$, the left child reduces to C and the right child reduces to 1 as shown in Figure 3.17.

Now consider the intersection operation applied to events A and B as shown in Figure 3.18. Note that $0 \cdot X = 0$ and that $1 \cdot X = X$. Thus, the reduced BDD for event $A \cdot B$ is shown in Figure 3.18.

Consider the Boolean logic $A \cdot B + C$ which is the union of the AND gate operation in Figure 3.18 with event C. The BDD construction and reduction is shown in Figure 3.19. Since A comes before C, A is considered as the root node and the union operation is applied to A's children.

The reduced BDD for $A \cdot B + C$ can further be reduced because node C appears two times. Each path from the root node A to the lowest terminal node 1 represents a disjoint combination of events that constitute failure (occurrence) of the root node. Thus in Figure 3.19, the failure paths (those leading to a bottom node 1) would be $\bar{A} \cdot C + \bar{A} \cdot \bar{B} + A \cdot \bar{B} \cdot C$. Since the paths are mutually exclusive, the calculation of the probability of failure is simple and similar to the truth table approach.

In real practice, a fault tree is transformed to its equivalent BDD often using software tools such as QRAS [10]. The failure paths leading to all end values of 1 would be identified and quantified by summing their probabilities. In developing the BDD for a fault tree, various reduction techniques are used to simplify the BDD.

The BDD approach is similar to the truth table method and yields an exact value of the top event probability. The exact probability is useful when many high-probability events appear in the model. The BDD approach is also the most efficient computer based approach for calculating probabilities. Because the minimal paths generated in the BDD approach are

FIGURE 3.16 BDD for basic event B.

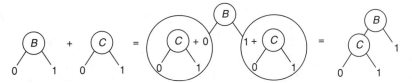

FIGURE 3.17 BDDs for the OR gate involving events B and C (Step 1).

disjoint, importance measures and sensitivities can be calculated more efficiently. It is important to note that generation of the minimal cut sets obtained, for example, through the substitution method provides important qualitative information as well as quantitative information. Examples are the use of minimal cut sets to highlight the most significant failure combinations (cut sets of the systems or scenarios) and show where design changes can remove or reduce certain failure combinations. Minimal cut sets also help validating fault tree modeling of the system or subsystem by determining whether such combinations are physically meaningful, and by examining the tree logic to see whether they would actually cause the top event to occur. Minimal cut sets are also useful in investigating the effect or dependencies among basic events modeled in the fault tree.

The most information is provided by using both approaches. The use of BDD approach, while faster and more exact, does not preclude generation of minimal cut sets, as these provide additional critical information in the PRA.

Example 3.2
Find the cut sets for the following fault tree through the BDD approach, and compare the results with the truth table method.

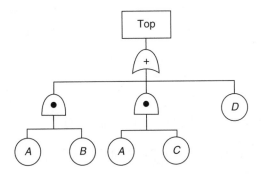

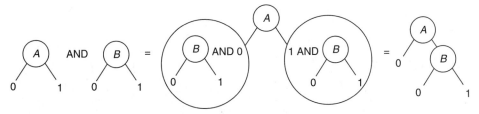

FIGURE 3.18 BDDs for the AND gate involving A and B (Step 2).

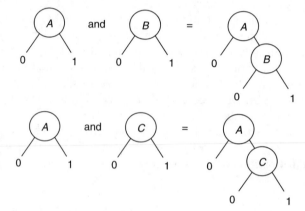

FIGURE 3.19 BDD construction and reduction for $A \cdot B + C$.

Solution
First, consider the basic events "*A* and *B*"; and "*A* and *C*."

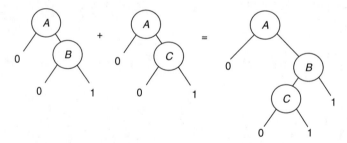

Now consider the union operation.

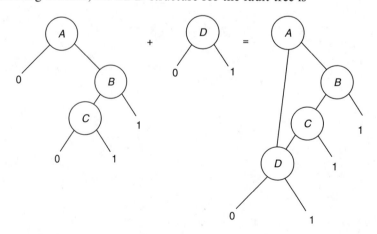

Finally considering event *D*, the BDD structure for the fault tree is

Each path from the root node A, to a terminal node with value 1 represents a disjoint combination of events that causes system failure.

Assuming that the probability of failure for A, B, C, and D is 0.2,

Table for Example 3.2

Disjoint Path	Probability of Failure	
AB	$\Pr(A)\Pr(B)$	0.0400
$A\bar{B}C$	$\Pr(A)(1-\Pr(B))\Pr(C)$	0.0320
$A\bar{B}\bar{C}D$	$\Pr(A)(1-\Pr(B))(1-\Pr(C))\Pr(D)$	0.0256
$\bar{A}D$	$(1-\Pr(A))\Pr(D)$	0.1600
		0.2576

The system failure probability for the top event is 0.2576. Table 3.2 illustrates the results for the truth table method.

Combinations 2, 4, 6, 8, 10, 11, 12, 13, 14, 15, and 16 lead to the occurrence of the top event T which results in system failure probability of $\Pr(T) = 0.2576$. This result is similar to the BDD approach.

3.4.3 EVENT TREES METHOD

If successful operation of a facility or complex system depends on an approximately chronological, but discrete, operation of its units (i.e., hazard barriers, subsystems, etc.), then an event tree is appropriate. Generally, the event trees model scenarios of successive events that

TABLE 3.2
Truth Table of Evaluating Fault Tree

Combination Number	Combination Definition (System States)	Probability of C_i	System Operation T
1	$A_S\ B_S\ C_S\ D_S$	0.4096	S
2	$A_S\ B_S\ C_S\ D_F$	0.1024	F
3	$A_S\ B_S\ C_F\ D_S$	0.1024	S
4	$A_S\ B_S\ C_F\ D_F$	0.0256	F
5	$A_S\ B_F\ C_S\ D_S$	0.1024	F
6	$A_S\ B_F\ C_S\ D_F$	0.0256	F
7	$A_S\ B_F\ C_F\ D_S$	0.0256	S
8	$A_S\ B_F\ C_F\ D_F$	0.0064	F
9	$A_F\ B_S\ C_S\ D_S$	0.1024	S
10	$A_F\ B_S\ C_S\ D_F$	0.0256	F
11	$A_F\ B_S\ C_F\ D_S$	0.0256	F
12	$A_F\ B_S\ C_F\ D_F$	0.0064	F
13	$A_F\ B_F\ C_S\ D_S$	0.0256	F
14	$A_F\ B_F\ C_S\ D_F$	0.0064	F
15	$A_F\ B_F\ C_F\ D_S$	0.0064	F
16	$A_F\ B_F\ C_F\ D_F$	0.0016	F

$\sum C_i = 1.0000$

S = success; F = failure.

lead to exposure of hazards and ultimately to the undesirable consequences. This may not always be the case for a simple system, but it is often the case for complex systems, such as nuclear power plants, where the subsystems (hazard barriers) should work according to a specific sequence of events to avert an undesirable initiating event and achieve a desirable outcome. The event tree is the primary technique uses in PRAs for modeling risk scenarios discussed in Chapters 1 and 2.

Let us consider the event tree built for a specific initiating event of a nuclear power plant as shown in Figure 3.20. The event trees are horizontally built structures that start on the left, where the initiating event is modeled. This event describes an abnormal situation or event that requires legitimate demand for the operation of subsequent subsystems (hazard barriers) of the facility or overall system. Development of the tree proceeds chronologically, with the demand on each unit (or subsystem) being postulated. The first "event heading" represents the first barrier to prevent, protect, or mitigate exposure of hazards. They include specific subsystem operation, software operation, or human actions needed, as shown on the top of the tree structure. In Figure 3.20, the barriers (referred to as event tree headings) are as follows:

RP   — operation of the reactor-protection system to shutdown the reactor
ECA — injection of emergency coolant water by subsystem A to cool the reactor in the short term
ECB — injection of emergency coolant water by subsystem B to cool the reactor in the short term
LHR — injection of coolant water to cool the reactor for long term.

Note that in this book, the convention used to represent the upper (success) branches is to use a complement event (e.g., $\bar{B}$) and to use a regular event (e.g., B) for the lower (failure) branches.

At a branch point, the upper branches of an event show the success of the event headings (barriers) and the lower branches show their failures. In Figure 3.20, following the occurrence of the initiating event A (which is loss of normal water due to a pipe break), RP needs to work (event B). If RP does not work, the overall system will fail, because the system is not shutdown (as shown by the lower branches of event B). If RP works, then it is important to know whether removal of the heat generated by the reactor due to postshutdown radioactive decay is possible. This function is performed by ECA. If ECA does not function, even though

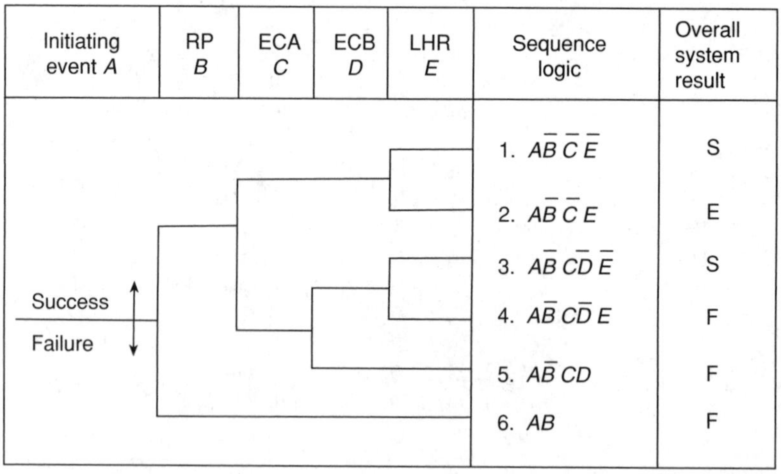

Initiating event A	RP B	ECA C	ECB D	LHR E	Sequence logic	Overall system result
					1. $A\bar{B}\,\bar{C}\,\bar{E}$	S
					2. $A\bar{B}\,\bar{C}\,E$	E
					3. $A\bar{B}\,CD\,\bar{E}$	S
					4. $A\bar{B}\,C\bar{D}\,E$	F
					5. $A\bar{B}\,CD$	F
					6. AB	F

FIGURE 3.20 Example of an event tree.

RP has worked, the overall system would still fail, unless the other heat removal option, ECB functions. However, if ECA operates properly, the cooling function is achieved, but it is important for LHR to function for long-term cooling of the reactor. Successful operation of LHR leads the system to a successful operating state, and failure of LHR (event E) leads the overall system to a failed state. Note that if ECA operates, it makes no difference whether ECB works, because the short-term cooling function would be realized. If LHR operates successfully, the overall system would be in a success state. It is obvious that operation of certain subsystems may not be necessarily dependent on the occurrence of some preceding events.

The overall outcome of each of the sequence or scenario of events is shown at the end of each sequence. This outcome, in essence, describes the final outcome of each sequence, whether the overall system succeeds, fails, initially succeeds but fails at a later time, or *vice versa*. The logical representation of each sequence can also be shown in the form of a Boolean expression. For example, for sequence 5 in Figure 3.20, events A, C, and D have occurred (for example, the corresponding hazard barrier subsystems have failed), but event B has not occurred (i.e., B properly works). Clearly, the event tree scenarios are mutually exclusive.

The event trees are usually developed in a binary format, i.e., the heading events are assumed to either occur or not. In cases where a spectrum of outcomes is possible, the branching process can proceed with more than two outcomes. In these cases, the qualitative representation of the event tree branches in a Boolean sense would not be possible.

The development of an event tree, although inductive in nature, in principle, requires a good deal of deductive thinking by the analyst. To demonstrate this issue and further understand the concept of event tree development, consider the system shown in Figure 3.21. One can think of a situation where the sensing and control system device S initiates one of the two pumps. At the same time, the ac power source AC should always exist to allow S and pumps P_1 and P_2 to operate. Thus, if we define three distinct events S, AC, and pumping system PS, for a scenario of events starting with the initiating event, an event tree that includes these three events can be constructed. Clearly if AC fails, both PS and S fail; if S fails only PS fails. This would lead to placing AC as the first event tree heading followed by S and PS. This event tree is illustrated in Figure 3.22. Note that Figure 3.22(a) and 3.22(b) are two logically equivalent event trees using two different graphical convensions. In this book we used the convensions displayed in Figure 3.22(a).

Events represent discrete states. The logic of these states can be modeled by other logic modeling methods such as fault trees. This way the event tree scenarios and the logical

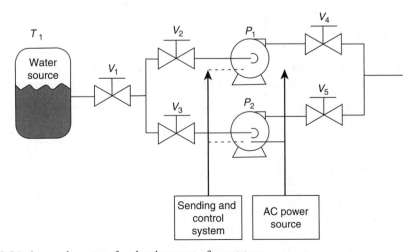

FIGURE 3.21 A sample system for development of event tree.

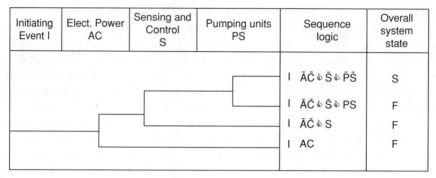

Initiating Event I	Elect. Power AC	Sensing and Control S	Pumping units PS	Sequence logic	Overall system state
				I $\bar{A}\bar{C}$⬥$\bar{S}$⬥$\bar{P}\bar{S}$	S
				I $\bar{A}\bar{C}$⬥$\bar{S}$⬥PS	F
				I $\bar{A}\bar{C}$⬥S	F
				I AC	F

FIGURE 3.22 Event tree for system in Figure 3.21.

combinations of basic failure events can be considered together. This is a powerful aspect of the event tree technique. If the event tree headings represent complex subsystems or units, the event tree heading can be represented by a logic modeling technique such as fault tree. Clearly, other system analysis models such as reliability block diagrams and logical representations in terms of cut sets or path sets, can also be used to model the event tree headings.

3.4.3.1 Evaluation of Event Trees

Qualitative evaluation of event trees is straightforward. The logical (e.g., cut sets) representation of each event tree heading, and ultimately each event tree scenario, is obtained and then reduced through the use of Boolean algebra rules. For example, in scenario 5 in Figure 3.20, if events B, C, and D are represented by the following Boolean expressions (minimal cut sets) shown below (e.g., obtained from fault tree evaluation), the reduced Boolean expression of the sequence can be obtained.

$$A = a,$$
$$B = b + c \cdot d,$$
$$C = e + d,$$
$$D = c + e \cdot h$$

The simultaneous Boolean expression and reduction proceeds as follows:

$$A \cdot \bar{B} \cdot C \cdot D = a \cdot \overline{(b + c \cdot d)} \cdot (e + d) \cdot (c + e \cdot h)$$
$$= a \cdot (\bar{b} \cdot \bar{c} + \bar{b} \cdot \bar{d}) \cdot (e \cdot c + e \cdot h + d \cdot c)$$
$$= a \cdot \bar{b} \cdot \bar{c} \cdot e \cdot h + a \cdot \bar{b} \cdot c \cdot \bar{d} \cdot e + a \cdot \bar{b} \cdot \bar{d} \cdot e \cdot h$$

If an expression explaining all failed states is desired, the union of the reduced Boolean equations of each scenario that leads to failure or hazard exposure should be obtained and reduced.

Quantitative evaluation of event trees is similar to the quantitative evaluation of fault trees. For example, to determine the probability associated with the $A \cdot \bar{B} \cdot C \cdot D$ sequence, one would consider:

$$\Pr(A \cdot \bar{B} \cdot C \cdot D) = \Pr(a \cdot \bar{b} \cdot \bar{c} \cdot e \cdot h + a \cdot \bar{b} \cdot c \cdot \bar{d} \cdot e + a \cdot \bar{b} \cdot \bar{d} \cdot e \cdot h)$$
$$= \Pr(a \cdot \bar{b} \cdot \bar{c} \cdot e \cdot h) + \Pr(a \cdot \bar{b} \cdot c \cdot \bar{d} \cdot e) + \Pr(a \cdot \bar{b} \cdot \bar{d} \cdot e \cdot h)$$
$$= \Pr(a) \cdot [1 - \Pr(b)][1 - \Pr(c)] \Pr(e) \cdot \Pr(h)$$
$$\quad + \Pr(a) \cdot [1 - \Pr(b)] \Pr(c) \cdot [1 - \Pr(d)] \Pr(e)$$
$$\quad + \Pr(a)[1 - \Pr(b)][1 - \Pr(d)] \Pr(e) \Pr(h)$$

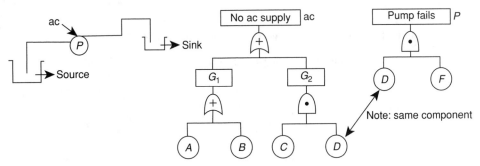

FIGURE 3.23 A simple pumping system and corresponding fault trees of its subsystems.

Since the two terms are disjoint, the above probability expression is exact. However, if the terms are not disjoint, the rare event approximation can be used here.

Example 3.3
Consider the pumping system shown at the left in Figure 3.23. Fluid from the source tank is moved to the sink tank assuming that ac power would be needed to start and run the pump when the sink tank level is low. No other failures or human errors are important to consider. Develop an event tree for this system. Assuming that the events of failing the ac power and pump failure can be represented by the following fault trees and the associated failure data shown, calculate the frequency of each scenario in the event tree and the total frequency of failure of the system. Note that failure of component "D" appears in both fault trees. That is, it plays two different roles and is called a "replicated event." For example, it could be failure of a signal that turns on the ac power and starts the pump. The mean values of failure probabilities and frequency of the initiating event (i.e., sink is low in fluid) are also given in Table 3.3.

Solution
The event tree of Figure 3.24 shows the scenario of events that follows in the initiating event "sink is low." Note that, since ac is needed to operate the pump, its failure results in a scenario that bypasses the "pump" event tree heading.

Consider scenario $I \cdot \overline{ac} \cdot P$. The corresponding Boolean expression for this scenario is

$$ac = G_1 + G_2$$
$$G_1 = A + B, G_2 = C \cdot D$$
$$ac = A + B + C \cdot D$$
$$\overline{ac} = \overline{A + B + C \cdot D} = \left(\overline{A} \cdot \overline{B} \right) \cdot \left(\overline{C \cdot D} \right)$$
$$= \overline{A} \cdot \overline{B} \cdot \left(\overline{C} + \overline{D} \right) = \overline{A} \cdot \overline{B} \cdot \overline{C} + \overline{A} \cdot \overline{B} \cdot \overline{D}$$

$$P = D \cdot F, \overline{P} = \overline{D} + \overline{F}$$
$$I \cdot \overline{ac} \cdot P = I \cdot \left(\overline{A} \cdot \overline{B} \cdot \overline{C} + \overline{A} \cdot \overline{B} \cdot \overline{D} \right) \cdot P$$
$$= I \cdot \left(\overline{A} \cdot \overline{B} \cdot \overline{C} + \overline{A} \cdot \overline{B} \cdot \overline{D} \right) \cdot D \cdot F$$
$$= I \cdot \overline{A} \cdot \overline{B} \cdot \overline{C} \cdot D \cdot F + \underbrace{I \cdot \overline{A} \cdot \overline{B} \cdot \overline{D} \cdot D \cdot F}_{\emptyset}$$

Since $D \cdot \overline{D}$ is a null set, the last term can be ignored and

$$I \cdot \overline{ac} \cdot P = I \cdot \overline{A} \cdot \overline{B} \cdot \overline{C} \cdot D \cdot F$$

TABLE 3.3
Failure Data for the Events Shown in Figure 3.23

Item	Failure Probability or Frequencies	Success Probability
I	10 per month	—
A	0.01	0.99
B	0.01	0.99
C	0.02	0.98
D	0.05	0.95
F	0.01	0.99

Similarly, the second scenario can be represented by the Boolean expression:

$$I \cdot \mathrm{ac} = I \cdot A + I \cdot B + I \cdot C \cdot D$$

The frequency of both scenarios can be calculated as a function of the occurrence of the initiating event and probability of failure of individual components of the systems modeled in the fault trees:

$$\mathrm{Pr(system\ failure)} = \mathrm{Pr}(I \cdot \mathrm{ac} + I \cdot \overline{\mathrm{ac}} \cdot P) = \mathrm{Pr}(I \cdot \mathrm{ac}) + \mathrm{Pr}(I \cdot \overline{\mathrm{ac}} \cdot P)$$

$$= \underbrace{\mathrm{f}(I)\,\mathrm{Pr}(A) + \mathrm{f}(I)\,\mathrm{Pr}(B) + \mathrm{f}(I)\,\mathrm{Pr}(C)\,\mathrm{Pr}(D)}_{\text{mutually exclusive terms}}$$

$$+ \underbrace{\mathrm{f}(I)\,\mathrm{Pr}(\overline{A})\,\mathrm{Pr}(\overline{B})\,\mathrm{Pr}(C)\,\mathrm{Pr}(F)}_{\text{non-mutually exclusive}}$$

Since we are dealing with high-probability events, it is more accurate not to use the rare event approximation. Thus,

$$\begin{aligned}
\text{Frequency of system failure} =\ & 10\{(0.01) + (0.01) - (0.01)(0.01) + (0.02)(0.05) \\
& - 2(0.01)(0.02)(0.05) + (0.01)(0.01)(0.02)(0.05)\} \\
& + 10 \times (0.99)(0.99)(0.98)(0.05)(0.01) \\
=\ & 0.2136/\text{month}
\end{aligned}$$

The total frequency of risk (loss of the system) is the sum of these two scenarios, 0.2136 per month, or an average of 4.68 months (~140 days) to loss of the system.

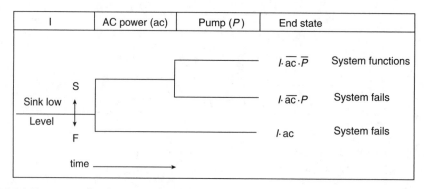

I	AC power (ac)	Pump (P)	End state

$I \cdot \overline{\mathrm{ac}} \cdot \overline{P}$ System functions

$I \cdot \overline{\mathrm{ac}} \cdot P$ System fails

$I \cdot \mathrm{ac}$ System fails

Sink low Level   S   F   time

FIGURE 3.24 Event tree for the system in Figure 3.23.

It is equally possible, but not customary or even practical in complex problems, to find the same result from the first scenario which represents the no-risk situation.

$$I \cdot \overline{ac} \cdot \overline{P} = I \cdot (\overline{A} \cdot \overline{B} \cdot \overline{C} + \overline{A} \cdot \overline{B} \cdot \overline{D}) \cdot (\overline{D} + \overline{F})$$

$$= I \cdot \overline{A} \cdot \overline{B} \cdot \overline{C} \cdot \overline{D} + I \cdot \overline{A} \cdot \overline{B} \cdot \overline{C} \cdot \overline{F} + I \cdot \overline{A} \cdot \overline{B} \cdot \overline{D} + I \cdot \overline{A} \cdot \overline{B} \cdot \overline{D} \cdot \overline{F}$$

$$= I \cdot \overline{A} \cdot \overline{B} \cdot \overline{C} \cdot \overline{F} + I \cdot \overline{A} \cdot \overline{B} \cdot \overline{D}$$

$$\Pr(I \cdot \overline{ac} \cdot \overline{P}) = \Pr(I \cdot \overline{A} \cdot \overline{B} \cdot \overline{C} \cdot \overline{F}) + \Pr(I \cdot \overline{A} \cdot \overline{B} \cdot \overline{D}) - \Pr(I \cdot \overline{A} \cdot \overline{B} \cdot \overline{C} \cdot \overline{D} \cdot \overline{F})$$

$$= 10\{(0.99 \times 0.99 \times 0.98 \times 0.99) + (0.99 \times 0.99 \times 0.95)$$

$$- (0.99 \times 0.99 \times 0.98 \times 0.95 \times 0.99)\}$$

$$9.7864$$

Clearly, the expected losses per month would be $10 - 9.7864 = 0.2136$.

3.4.4 MASTER LOGIC DIAGRAM

For complex systems and facilities such as a nuclear power plant, modeling for reliability analysis or risk assessment may become very difficult. In such systems and facilities, there are always several functionally separate subsystems (such as hazard barriers) that interact with each other, each of which can be modeled independently. However, it is necessary to find a logical representation of the overall system interactions with respect to the individual subsystems. The MLD is such a model.

Consider a functional block diagram of a complex system in which all of the functions modeled are necessary in one way or another to achieve a desired objective. For example, in the context of a nuclear power plant, the independent functions of heat generation, normal heat transport, emergency heat transport, reactor shutdown, thermal energy to mechanical energy conversion, and mechanical energy to electrical energy conversion collectively achieve the goal of safely generating electric power. Each of these functions, in turn, is achieved through the design and operating function from others. For example, emergency heat transport may require internal cooling, which is obtained from other so-called support functions.

The MLD clearly shows the interrelationships among the independent functions (or systems) and the independent support functions. The MLD (in success space) can show the manner in which various functions, subfunctions, and hardware interact to achieve the overall system objective. On the other hand, an MLD in a failure space can show the logical representation of the causes for failure of functions (or systems). The MLD (in success or failure space) can easily map the propagation of the effect of failures, i.e., establish the trajectories of basic event failure (or incipient failure) propagation.

In essence, the hierarchy of the MLD is displayed by a dependency matrix in form of a lattice. For each function, subfunction, subsystem, and hardware item shown on the MLD, the effect of failure or success of all combinations of items is established and explicitly shown by a solid round bullet symbol. Consider the MLD shown in a success space in Figure 3.25. In this diagram, there are two major protective or mitigative functions (or systems) F_1 and F_2.

Together, they achieve the system objective (e.g., assure safety). Each of these functions, because of modeling and assessing their performance, is further divided into two identical subfunctions, each of which can achieve the respective parent functions. This means that both subfunctions must be lost for F_1 or F_2 to be lost. Suppose the development of the subfunctions (or systems) can be represented by their respective hardware, software, or human actions which interface with other support functions (or support systems) S_1, S_2, and S_3. Support

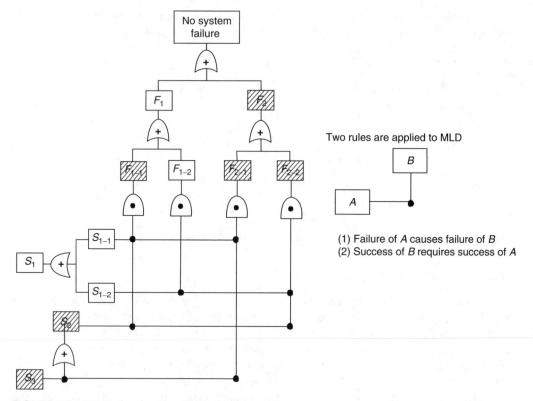

FIGURE 3.25 MLD showing the effect of failure of S_3.

functions are defined as those that provide the proper environment or operating condition for the main functions to be realized (e.g., electric, mechanical, thermal motive or power, cooling, lubrication, and control of the main components). For example, if a pump function is to "provide high pressure water," then functions "provide ac power," "cooling and lubrication," and "activation and control" are called support functions. However, function (or system) S_1 can be divided into two independent subfunctions (or systems) (S_{1-1} and S_{1-2}), so that each can interact independently with the subfunctions (or systems) of F_1 and F_2. The dependency matrix is established by reviewing the design specifications or operating manuals that describe the relationship between the items shown in the MLD, in which the dependencies are explicitly shown by a round solid bullet symbol. For instance, the dependency matrix shows that failure of S_3 leads directly to failure of S_2, which in turn results in failures of F_{1-1}, F_{2-2}, and F_{2-1}. This failure cause and effect relationship is highlighted on the MLD in Figure 3.25.

A key element in the development of an MLD is the assurance that the elements of the system or facility for which the dependency matrix is developed (e.g., S_{1-1}, S_{1-2}, S_3, F_{2-1}, F_{1-2}, and F_{2-2}) are all physically independent. "Physically independent" means they do not share any other system elements or parts. Each element may have other dependencies, such as CCF (discussed in Chapter 4). Sometimes it is difficult to distinguish, *a priori*, between main functions and supporting functions. In these situations, the dependency matrix can be developed irrespective of the main and supporting functions. Figure 3.26 shows an example of such a development. However, the main functions can be identified easily by examining the resulting MLD; they are those functions that appear hierarchically, at the top of the MLD model and do not support other items.

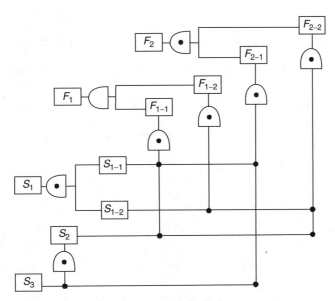

FIGURE 3.26 MLD with all system functions treated similarly.

The evaluation of the MLD is straightforward. One must determine all possible 2^n combinations of failures and successes of independent items (elements) map them onto the MLD and propagate their effects, using the MLD logic. The combinatorial approach discussed in Section 3.4.2 is the most appropriate method for that purpose, although the Boolean reduction method can also be applied. Table 3.4 shows the combinations for the example in Figure 3.25. For reliability calculations, one can combine those end-state effects that lead to the system success. Suppose independent items (here, systems or subsystems) S_{1-1}, S_{1-2}, S_2, and S_3 have a failure probability of 0.01 for a given mission, and the probability of independent failure of F_{1-1}, F_{1-2}, F_{2-1}, and F_{2-2} is also 0.01. Table 3.4 shows the resulting probability of the end-state effects. If needed, calculation of failure probabilities for the MLD items (e.g., subsystems) can be completed independent of the MLD, through one of the conventional system reliability analysis methods (e.g., fault free analysis).

Table 3.4 shows all possible mutually exclusive combinations of items modeled in the MLD (i.e., S_{1-1}, S_{1-2}, S_2, S_3, F_{1-1}, F_{1-2}, F_{2-1}, and F_{2-2}). Those combinations that lead to a failure of the system are mutually exclusive cut sets.

One may only select those combinations that lead to a complete failure of the system. The sum of the probabilities of occurrence of these combinations determines the failure probability of the system. If one selects the combinations that lead to the system's success, then the sum of the probabilities of occurrence of these combinations determines the reliability of the system. Table 3.5, for example, shows dominant combinations that lead to the failure of the system.

Another useful analysis that may be performed via MLD is the calculation of the conditional system probability of failure. In this case, a particular element of the system is set to failure, and all other combinations that lead to the system's failure may be identified. Table 3.6 shows all combinations within the MLD that lead to the system's failure, when element S_{1-2} is known to be in the failed state.

Example 3.4
Consider the simplified diagram of the H-coal process facility shown in Figure 3.27. In case of an emergency, a shutdown device (SDD) is used to shutdown the hydrogen flow. If the

TABLE 3.4
Most Likely Combinations of Failure and Their Respective Probabilities

Combination No. (i)	Failed Items	Probability of Failed Items (and Success of Other Items)	End State (Actually Failed and Casually Failed Elements)
1	None	9.3×10^{-1}	Success
2	S_3	9.3×10^{-3}	$F_{1-1}, F_2, F_{2-1}, F_{2-2}, S_2, S_3$
3	F_{2-2}, S_3	9.4×10^{-5}	$F_{1-1}, F_2, F_{2-1}, F_{2-2}, S_2, S_3$
4	S_2, S_3	9.4×10^{-5}	$F_{1-1}, F_2, F_{2-1}, F_{2-2}, S_2, S_3$
5	F_{1-1}, S_3	9.4×10^{-5}	$F_{1-1}, F_2, F_{2-1}, F_{2-2}, S_2, S_3$
6	F_{2-1}, S_3	9.4×10^{-5}	$F_{1-1}, F_2, F_{2-1}, F_{2-2}, S_2, S_3$
7	S_{1-2}	9.3×10^{-3}	$F_{2-1}, F_{2-2}, S_{1-2}, S_1$
8	F_{1-2}, S_{1-2}	9.4×10^{-5}	$F_{1-2}, F_{2-2}, S_{1-2}, S_1$
9	F_{2-2}, S_{1-2}	9.4×10^{-5}	$F_{1-2}, F_{2-2}, S_{1-2}, S_1$
10	S_{1-1}	9.3×10^{-3}	$F_{1-2}, F_{2-2}, S_{1-2}, S_1$
11	F_{2-1}, S_{1-1}	9.4×10^{-5}	$F_{1-1}, F_{2-1}, S_{1-1}, S_1$
12	F_{1-1}, S_{1-1}	9.4×10^{-5}	$F_{1-1}, F_{2-1}, S_{1-1}, S_1$
13	S_2	9.3×10^{-3}	F_{1-1}, F_{2-2}, S_2
14	F_{1-1}, S_2	9.4×10^{-5}	F_{1-1}, F_{2-2}, S_2
15	F_{2-2}, S_2	9.4×10^{-5}	F_{1-1}, F_{2-2}, S_2
16	F_{2-2}	9.3×10^{-3}	F_{2-2}
17	F_{2-1}	9.3×10^{-3}	F_{2-1}
18	F_{1-2}	9.3×10^{-3}	F_{1-2}
19	F_{1-1}	9.3×10^{-3}	F_{1-1}
20	S_{1-2}, S_3	9.4×10^{-5}	$F_1, F_{1-1}, F_{1-2}, F_2, F_{2-1}, F_{2-2}, S_{1-2}, S_2, S_3, S_1$

reactor temperature is too high, an emergency cooling system (ECS) is also needed to reduce the reactor temperature. To protect the process plant when the reactor temperature becomes too high, both ECS and SDD must succeed. The SDD and ECS are actuated by a control device. If the control device fails, the ECS will not be able to work. However, an operator can manually operate (OA) the SDD and terminate the hydrogen flow. The power for the SDD,

TABLE 3.5
Combinations that Lead to Failure of the System

Combination No. (i)	Failure Combination	Probability of State
1	F_{1-2}, S_3	9.4×10^{-5}
2	S_{1-1}, S_{1-2}	9.4×10^{-5}
3	S_{1-2}, S_3	9.4×10^{-5}
4	F_{2-1}, S_{1-2}, S_3	9.5×10^{-7}
5	$F_{2-2}, S_{1-1}, S_{1-2}$	9.5×10^{-7}
6	F_{2-1}, S_{1-2}, S_2	9.5×10^{-7}
7	$F_{2-1}, S_{1-1}, S_{1-2}$	9.5×10^{-7}
8	$F_{1-1}, F_{2-1}, S_{1-2}$	9.5×10^{-7}
9	S_{1-1}, S_{1-2}, S_3	9.5×10^{-7}
10	S_{1-2}, S_2, S_3	9.5×10^{-7}
Total probability of system failure		2.9×10^{-4}

TABLE 3.6
Combinations Leading to the System Failure When S_{1-2} Is Known to Have Failed (Conditional Probabilities)

Combination No. (i)	Failure Combination	Probability of State
1	S_{1-2}	9.3×10^{-3}
2	S_{1-1}	9.3×10^{-3}
3	S_2, S_3	9.4×10^{-5}
4	F_{2-1}, S_3	9.4×10^{-5}
5	F_{1-1}, F_{2-1}	9.4×10^{-5}
6	F_{2-1}, S_2	9.4×10^{-5}
7	F_{2-1}, S_{1-1}	9.4×10^{-5}
8	S_{1-1}, S_2	9.4×10^{-5}
9	F_{1-1}, S_3	9.4×10^{-5}
10	S_{1-1}, S_2	9.4×10^{-5}
11	S_{1-1}, S_{1-2}	9.4×10^{-5}
Total probability of system failure *given* failure of S_{1-2}		1.95×10^{-2}

ECS, and control device comes from an outside electric company (off-site power OSP). The failure data for these systems are listed in Table 3.7. Draw an MLD and use it to find the probability of losing both the SDD and ECS.

Solution
The MLD is shown in Figure 3.28. Important combinations of independent failures are listed in Table 3.8. The probability of losing both the ECS and SDD for each end state is calculated and listed in the third column of Table 3.8. Combinations that exceed 1×10^{-10} are included in

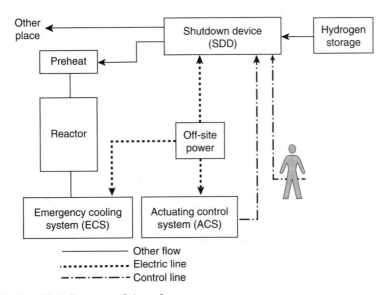

FIGURE 3.27 Simplified diagrams of the safety systems.

TABLE 3.7
Failure Probability of Each
System

	System Failure
OSP	2.0×10^{-2}
OA	1.0×10^{-2}
ACS	1.0×10^{-3}
SDD	1.0×10^{-3}
ECS	1.0×10^{-3}

Table 3.8. The combinations that lead to failure of both SDD and ECS are shown in Table 3.9. The probability of losing both systems is calculated as 2.29×10^{-2}.

Example 3.5
The simple event tree shown in Figure 3.29 has five event tree headings (A, B, C, D, and E). The initiating event is labeled I. Consider scenario No. 5, the logical equivalent of the scenario is

$$S_5 = I \cdot \overline{A} \cdot B \cdot \overline{C} \cdot D$$

where I is the initiating event. Develop an equivalent MLD representation of this event tree.

Solution
Sequence 5 occurs when the expression $\overline{A} \cdot B \cdot \overline{C} \cdot D$ is true. Note that the above Boolean expression involves two failed elements (i.e., B and D). We can express these terms, in the success space, through the complement of $\overline{A} \cdot B \cdot \overline{C} \cdot D$, which is

$$\overline{\overline{A} \cdot B \cdot \overline{C} \cdot D} = \overline{\left(\overline{A} \cdot \overline{C}\right) \cdot (B \cdot D)}$$
$$= \overline{\left(\overline{A} \cdot \overline{C}\right)} + \overline{(B \cdot D)}$$
$$= \overline{\left(\overline{A} \cdot \overline{C}\right)} + \left(\overline{B} + \overline{D}\right)$$

The last expression represents every event in a success space (e.g., $\overline{A}, \overline{B}, \overline{C}, \overline{D}$) and its equivalent MLD logic is shown in Figure 3.30.

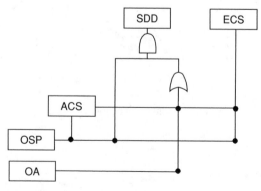

FIGURE 3.28 MLD for the safety system in Figure 3.26.

TABLE 3.8
Leading Combination of Unit Failures in the System

State No. (i)	Failed Units	Probability*	Failed and inoperable Units	System State
1	None	9.67×10^{-1}	None	Success
2	ECS	9.68×10^{-4}	ECS	Success
3	SDD	9.68×10^{-4}	SDD	Success
4	ACS	9.68×10^{-4}	ECS, ACS	Success
5	OSP	1.97×10^{-2}	SDD, ECS, ACS, OSP	Failure
6	OA	9.77×10^{-3}	OA	Success
7	ECS, SDD	9.69×10^{-7}	SDD, ECS	Failure
8	ECS, ACS	9.69×10^{-7}	ECS, ACS	Success
9	ECS, OSP	1.98×10^{-5}	SDD, ECS, ACS, OSP	Failure
10	ECS, OA	9.78×10^{-6}	ECS, OA	Success
11	SDD, ACS	9.96×10^{-7}	ACS, ECS, SDD	Failure
12	SDD, OSP	1.98×10^{-5}	SDD, ECS, ACS, OSP	Failure
13	SDD, OA	9.78×10^{-6}	OA, SDD	Success
14	ACS, OSP	1.98×10^{-5}	SDD, ECS, ACS, OSP	Failure
15	ACS, OA	9.79×10^{-6}	ECS, ACS, OA	Success
16	OSP, OA	1.99×10^{-4}	SDD, ECS, ACS, OSP,OA	Failure
17	ECS, SDD, ACS	9.70×10^{-10}	ACS, ECS, SDD	Failure
18	ECS, SDD, OSP	1.9×10^{-8}	SDD, ECS, ACS, OSP	Failure
19	ECS, SDD, OA	9.79×10^{-9}	OA, ECS, SDD	Failure
20	SDD, ACS, OSP	1.98×10^{-8}	SDD, ECS, ACS, OSP	Failure
21	SDD, ACS, OA	9.79×10^{-9}	OA, ACS, ECS, SDD	Failure
22	ACS, OSP, OA	2.00×10^{-7}	SDD, ECS, ACS, OSP,OA	Failure

* Includes probability of success of elements not affected.

3.5 MODELING OF DEPENDENT FAILURES IN RISK ASSESSMENT

Dependent failures are extremely important in risk assessment and must be given adequate treatment so as to minimize gross overestimation of performance. In general, dependent failures are defined as events in which the probability of each failure is dependent on the occurrence of other failures. According to (A.14), if a set of dependent events $\{E_1, E_2, \ldots, E_n\}$

TABLE 3.9
Failure Contributions from Failure of One and Two Units

Combination No.	Units Failed	Probability*	Contribution to Total Failure Prob. (%)
1	OSP	1.97×10^{-2}	98.69
2	ECS, SDD	9.69×10^{-7}	0.00
3	ECS, OSP	1.98×10^{-5}	0.10
4	SDD, ACS	9.96×10^{-7}	0.00
5	SDD, OSP	1.98×10^{-5}	0.10
6	ACS, OSP	1.98×10^{-5}	0.10
7	OSP, OA	1.99×10^{-4}	1.00
8	Sum of all others	2.60×10^{-7}	0.01

* Includes probability of success of elements not affected.

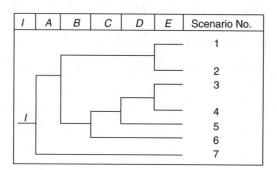

FIGURE 3.29 Simple event tree.

exists, then the probability of each failure in the set depends on the occurrence of other failures in the set.

The probabilities of dependent events in the left-hand side of (A.14) are usually, but not always, greater than the corresponding independent probabilities. Determining the conditional probabilities in (A.14) is generally difficult. However, there are parametric methods that can take into account the conditionality and generate the probabilities directly. These methods are discussed later in this section.

Generally, dependence among various events, e.g., failure events of two items, is either due to the internal environment or external environment (or events). The internal aspects can be divided into three categories: internal challenges, intersystem dependencies, and intercomponent dependencies. The external aspects are natural or human-made environmental events that make failures dependent. For example, the failure rates for items exposed to extreme heat, earthquakes, moisture, and flood will increase. The intersystem and intercomponent dependencies can be categorized into four broad categories: functional, shared equipment, physical, and human-caused dependencies. CCFs are considered as the collection of all sources of dependencies, especially between components, that are not known, or are difficult to explicitly model in the system or component risk assessment models. For example, functional and shared equipment dependencies are often handled by explicitly modeling

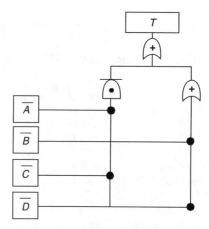

FIGURE 3.30 MLD equivalent of event tree shown in Figure 3.29.

them in the system analysis part of risk assessment, but other dependencies are considered collectively using CCF.

CCFs have been shown by many studies to contribute significantly to the overall performance (unavailability or unreliability) of complex systems. There is no unique and universal definition for CCFs. However, a fairly general definition of CCF is given by Mosleh et al. [12] as "…a subset of dependent events in which two or more component fault states exist at the same time, or in a short-time interval, and are direct results of a shared cause."

To better understand CCFs, consider a hazard barrier (e.g., a cooling system) with three redundant components A, B, and C. The total failure probability of A can be expressed in terms of its independent failure A_I and dependent failures as follows:

C_{AB} is the failure of A and B (and not C) from common causes
C_{AC} is the failure of A and C (and not B) from common causes
C_{ABC} is the failure of A, B, and C from common causes.

Component A fails if any of the above events occur. The equivalent Boolean representation of total failure of component A is $A_T = A_I + C_{AB} + C_{AC} + C_{ABC}$. Similar expressions can be developed for components B and C.

Now suppose that the success criterion for the system is 2-out-of-3 for components A, B, and C. Accordingly, the failure of the system can be represented by the following events (cut sets):

$$(A_I \cdot B_I), (A_I \cdot C_I), (B_I \cdot C_I), C_{AB}, C_{AC}, C_{BC}, C_{ABC}$$

Thus, the Boolean representation of the system failure will be

$$S = (A_I \cdot B_I) + (A_I \cdot C_I) + (B_I \cdot C_I) + C_{AB} + C_{AC} + C_{BC} + C_{ABC}$$

It is evident that if only independence is assumed, the first three terms of the above Boolean expression are used, and the remaining terms are neglected. Applying the rare event approximation, the system failure probability Q_S is given by

$$Q_S \approx \Pr(A_I)\Pr(B_I) + \Pr(A_I)\Pr(C_I) + \Pr(B_I)\Pr(C_I)$$
$$+ \Pr(C_{AB}) + \Pr(C_{AC}) + \Pr(C_{BC}) + \Pr(C_{ABC})$$

If components A, B, and C are similar (which is often the case since common causes among different components have a much lower probability), then

$$\Pr(A_I) = \Pr(B_I) = \Pr(C_I) = Q_1$$
$$\Pr(C_{AB}) = \Pr(C_{AC}) = \Pr(C_{BC}) = Q_2$$

and

$$\Pr(C_{ABC}) = Q_3$$

Therefore,

$$Q_s = 3(Q_1)^2 + 3Q_2 + Q_3$$

TABLE 3.10
Key Characteristics of the CCF Parametric Models [15]

Estimation Approach	Model	Model Parameters	General Form for Multiple Component Failure Probabilities
Nonshock models Single parameter	β-Factor	β	$Q_k = \begin{cases} (1-\beta)Q_t & k=1 \\ 0 & l < k < m \\ \beta Q_t & k = m \end{cases}$
Nonshock models Multiparameter	Multiple Greek letters	β, γ, δ	$Q_k = \dfrac{1}{\binom{m-1}{k-1}}\left(1-\rho_{k+1}\right)\left(\prod_{i=1}^{k}\rho_i\right)Q_t$ $k = 1, 2, \ldots, m \; \rho_1 = 1,$ $\rho_2 = \beta, \ldots, \rho_{k+1} = 0$
	α-Factor	$\alpha_1, \alpha_2, \ldots, \alpha_m$	$Q_k = \dfrac{k}{\binom{m-1}{k-1}}\dfrac{\alpha_k}{\alpha_t}Q_t$ $k = 1, 2, \ldots, m$ $\alpha_t = \sum_{k=1}^{m}k\alpha_k$
Shock models	Binomial failure rate	μ, ρ, ω	$Q_k = \begin{cases} \mu\rho^k(1-\rho)^{m}-k & k=1 \\ \mu\rho^m + \omega & k = m \end{cases}$

In general, one can introduce the probability Q_k representing the probability of CCF among k-specific components in a component group of size m, such that $1 \le k \le m$. The CCF models for calculating Q_k are summarized in Table 3.10. In this table, Q_t is the total probability of failure accounting both for common cause and independent failures, and α, β, γ, δ, μ, and ω are the model parameters, estimated from the failure data on these components.

The CCF parametric models can be divided into two categories: single parameter models and multiple parameter models. The remainder of this section discusses these two categories in more detail as well as elaborates on the parameter estimation of the CCF models.

Single parameter models are those that use one parameter in addition to the total component failure probability to calculate the CCF probabilities. One of the most commonly used single parameter models defined by Fleming [11] is called the β-factor model. It is the first parametric model applied to CCF events in risk and reliability analysis. The sole parameter of the model, β, can be associated with that fraction of the component failure rate that is due to the CCFs experienced by the other components in the system. That is,

$$\beta = \frac{\lambda_c}{\lambda_c + \lambda_i} = \frac{\lambda_c}{\lambda_t} \qquad (3.5)$$

where λ_c is a failure rate due to CCFs, λ_I is a failure rate due to independent failures, and $\lambda_t = \lambda_c + \lambda_i$.

An important assumption of this model is that whenever a common cause event occurs, all components of a redundant component system fail. In other words, if a CCF "shock" strikes a redundant system; all components are assumed to fail instantaneously.

Based on the β-factor model, for a system of m components, the probabilities of basic events involving k-specific components (Q_k), where $1 \leq k \leq m$ are equal to zero, except Q_1 and Q_m. These quantities are given as

$$Q_1 = (1 - \beta)Q_t$$
$$Q_2 = 0$$
$$\vdots$$
$$Q_{m-1} = 0$$
$$Q_m = \beta Q_t$$

In general, the estimate for the total component failure rate is generated from generic sources of failure data, while the estimators of the corresponding β-factor do not explicitly depend on generic failure data, but rather rely on specific assumptions concerning data interpretation. Some recommended values of β are given by Mosleh et al. [12]. It should be noted that although this model can be used with a certain degree of accuracy for two component redundancy, the results tend to be conservative for a higher level of redundancy. However, due to its simplicity, this model has been widely used in risk assessment studies. To get more reasonable results for a higher level of redundancy, more generic parametric models should be used.

Example 3.6
Consider the following cooling system used as a hazard barrier in a facility. The system has two redundant trains. Suppose each train is composed of a valve and a pump (each driven by an electric motor). The pump failure modes are "failure to start" (PS) and "failure to run following a successful start" (PR). The valve failure mode is "failure to open" (VO). Develop an expression for the probability of system failure.

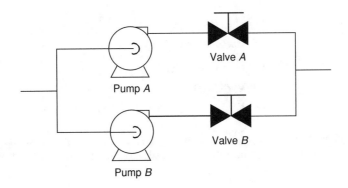

Solution
Develop a system fault tree to include both independent and CCFs of the components, where P_A is the independent failure of pump A, P_B is the independent failure of pump B, P_{AB} is the dependent failure of pumps A and B, V_A is the independent failure of valve A, V_B is the independent failure of valve B, V_{AB} is the dependent failure of valves A and B.

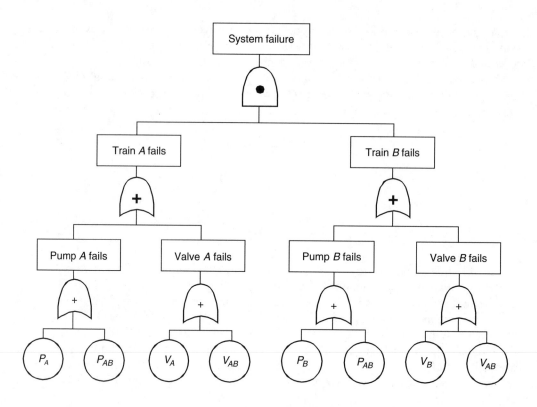

By solving the fault tree, the following cut sets can be identified:

$$C_1 = (P_A, P_B), \ C_2 = (P_{AB})$$
$$C_3 = (V_A, V_B), \ C_4 = (V_{AB})$$
$$C_5 = (P_A, V_B), \ C_6 = (P_B, V_A)$$

Use the β-factor method to calculate the probability of each cut set.

$$\Pr(C_1) = (1 - \beta_{PS})^2 (q_{PS})^2 + (1 - \beta^{PR})^2 (\lambda_{PR} t)^2$$
$$\Pr(C_2) = \beta_{PS}(q_{PS}) + \beta_{PR}(\lambda_{PR} t)$$
$$\Pr(C_3) = (1 - \beta_{VO})^2 (q_{VO})^2$$
$$\Pr(C_4) = \beta_{VO}(q_{VO})$$
$$\Pr(C_5) = \Pr(C_6) = (q_{PS} + \lambda_{PR} t)(q_{VO})$$

where q is the probability of failure rate on demand, λ is the failure rate to run, and t is mission time. System failure probability is calculated using rare event approximation, as follows:

$$Q_s = \sum_{i=1}^{6} \Pr(C_i)$$

Multiple parameter models are used to get a more accurate assessment of CCF probabilities in systems with a higher level of redundancy. These models have several parameters that are usually associated with different event characteristics. This category of models can be further divided into two subcategories namely, shock and nonshock models. Multiple Greek model and α-factor model are nonshock models, whereas binomial failure rate (BFR) model is a shock model. These models are further discussed below.

3.5.1 MULTIPLE GREEK LETTER MODEL

The multiple Greek letter (MGL) model introduced by Fleming et al. [13] is a generalization of the β-factor model. New parameters such as γ, δ, etc., are used in addition to β to distinguish among common cause events affecting different numbers of components in a higher level of redundancy. For a system of m redundant components, $m-1$, different parameters are defined. For example, for $m=4$, the model includes the following three parameters (see Table 3.10):

- Conditional probability that the common cause of an item failure will be shared by one or more additional items, β
- Conditional probability that the common cause of an item failure that is shared by one or more items will be shared by two or more items in addition to the first, γ
- Conditional probability that the common cause of an item failure shared by two or more items will be shared by three or more items in addition to the first, δ.

It should be noted that the β-factor model is a special case of the MGL model in which all other parameters excluding β are equal to 1.

The following estimates of the MGL model parameters are used as conservative generic values:

Number of Components (m)	MGL Parameters		
	β	γ	δ
2	0.10	—	—
3	0.10	0.27	—
4	0.11	0.42	0.4

Consider the 2-out-of-3 success model described before. If we were to use the MGL model, then

$$Q_k = \frac{1}{\binom{3-1}{1-1}}(1 - \rho_{1+1})\left(\prod_{i=1}^{1}\rho_i\right)Q_t = \frac{1}{2}(1 - \rho_2)(\rho_1)Q_t$$

since $\rho_1 = 1$ and $\rho_2 = \beta$, then

$$Q_1 = \frac{1}{2}(1 - \beta)Q_t$$

Similarly,

$$Q_2 = \frac{1}{2}(1 - \rho_3)(\rho_1\rho_2)Q_t$$

with $\rho_1 = 1$, $\rho_2 = \beta$, and $\rho_3 = \gamma$,

$$Q_2 = \frac{1}{2}\beta(1 - \gamma)Q_t$$

Also,

$$Q_3 = \frac{1}{2}(1 - \rho_4)(\rho_1\rho_2\rho_3)Q_t$$

with $\rho_1 = 1$, $\rho_2 = \beta$, $\rho_3 = \gamma$, and $\rho_4 = 0$,

$$Q_3 = \beta\gamma Q_t$$

To compare the result of the β-factor and MGL, consider a case where the total failure probability of each component (accounting for both dependent and independent failures) is 8×10^{-3}. According to the β-factor model, failure probability of the system including CCFs, if $\beta = 0.1$, would be

$$Q_s = 3(1 - \beta)^2 Q_t^2 + \beta Q_t$$
$$Q_s = 3(1 - 0.1)^2 (8 \times 10^{-3})^2 + 0.1(8 \times 10^{-3})$$
$$Q_s = 9.6 \times 10^{-4}$$

However, MGL model with $\beta = 0.1$ and $\gamma = 0.27$ will predict the system failure probability as

$$Q_s = \frac{3}{4}(1 - \beta)^2 Q_t^2 + \frac{3}{2}\beta(1 - \gamma)^2 Q_t + \beta\gamma Q_t$$
$$Q_s = \frac{3}{4}(1 - 0.1)^2 (8 \times 10^{-3}) + \frac{3}{2}0.1(1 - 0.27)^2 (8 \times 10^{-3}) + 0.1(0.27)(8 \times 10^{-3})$$
$$Q_s = 1.1 \times 10^{-3}$$

The difference is obviously small, but MGL model is more accurate than the β-factor model.

3.5.2 α-FACTOR MODEL

The α-factor model discussed by Mosleh and Siu [14] develops CCF failure probabilities from a set of failure ratios and the total component failure rate. The parameters of the model are the fractions of the total probability of failure in the system that involves the failure of k components due to a common cause, a_k.

The probability of a common cause basic event involving failure of k components in a system of m components is calculated, according to the equation given in Table 3.10. For example, the probabilities of the basic events of the three-component system described earlier will be

$$Q_1 = (\alpha_1/\alpha_t)Q_t$$
$$Q_2 = (\alpha_2/\alpha_t)Q_t$$
$$Q_3 = (3\alpha_3/\alpha_t)Q_t$$

where $\alpha_t = \alpha_1 + 2\alpha_2 + 3\alpha_3$. Table below from Mosleh [15] provides conservative generic values of α-factors.

Number of Items (m)	α-Factor			
	α_1	α_2	α_3	α_4
2	0.95	0.050	—	—
3	0.95	0.040	0.01	—
4	0.95	0.035	0.01	0.005

Therefore, the system failure probability for the three redundant components discussed earlier can now be written as

$$Q_s = 3\left(\frac{\alpha_1}{\alpha_t}Q_t\right)^2 + 3\left(\frac{\alpha_2}{\alpha_t}Q_t\right)^2 + 3\left(\frac{\alpha_3}{\alpha_t}Q_t\right)^2$$

Accordingly, using the generic α values for the 2-out-of-3 success, $\alpha_t = 0.95 + 0.08 + 0.03 = 1.06$. Thus,

$$Q_s = 3\left(\frac{\alpha_1}{\alpha_t}Q_t\right)^2 + 3\left(\frac{\alpha_2}{\alpha_t}Q_t\right)^2 + 3\left(\frac{\alpha_3}{\alpha_t}Q_t\right)^2$$

$$Q_s = 3\left[\frac{0.95}{1.06}\left(8\times10^{-3}\right)\right]^2 + 3\left[\frac{0.04}{1.06}\left(8\times10^{-3}\right)\right]^2 + 3\left[\frac{0.01}{1.06}\left(8\times10^{-3}\right)\right]^2$$

$$Q_s = 1.3 \times 10^{-3}$$

this is closely consistent with the MGL model results.

3.5.3 BINOMIAL FAILURE RATE MODEL

The BFR model discussed by Atwood [16], unlike the α-factor model and MGL model, is a shock-dependent model. It estimates the failure frequency of two or more components in a redundant system as the product of the CCF shock arrival rate and the conditional failure probability of components given the shock has occurred. This model considers two types of shock: lethal and nonlethal. The assumption is that, given a nonlethal shock, components fail independently, each with a probability of ρ, whereas in the case of a lethal shock, all components fail with a probability of 1. It should be noted that due to the BFR model complexity and the lack of data to estimate its parameters, it is not very widely used in practice.

3.5.4 ASSESSMENT OF PARAMETERS OF CCF MODELS

Despite the difference among the CCF models described, they all have similar data requirements in terms of parameter estimation. The most important steps in the quantification of CCFs are collecting information from the raw data and selecting a model that can use most of this information. Table 3.11 summarizes simple point estimators for parameters of various nonshock CCF models. In this table, n_k is the total number of observed failure events involving failure of k similar items due to a common cause, m is the total number of redundant items considered; and N_D is the total number of item operation demands. If the item is normally operating (not on a standby), then N_D can be replaced by the total test

TABLE 3.11
Simple Point Estimators for Various Parametric Models

Model	Point Estimator
β-Factor	$\hat{Q}_t = \dfrac{1}{mN_D} \displaystyle\sum_{i=1}^{m} kn_k$ $\hat{\beta} = \displaystyle\sum_{i=2}^{m} kn_k \Big/ \sum_{i=1}^{m} kn_k$
Multiple Greek letters	$\hat{Q} = \dfrac{1}{m\,N_D} \displaystyle\sum_{k=1}^{m} kn_k$ $\hat{\beta} = \left(\displaystyle\sum_{k=2}^{m} kn_k\right)\Big/\left(\sum_{k=1}^{m} kn_k\right)$ $\hat{\gamma} = \left(\displaystyle\sum_{k=3}^{m} kn_k\right)\Big/\left(\sum_{k=1}^{m} kn_k\right)$ $\hat{\delta} = \left(\displaystyle\sum_{k=4}^{m} kn_k\right)\Big/\left(\sum_{k=1}^{m} kn_k\right)$ $\hat{\alpha}_k = n_k \Big/ \left(\displaystyle\sum_{k=1}^{m} kn_k\right)$
α-Factor	$\hat{Q}_t = \dfrac{1}{mN_D} \displaystyle\sum_{i=1}^{m} kn_k$ $\hat{\alpha}_k = n_k \Big/ \displaystyle\sum_{i=1}^{m} kn_k \quad (k = 1, 2, \dots, m)$

(or operation) time T. The estimators in Table 3.11 are based on the assumption that in every system (or hazard barrier) demand, all components and possible combinations of components may be challenged. Therefore, the estimators apply to systems whose tests are nonstaggered.

Example 3.7
For the system described in Example 3.6, estimate the β parameters, λ and q, for the valves and pumps based on the following failure data:

	Event Statistic		
Failure Mode	n_1	n_2	T (h) or N_D
Pump fails to start (PS)	10	1	500 (demands)
Pump fails to run (PR)	50	2	10,000 (h)
Valve fails to open (VO)	10	1	10,000 (demands)

In the above table, n_1 is the number of observed independent failures, n_2 is the number of observed events involving double CCF. Calculate the β parameter for the pump and valve failure modes.

Solution
From Table 3.11,

$$\hat{\beta} = \frac{2n_2}{n_1 + 2n_2}$$

Apply this formula to β_{PR}, β_{PS}, and β_{VO}, by using appropriate values for n_1 and n_2:

$$n_{PS} = n_1 + 2n_2 = 12$$
$$n_{PR} = n_1 + 2n_2 = 54$$
$$n_{VO} = n_1 + 2n_2 = 17$$

Accordingly,

$$\beta_{PS} = \frac{2}{12} = 0.17$$
$$\beta_{PR} = \frac{4}{54} = 0.07$$
$$\beta_{VO} = \frac{2}{17} = 0.13$$

3.6 A SIMPLE EXAMPLE OF PRA

Consider the fire protection system shown in Figure 3.31.

This system is designed to extinguish all possible fires in a plant with toxic chemicals. Two physically independent water extinguishing nozzles are designed such that each is capable of controlling all types of fires in the plant. Extinguishing nozzle 1 is the primary method of injection. Upon receiving a signal from the detector/alarm/actuator device, pump 1 starts

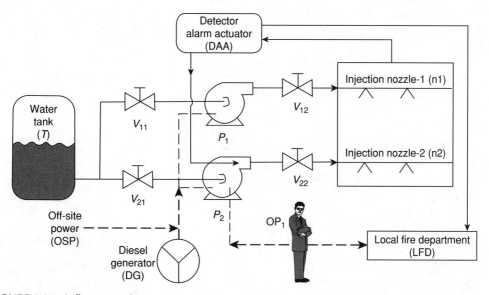

FIGURE 3.31 A fire protection system.

automatically, drawing water from the reservoir tank and injecting it into the fire area in the plant. If this pump injection path is not actuated, plant operators can start a second injection path manually. If the second path is not available, the operators will call for help from the local fire department, although the detector also sends a signal directly to the fire department. However, due to the delay in the arrival of the local fire department, the magnitude of damage would be higher than it would be if the local fire extinguishing nozzles were available to extinguish the fire. Under all conditions, if the normal OSP is not available due to the fire or other reasons, a local diesel generator which is normally on standby would provide electric power to the pumps. The power to the detector/alarm/actuator system is provided through the batteries, which are constantly charged by the OSP. Even if the ac power is not available, the dc power provided through the battery is expected to be available at all times. The manual valves on the two sides of pumps 1 and 2 are normally opened, and only remain closed when they are being repaired. The entire fire system and the generator are located outside of the main chemical reactor compartment, and are therefore not affected by an internal fire. Assuming all the preliminary steps of familiarization and choice of the PRA methodology that have been accomplished, the remaining steps of the PRA for this situation have been explained below:

1. Identification of Initiating Events. In this step, all initiating events and scenarios that lead to or promote the hazard of interest, which is a fire in the main chemical reactor compartment, must be identified. These should include equipment malfunctions, human errors, and facility conditions. The frequency of each event should be estimated. Assuming that all events would lead to the same magnitude of fire, the ultimate initiating event is occurrence of a fire, the frequency of which is the sum of the frequencies of the individual fire-causing events. Assume in this example, the frequency of fire is estimated at 1×10^{-6} per year (for example by using an initiating event logic diagram). Since fire is the only challenge to the plant in this example, we end up with only one initiating event. However, in more complex situations, a large set of initiating events can be identified, each posing a different challenge to the plant.

2. Scenario Development. In this step, we should explain the cause and effect relationship between the fire and the progression of events following the occurrence of fire. We will use the event tree method to depict this relationship. Generally, this is done inductively, and the level of detail considered in the event tree is somewhat dependent on the analyst. Two protective measures as hazard barriers have been considered in the event tree shown in Figure 3.32: on-site protective measures (on-site pumps, tanks, etc.) and off-site fire department measures.

The selection of these protective barriers is based on the fact that availability or unavailability of the on-site or off-site, and these measures would lead to different damage states.

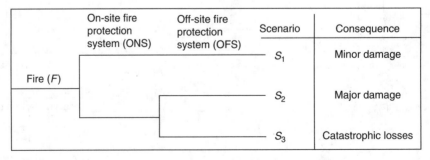

FIGURE 3.32 Scenarios of events following a fire using the event tree methods.

3. Logic Modeling. In this step, we should identify all failures (equipment or human) that lead to failure of the event tree headings (on-site or off-site protective measures).

For example, Figure 3.33 shows the fault tree developed for the hazard barrier on-site fire protection system. In this fault tree, all basic events that lead to the failure of the two independent paths of this system are described. Note that DAA, electric power to the pumps, and the water tank are shared by the two water injection paths.

Clearly, these are considered as physical dependencies. This is taken into account in the quantification step of the PRA. In this fault tree, all external event failures and passive failures are neglected.

This tree is simple since it only includes all failures not leading to an on-time response from the local fire department. Similarly an off-site fault tree model can be developed as shown in Figure 3.34.

It is also possible to use the MLD for logic tree analysis. An example of the MLD for this problem is shown in Figure 3.35 (manual values have not been included). However for this example, only the fault trees are used in the PRA, although MLD can also be used.

4. Failure Data Analysis. It is important at this point to calculate the probabilities of the basic failure events described in the event trees and fault trees. As indicated earlier, this can be done by using system-specific data, generic data, or expert judgment. Table 3.12 describes the generic and facility-specific failure data used and their sources. Chapter 4 describes more detailed methods for the analysis of failure and other event data. It is assumed that at least 10 h of operation is needed for the fire to be completely extinguished.

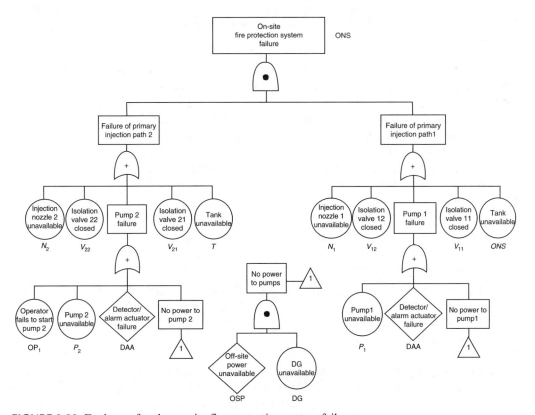

FIGURE 3.33 Fault tree for the on-site fire protection system failure.

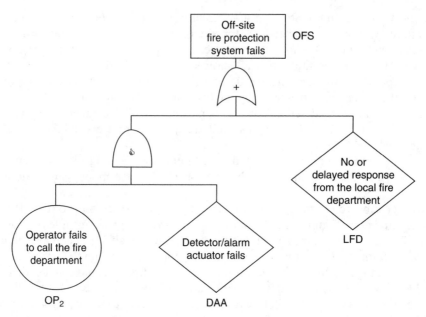

FIGURE 3.34 Fault tree for the off-site fire protection system failure.

   5. Quantification. To calculate the frequency of each scenario defined in Figure 3.32, we must first determine the cut sets of the two fault trees shown in Figure 3.33 and Figure 3.34. From this, the cut sets of each scenario are determined, followed by calculation of the probabilities of each scenario based on the occurrence of one of its cut sets.
   These steps are described below:

1. The cut sets of the on-site fire protection system failure are obtained using the technique described in Section 3.3. These cut sets are listed in Table 3.13. Only cut set number 22, which is failure of both pumps is subjected to a CCF. This is shown by adding a new cut set (cut set number 24), which represents this CCF.
2. The cut sets of the off-site fire protection system failure are similarly obtained and listed in Table 3.14.
3. The cut sets of the three scenarios are obtained using the following Boolean equations representing each scenario:

$$\text{Scenario-1} = F \cdot \overline{\text{ONS}}$$
$$\text{Scenario-2} = F \cdot \text{ONS} \cdot \overline{\text{OFS}}$$
$$\text{Scenario-3} = F \cdot \text{ONS} \cdot \text{OFS}$$

4. The frequency of each scenario is obtained using data listed in Table 3.13 and Table 3.14. These frequencies are shown in Table 3.15.
5. The total frequency of each scenario is calculated using the rare event approximation. These are also shown in Table 3.15.

   6. Consequences. In the scenario development and quantification tasks, we identified three distinct scenarios of interest, each with different outcomes and frequencies. The consequences associated with each scenario should be specified in terms of both economic and human losses. This part of the analysis is one of the most difficult for several reasons:

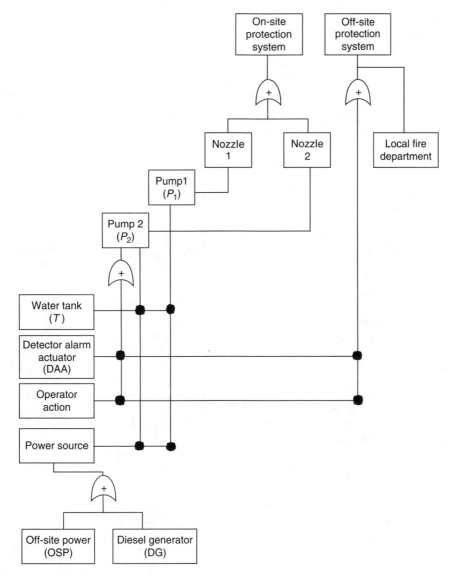

FIGURE 3.35 MLD for the entire fire protection system.

(1) Each scenario poses different hazards and methods of hazard exposure. In this case, the model should include the ways the fire can spread through the plant, how people can be exposed, evacuation procedures, the availability of protective clothing, etc.

(2) The consequence of the scenario can be measured in terms of human losses. It can also be measured in terms of economic losses, i.e., the total cost associated with the scenario. This involves assigning a dollar value to human life or casualties, which is a source of controversy (this subject is discussed in more detail in Section 7.4.6).

Suppose a careful analysis of the spread of fire and fire exposure is performed, with consideration of the above issues, in which consequences are measured only in terms of economic losses. These results are shown in Table 3.16.

TABLE 3.12
Sources of Data and Failure Probabilities

Failure Event	Plant-Specific Experience	Generic Data (see Chapter 4 for more detail)	Probability Used (see Chapter 4 for more detail)	Comments
Fire initiation frequency	No such experience in 10 years of operation	Five fires in similar plants. There are 70,000 plant-years of experience	$F = 5/70,000 = 7.1E\text{-}4/\text{year}$	Use generic data
Pumps 1 and 2 failure	Four failures of two pumps to start per year each having an average of 10 demands (tests) per month. Repair time takes about 2.5 hour. No experience of failure to run	Failure to run $= 1 \times 10^{-5}/\text{h}$	$\dfrac{4}{2(12)(10)} = 1.7 \times 10^{-2}/$ demand Unavailability $= 1.7 \times 10^{-2} + \dfrac{2.5(4)}{8760}$ $= 1.8 \times 10^{-2}/\text{demand}$ $P_1 = P_2 = 1.8 \times 10^{-2}$	Failure to start is facility-specific. But failure to run is generic and insignificant comparing to demand failure probability
CCF between pumps 1 and 2	No such experience	Using the β-factor method, $\beta = 0.1$ for failure of pumps to start	Unavailability due to CCF $CCF = 0.1 \times 1.8 \times 10^{-2}$ $= 1.8 \times 10^{-3}$ demand	Assume no significant CCF exists between valves and nozzles
Failure of isolation valves	Two failures to leave the valve in open position following 10 pump tests in 1 year	Not used	$V_{11} = V_{12} = V_{21} = V_{22}$ $= \dfrac{2}{10(12)(4)}$ $= 4.2 \times 10^{-3}/\text{demand}$	Facility-specific data used
Failure of nozzles	No such experience	$1 \times 10^{-5}/\text{demand}$	$N_1 = N_2$ $= 1.0 \times 10^{-5}/\text{demand}$	Generic data used
Diesel generator failure	Three failures in tests per year. 10 hours of operation is required	$3.0 \times 10^{-3}/\text{h}$	Failure on demand $= \dfrac{3}{12(10)}$; failure on demand $= 2.5 \times 10^{-2}/\text{demand}$; failure on run $= 3.0 \times 10^{-3}/\text{h}$; total failure of DG $= 2.5 \times 10^{-2} + 3.0 \times 10^{-3}$ $= 5.5 \times 10^{-2}$	Facility-specific data used for demand failure
Loss of off-site power	No experience	0.1/year	$OSP = 0.1 \times \dfrac{10}{8760}$ $= 1.1 \times 10^{-4}/\text{demand}$	Assume 10 hours of operation for fire extinguisher and use generic data
Failure of DAA	No experience	No data available	$DAA = 1 \times 10^{-4}/\text{demand}$	This estimate is based on expert judgment

continued

TABLE 3.12
Sources of Data and Failure Probabilities — *continued*

Failure Event	Plant-Specific Experience	Generic Data (see Chapter 4 for more detail)	Probability Used (see Chapter 4 for more detail)	Comments
Failure of operator to start pump 2	No such experience	Using the THERP method	$OP1 = 1 \times 10^{-2}$/demand	The method is discussed in Chapter 4
Failure of operator to call the fire department	No such experience	1×10^{-3}	$OP2 = 1 \times 10^{-3}$/demand	This is based on experience from no response to similar situations. Generic probability is used
No or delayed response from fire department	No such experience	1×10^{-4}	$LFD = 1 \times 10^{-4}$/demand	This is based on response to similar cases from the fire department. Delayed/no arrival is due to accidents, traffic, communication problems, etc.
Tank failure	No such experience	1×10^{-5}	$T = 1 \times 10^{-5}$/demand	This is based on data obtained from rupture of the tank or insufficient water content

7. *Risk Value Calculation and Evaluation.* Using values from Table 3.16, we can calculate the risk associated with each scenario. These risks are shown in Table 3.17.

Since this analysis shows that the mean risk value due to fire is rather low, uncertainty analysis is not very important. However, one of the methods described in Chapter 5 could be used to estimate the uncertainty associated with models and parameters (e.g., failure rates) of each component and the fire-initiating event if necessary. The uncertainties should be propagated through the cut sets of each scenario to obtain the uncertainty associated with the frequency estimation of each scenario. The uncertainty associated with the consequence estimates can also be obtained. When uncertainty associated with the consequence values are combined with the scenario frequencies and their uncertainties, the uncertainty associated with the estimated risk can be calculated. Figure 3.36 shows the risk profile based on the mean values of Table 3.17.

Example 3.8

We are interested to assess societal and individual risks of influenza shots to protect individuals against the flu virus. Influenza is a contagious disease and the entire population may be exposed to it, especially in the late fall and the winter seasons. Many factors influence the decision to take a shot, namely, age, severity of flu epidemic, and recipient's state of health. Data on influenza go back to more than 85 years. Once infected, the severity can range from

TABLE 3.13
Cut Sets of the On-Site Fire Protection System Failure

Cut Set No.	Cut Set	Probability (% Contribution to the Total Probability)
1	T	1.0×10^{-5} (0.35)
2	DAA	1.0×10^{-4} (3.5)
3	OSP $\cdot$ DG	6.0×10^{-6} (0.21)
4	$N_2 \cdot N_1$	1.0×10^{-10} (~0)
5	$N_2 \cdot V_{12}$	4.2×10^{-8} (~0)
6	$N_2 \cdot P_1$	1.8×10^{-7} (~0)
7	$N_2 \cdot V_{11}$	4.2×10^{-8} (~0)
8	$V_{22} \cdot N_1$	4.2×10^{-8} (~0)
9	$V_{22} \cdot V_{12}$	1.8×10^{-5} (0.64)
10	$V_{22} \cdot P_1$	7.6×10^{-5} (2.5)
11	$V_{22} \cdot V_{11}$	1.8×10^{-5} (0.64)
12	$V_{21} \cdot N_1$	4.2×10^{-8} (~0)
13	$V_{21} \cdot V_{12}$	1.8×10^{-5} (0.35)
14	$V_{22} \cdot P_1$	7.6×10^{-5} (2.5)
15	$V_{21} \cdot V_{11}$	1.8×10^{-5} (0.64)
16	$OP_1 \cdot N_1$	1.0×10^{-7} (~0)
17	$OP_1 \cdot V_{12}$	4.2×10^{-5} (1.5)
18	$OP_1 \cdot P_1$	1.8×10^{-4} (6.0)
19	$OP_1 \cdot V_{11}$	4.2×10^{-5} (1.5)
20	$P_2 \cdot N_1$	1.8×10^{-7} (~0)
21	$P_2 \cdot V_{12}$	7.6×10^{-5} (2.5)
22	$P_2 \cdot P_1$	3.2×10^{-4} (0.3)
23	$P_2 \cdot V_{11}$	7.6×10^{-5} (2.5)
24	CCF	1.8×10^{-3} (63.8)

$\Pr(ON) = \Sigma_i C_i = 2.8 \times 10^{-3}$

mildly sick to extremely sick and in rare cases, even death. Young and healthy population have much less excess deaths due to influenza than older and chronically ill population. Rowe [17] reports that a risk assessment can be carried out for subgroups of the population or the entire population of the U.S. (e.g., for the elderly and chronically ill, young and less sensitive). Table 3.18 shows the initiating cause of sickness in terms of the annual frequency of specific epidemics. Once the epidemic occurs, one may assume that the general population is susceptible to it, with the probabilities provided in Table 3.19. If someone is exposed to the virus, the severity of the flu that he/she experiences would be different. Table 3.20 lists the probability of

TABLE 3.14
Cut Sets of the Off-Site Fire Protection System

Cut Set No.	Cut Set	Probability (% Contribution to the Total Probability)
1	LFD	1×10^{-4} (100)
2	$OP_2 \cdot DAA$	1×10^{-7} (~0)

Total $\Pr^{(OFF)} \approx 1 \times 10^{-4}$

TABLE 3.15
Dominant Minimal Cut Sets of the Scenarios

Scenario No.	Cut Sets	Frequency	Comment
1	$F \cdot \overline{ON}$	$7.1 \times 10^{-4} \left(1 - 2.8 \times 10^{-2}\right)$ $= 7.1 \times 10^{-4}$	Since the probability can be directly evaluated for $\overline{ON}$ without evaluating the need to generate cut sets, only the probability is calculated
2	$F \cdot DAA \cdot LFD \cdot OP_2$	7.0×10^{-8}	1. Only cut sets from Table 3.13
	$F \cdot V_{22} \cdot P_1 \cdot \overline{LFD} \cdot \overline{OP_2}$	5.0×10^{-8}	that have a contribution
	$F \cdot V_{22} \cdot P_1 \cdot \overline{LFD} \cdot \overline{DAA}$	5.0×10^{-8}	greater than 1% are shown
	$F \cdot V_{21} \cdot P_1 \cdot \overline{LFD} \cdot \overline{OP_2}$	5.0×10^{-8}	2. Cut set
	$F \cdot V_{22} \cdot P_1 \cdot \overline{LFD} \cdot \overline{DAA}$	5.0×10^{-8}	$F \cdot DAA \cdot \overline{LFD} \cdot \overline{DAA}$
	$F \cdot OP_1 \cdot V_{12} \cdot \overline{LFD} \cdot \overline{OP_2}$	2.9×10^{-9}	is eliminated since
	$F \cdot OP_1 \cdot V_{12} \cdot \overline{LFD} \cdot \overline{DAA}$	2.9×10^{-9}	$DAA \cdot \overline{DAA} = \emptyset$
	$F \cdot OP_1 \cdot P_1 \cdot \overline{LFD} \cdot \overline{OP_2}$	1.1×10^{-7}	
	$F \cdot OP_1 \cdot P_1 \cdot \overline{LFD} \cdot \overline{DAA}$	1.1×10^{-7}	
	$F \cdot OP_1 \cdot V_{11} \cdot \overline{LFD} \cdot \overline{OP_2}$	2.9×10^{-9}	
	$F \cdot OP_1 \cdot V_{11} \cdot \overline{LFD} \cdot \overline{DAA}$	2.9×10^{-9}	
	$F \cdot P_2 \cdot V_{12} \cdot \overline{LFD} \cdot \overline{OP_2}$	5.0×10^{-8}	
	$F \cdot P_2 \cdot V_{12} \cdot \overline{LFD} \cdot \overline{DAA}$	5.0×10^{-8}	
	$F \cdot P_2 \cdot P_1 \cdot \overline{LFD} \cdot \overline{OP_2}$	5.0×10^{-8}	
	$F \cdot P_2 \cdot P_1 \cdot \overline{LFD} \cdot \overline{DAA}$	2.0×10^{-7}	
	$F \cdot P_2 \cdot V_{11} \cdot \overline{LFD} \cdot \overline{OP_2}$	2.0×10^{-7}	
	$F \cdot P_2 \cdot V_{11} \cdot \overline{LFD} \cdot \overline{DAA}$	5.0×10^{-8}	
	$F \cdot CCF \cdot \overline{LFD} \cdot \overline{OP_2}$	5.0×10^{-8}	
	$F \cdot CCF \cdot \overline{LFD} \cdot \overline{DAA}$	1.3×10^{-6}	

$\Sigma_i = 2.5 \times 10^{-6}$

Scenario No.	Cut Sets	Frequency	Comment
3	$F \cdot DAA \cdot LFD$	7.1×10^{-12}	1. Only cut sets from
	$F \cdot V_{22} \cdot P_1 \cdot LFD$	5.0×10^{-12}	Table 3.11 and
	$F \cdot V_{21} \cdot P_1 \cdot LFD$	5.0×10^{-12}	Table 3.12 that have
	$F \cdot OP_1 \cdot V_{12} \cdot LFD$	2.9×10^{-12}	contribution to the
	$F \cdot OP_1 \cdot P_1 \cdot LFD$	2.8×10^{-12}	scenario are shown
	$F \cdot OP_1 \cdot V_{11} \cdot LFD$	2.9×10^{-12}	
	$F \cdot P_2 \cdot P_{12} \cdot LFD$	5.0×10^{-12}	
	$F \cdot P_2 \cdot P_1 \cdot LFD$	2.0×10^{-11}	
	$F \cdot P_2 \cdot V_{11} \cdot LFD$	5.0×10^{-12}	
	$F \cdot CCF \cdot LFD$	3.0×10^{-11}	

$\Sigma_i = 8.6 \times 10^{-11}$

all the possible consequences once exposed to the virus. It is expected that the consequences from becoming ill or dying would determine a basis for measuring the risk. Table 3.21 shows the consequences of each possible outcome. To be able to compare all risks, according to Rowe [17], it is possible to define a unified utility loss measure called FLU to quantify relatively the amount of loss (i.e., consequence) experienced by the flu sufferers. If one FLU is representative of an economic loss of approximately 10 days of lost workdays (i.e., a severe case of flu), then other consequences can be relatively measured on this basis in terms of FLU units as listed in Table 3.21.

TABLE 3.16
Economic Consequences of Fire Scenarios

Scenario Number	Economic Consequence
1	$1,000,000
2	$92,000,000
3	$210,000,000

TABLE 3.17
Risk Associated with Each Scenario

Scenario Number	Economic Consequence (Expected Loss)
1	(7.1×10^{-4}) ($1,000,000) = $710.000
2	(2.5×10^{-6}) ($92,000,000) = $230.000
3	(8.6×10^{-11}) ($210,000,000) = $0.018

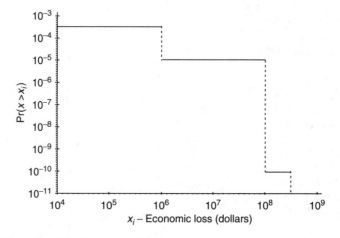

FIGURE 3.36 Risk profile of the fire protection system.

TABLE 3.18
Initiating Event Frequency

Type of Epidemic	Frequency/Year
Severe epidemic	0.1
Mild epidemic	0.4
Nonepidemic	0.5

TABLE 3.19
Probability of Developing Flu If Exposed to the Virus in Various Seasons

Type of Epidemic Season	Probability of Exposure and Flu Development
Severe epidemic	0.40
Mild epidemic	0.15
Nonepidemic	0.05

TABLE 3.20
Probability of Severe and Mild Flu Cases and Subsequent Mortality Probability

	Severe Case of Flu	Mild Case of Flu	Mortality Given a Flu Case
Severe epidemic seasons	0.500	0.499	0.001
Mild epidemic seasons	0.330	0.669	0.001
Nonepidemic seasons	0.100	0.899	0.001

TABLE 3.21
Consequence Values of Developing Flu

Consequence	Days Lost	FLU* Utility
Mortality (in severe cases of flu)	500–2000	100
Morbidity (state of disease)	2–20	1
Mild case of morbidity	3–9	0.5

*FLU is a unit measure of the degree of suffering from the flu.

We are interested to measure and compare the risk of vaccinating against the flu for the general population. The data on the flu shot show that while it could cause some reactions and thus mild consequences, it substantially reduces the likelihood of developing a subsequent flu, when exposed to the virus. Table 3.22 shows the data on the effectiveness and consequences of flu shots. Flu shots reduce the chance of flu, if exposed to the virus, to 8%. That is, 92% of flu shot recipients will become protected.

Repeat the calculation for the subgroup of very young, and chronically ill and elderly population for the corresponding data shown in Table 3.23 and Table 3.24. As noted, the only difference between the subgroups and the general population is in the exposure potential and chances of developing flu, once exposed to the virus.

Solution
Scenario development proceeds with the assumption that the initiating event (exposure to flu virus) occurs, and followed by the events developing flu of a given type and probability of

TABLE 3.22
Consequences of Vaccination

Consequence	Days Lost	FLU Utility	Frequency
Chills and fever	3–10	0.9	0.3
Mild reaction	0–1	0.2	0.3
Sore arm	—	0.1	0.4

TABLE 3.23
Flu Exposure Data (Initiation and Exposure) for Subgroups of Recipients

Type of Year	Without Vaccination	With Vaccination
Epidemic	0.03301	0.00268
Mild epidemic	0.04190	0.00263
Sore arm	0.04105	0.00265

subsequent mortality. Multiply frequency of each scenario by the sum consequences associated with that scenario and sum over all scenarios to calculate the risk values. Figure 3.37 shows the scenario of events for the "general population" involving no vaccination case. Moreover, Figure 3.38 illustrates the scenario of events for the "young population" and mild epidemic year case. Figure 3.38 is just one example of the similar event trees that should be developed (an example for several epidemic and nonepidemic seasons and for all population groups).

According to the results summarized in Table 3.25, the benefits of a flu shot for the general population (generally the healthy) is not significant. Therefore, it would be a personal choice. However, the benefit of the vaccine is sizable for the young (primarily those between the ages 6 months and 7 years) and the elderly.

The mean risk values indicate that the general population risk is not significantly different with or without flu vaccination (especially in the nonepidemic years). The very young group show somewhat of a difference. By far, the highest impact can be seen in the elderly and chronically ill population who can see a risk reduction of as much as fivefold, if vaccinated. In Chapter 5, we will use the same example and consider uncertainties associated with the data in this example.

TABLE 3.24
Probability of Outcome of Flu on Subgroup Populations

Year	Young			Elderly and Chronically Ill		
	Severe Case	Mild Case	Mortality	Severe Case	Mild Case	Mortality
Epidemic	0.499	0.498	0.003	0.496	0.495	0.009
Mild epidemic	0.399	0.598	0.003	0.446	0.545	0.009
Nonepidemic	0.199	0.498	0.003	0.296	0.695	0.009

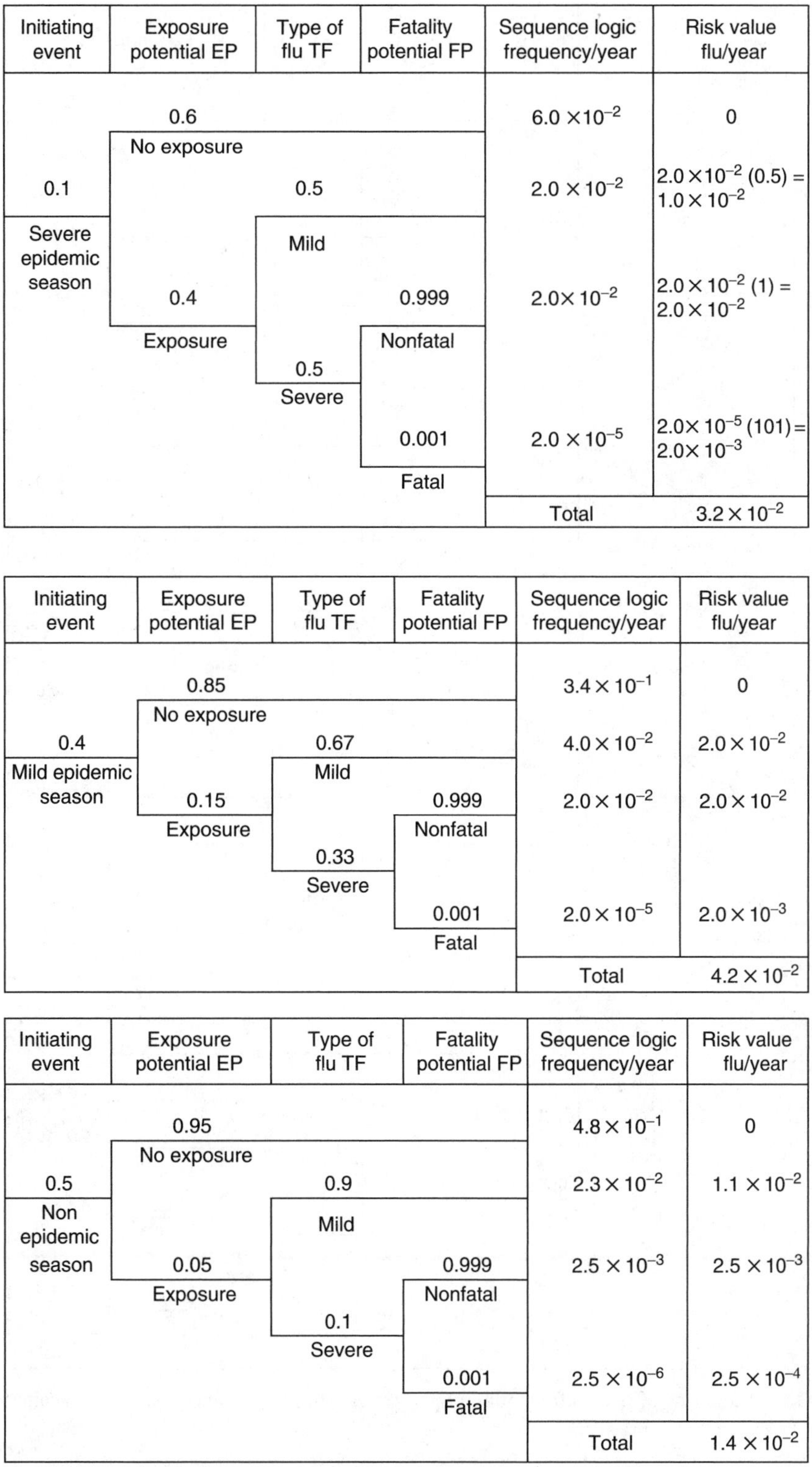

Initiating event	Exposure potential EP	Type of flu TF	Fatality potential FP	Sequence logic frequency/year	Risk value flu/year
	0.6			6.0×10^{-2}	0
	No exposure				
0.1		0.5		2.0×10^{-2}	$2.0 \times 10^{-2} (0.5) = 1.0 \times 10^{-2}$
Severe epidemic season		Mild			
	0.4		0.999	2.0×10^{-2}	$2.0 \times 10^{-2} (1) = 2.0 \times 10^{-2}$
	Exposure		Nonfatal		
		0.5			
		Severe			
			0.001	2.0×10^{-5}	$2.0 \times 10^{-5} (101) = 2.0 \times 10^{-3}$
			Fatal		
				Total	3.2×10^{-2}

Initiating event	Exposure potential EP	Type of flu TF	Fatality potential FP	Sequence logic frequency/year	Risk value flu/year
	0.85			3.4×10^{-1}	0
	No exposure				
0.4		0.67		4.0×10^{-2}	2.0×10^{-2}
Mild epidemic season		Mild			
	0.15		0.999	2.0×10^{-2}	2.0×10^{-2}
	Exposure		Nonfatal		
		0.33			
		Severe			
			0.001	2.0×10^{-5}	2.0×10^{-3}
			Fatal		
				Total	4.2×10^{-2}

Initiating event	Exposure potential EP	Type of flu TF	Fatality potential FP	Sequence logic frequency/year	Risk value flu/year
	0.95			4.8×10^{-1}	0
	No exposure				
0.5		0.9		2.3×10^{-2}	1.1×10^{-2}
Non epidemic season		Mild			
	0.05		0.999	2.5×10^{-3}	2.5×10^{-3}
	Exposure		Nonfatal		
		0.1			
		Severe			
			0.001	2.5×10^{-6}	2.5×10^{-4}
			Fatal		
				Total	1.4×10^{-2}

FIGURE 3.37 Scenarios of events for general population and no vaccination case.

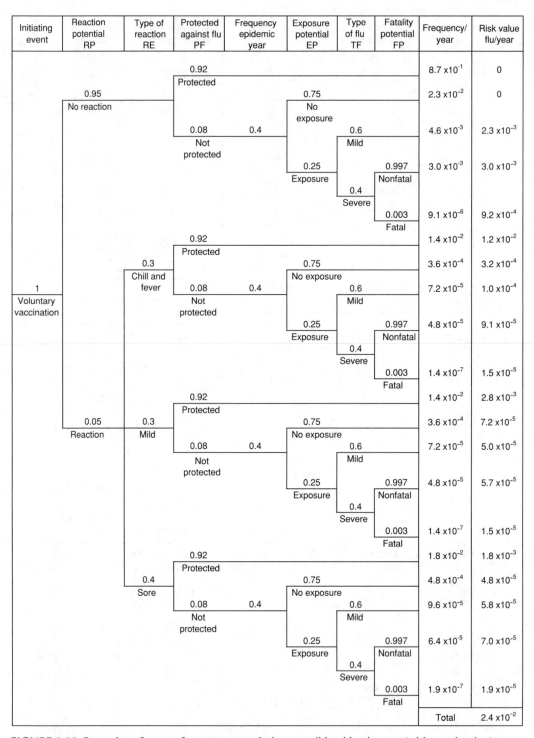

FIGURE 3.38 Scenarios of events for young population — mild epidemic year (with vaccination).

TABLE 3.25
Summary of Results

| Population | Type of Year | Total Risk Values (in FLU units) | |
		Without Vaccination	With Vaccination
General population	Epidemic year	3.2×10^{-2}	2.0×10^{-2}
	Mild epidemic year	4.2×10^{-2}	2.1×10^{-2}
	Nonepidemic year	1.4×10^{-2}	1.9×10^{-2}
Young	Epidemic year	4.9×10^{-2}	2.1×10^{-2}
	Mild epidemic year	8.2×10^{-2}	2.4×10^{-2}
	Nonepidemic year	4.9×10^{-2}	2.2×10^{-2}
Elderly and chronically ill	Epidemic year	7.4×10^{-2}	2.3×10^{-2}
	Mild epidemic year	1.2×10^{-1}	2.7×10^{-2}
	Nonepidemic year	8.3×10^{-2}	2.4×10^{-2}

Table for Example 3.7

| Failure Mode | Event Statistic | | |
	n_1	n_2	T (h) or N_D
Pump fails to start (PS)	10	1	500 (demands)
Pump fails to run (PR)	50	2	10,000 (h)
Valve fails to open (VO)	10	1	10,000 (demands)

Exercises

3.1 Consider the pumping system below. System cycles every hour. Ten minutes are required to fill the tank. Timer is set to open contact 10 min after switch is closed. Operator opens switch or the tank emergency valve if he/she notices an overpressure alarm. Develop a fault tree for this system with the top event "tank ruptures."

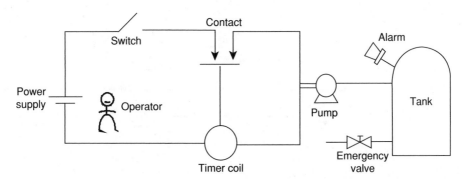

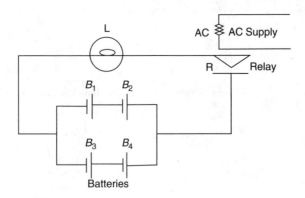

In this circuit, the relay is held open as long as AC power is available, and either of the four batteries is capable of supplying light power. Start with the top event "no light when needed."

    (a) Draw a fault tree for this system.
    (b) Find the minimal cut sets of the system.
    (c) Find the minimal path sets of the system.

  3.3 Consider the reliability diagram below.

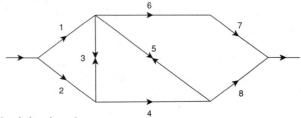

    (a) Find all minimal path sets.
    (b) Find all minimal cut sets.
    (c) Assuming each component has a reliability of 0.90 for a given mission time, compute the system reliability over mission time.

  3.4 Consider the fault tree below.

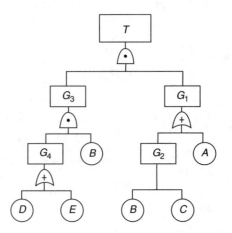

Find the following:

    (a) Minimal cut sets
    (b) Minimal path sets
    (c) Probability of the top event if the following probabilities apply:

$$\Pr(A) = \Pr(C) = \Pr(E) = 0.01$$
$$\Pr(B) = \Pr(D) = 0.0092$$

3.5 In the following fault tree,

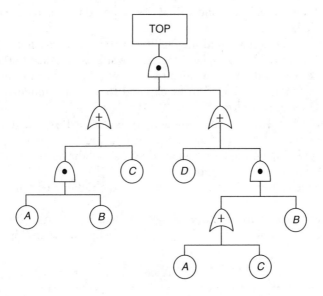

find all minimal cut sets and path sets. Assuming all component failure probabilities are 0.01, find the top event probability.

3.6 An event tree is used in reactor accident estimation as shown:

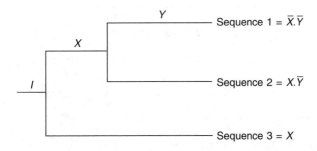

where sequence 1 is a success and sequences 2 and 3 are failures. The cut sets of system X and Y are

$$X = A \cdot B + A \cdot C + D$$

and

$$Y = B \cdot D + E + A$$

Find cut sets of sequences 2 and 3.

3.7 If a constant release of a gaseous low-level radioactive isotope from a medical facility is allowed, what would be the acceptable release rate (in terms of millirem/h) given the data below;

(a) Acceptable annual extra individual risk imposed by this release should not exceed 1-in-1,000,000.
(b) Population density around the facility 15,000 per square mile.
(c) Only a radius of I mile around the release site is considered affected.
(d) Because of variations in wind direction, weather, and movement of people, an average person is assumed to be exposed to a total cumulative exposure time of about 3 months per year.
(e) Due to a rapid decay, the radioactivity is decayed exponentially with a rate of 0.1 rem/mile as it moves from the release outward.
(f) The population dose relation which describes all the health effects of this exposure is 10,000 person-rem = 1 fatality.
(g) Ignore the change in the concentration as the radioactivity moves outward.

3.8 In a cement production factory, a system such as the one shown below is used to provide cooling water to the outside of the furnace. Develop an MLD for this system.

Table for Exercise 3.8

System bounds	$S_{12}, S_{11}, S_9, S_8, S_7, S_6, S_5, S_4, S_3, S_2, S_1$
Top event	Cooling from legs 1 and 2
Not allowed events	Passive failures and external failures
Assumptions	Only one of the pumps or legs is sufficient to provide the necessary cooling. Only one of the tanks is sufficient as a source

---

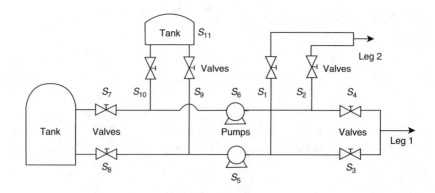

3.9 Develop a fault tree for the following block diagram system.

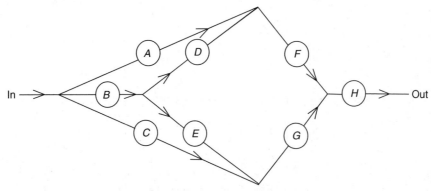

(a) Assume all component failure modes can be represented by a single basic event failure (for example, *A* means failure of "Component *A*" under all failure modes).
(b) Determine the cut sets of this fault tree using the Boolean substitution technique.
(c) Derive the path sets from the cut sets in question (b).

3.10 Develop an MLD model of the following system. Assume the following:

• One of the two product lines is sufficient for success.
• Control instruments feed the sensor values to the process-control computer, which calculates the position of the control valves.
• The plant computer controls the process-control computer.
(a) Develop a fault tree for the top event "inadequate product feed."
(b) Find all the cut sets of the top event.
(c) Find the probability of the top event.
(d) Determine which components are critical to the design.

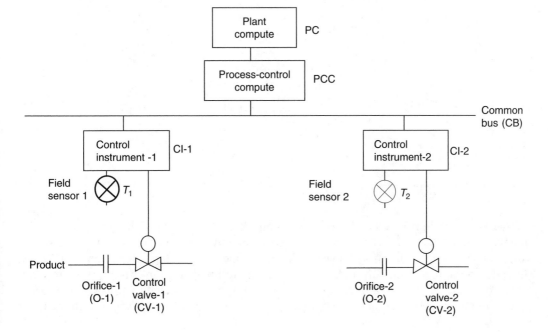

3.11 Consider the system shown below.

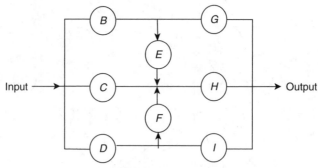

E and F are unidirectional

Develop a fault tree for this system with the top event of "No output from the system." Calculate the reliability of this system for a 100-h mission. Assume MTTF of 600 h for all components.

3.12 If a hazard should go through three independent barriers to be exposed, the probability of each barrier to fail is 0.001, and the consequence of hazard exposure is 3000 cancer-deaths, estimate the risk due to this hazard.

3.13 The system shown below discharges gas from a reservoir into a pressure tank.

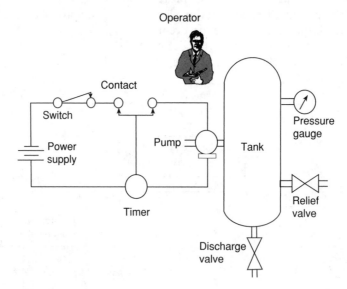

The switch is normally closed and the pumping cycle is initiated by an operator who manually resets the timer. The timer contact closes and the pumping starts. Later (well before any overpressure condition can exist) the timer times out and the timer contact opens. Current to the pump cuts off and pumping ceases, to prevent a tank rupture due to overpressure. If the timer contact does not open, the operator is instructed to observe the pressure gauge and to open the manual switch, thus causing the pump to stop. Even if both the timer and operator fail, overpressure can be relieved by the relief valve. An undesired event, from a risk viewpoint is a pressure tank ruptured by overpressure.

    (a) Develop scenarios leading to overpressure. The scenarios all started by the initiating event "pump overrun," and followed by functional requirements "operator shutdown," and "pressure protection." For, each scenario describe the end-effect (consequence) interims of tank rupture or no effect.

    (b) Develop three fault trees showing the ways that "pump overrun," "operator shutdown," and "pressure protection" could occur.

    (c) Calculate the risk of a tank rupture if probability of each basic event is 1×10^{-2}.

3.14 Consider the event tree and fault trees below.

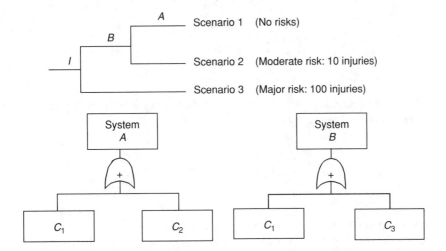

    (a) Determine a Boolean equation representing each of the event tree scenarios in terms of the fault tree basic events (C_1, C_2, and C_3).

    (b) If the frequency of the initiating event I is 10^{-3}/year, and $\Pr(C_1)=0.001$, $\Pr(C_2)=0.008$, and $\Pr(C_3)=0.005$, calculate the risk (injuries per year).

    (c) Plot the risk profile curve (Farmer's curve) for this problem.

3.15 A plant requires both systems A and B function so that an initiating event I is mitigated. If either A alone fails consequence x occurs. If B alone fails consequence $5x$ occurs, if both A and B fail consequence $100x$ occurs.

    (a) Draw an event tree representing this situation.

    (b) If $A = n + m \cdot h$, and $B = k + m \cdot n$ are cut sets of systems A and B, and if probability of occurrence of n, m, h, and k are equal (i.e., equal to 0.01), calculate probability of all sequences in the event tree (assume frequency of 1 is 0.1 per year).

    (c) Calculate and *draw* the risk profile as a function of x.

    (d) Calculate the expected loss (in terms of x).

3.16 Show an event tree representation of logic of a system composed of two parallel units.

3.17 Consider the event tree below:

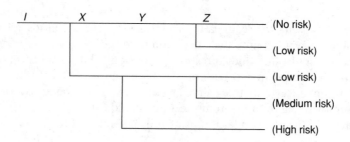

where systems X and Z are represented by the following Boolean expressions:

$$X = A + B \cdot C$$
$$Y = A \cdot B + D \cdot C$$
$$Z = C + E$$

(a) For the following frequencies and probabilities, determine the frequency of each scenario in the event tree above:

$$f(I) = 0.1/\text{year}$$
$$\Pr(A) = 0.001, \ \Pr(B) = 0.008, \ \Pr(C) = 0.01$$
$$\Pr(D) = 0.005, \ \Pr(E) = 0.008$$

(b) What is the total annual frequency of a risk?
(c) Plot the risk profile curve in complementary cumulative distribution form.

3.18 Consider the sequential series system below:

When a condition I (initiating event) exists, X is activated first, followed by Y and then Z. When this exact sequence successfully occurs, the sequential system properly responds to the initiating condition; otherwise, the system may be considered failed.

(a) Draw an event tree that depicts the minimum number of scenarios possible.
(b) If blocks X, Y, and Z are represented by the following reliability block diagrams, determine a Boolean representation of each event scenario in (a) in terms of A, B, C, D, and E.

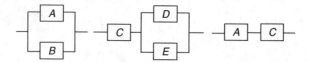

(c) If reliability of each block A, B, C, D, and E is expressed by

$$R(t) = \exp\left[-\left(\frac{t}{\alpha}\right)^{\beta}\right]$$

with $\beta = 1.5$, and $\alpha = 1000$ h, determine the unreliability of the system given the condition occurs at $t = 720$ h.

3.19 If an accident requires occurence of event I followed by events B or C so that a major consequence occurs:
(a) Develop an event tree to depict all scenarios that are possible.
(b) If frequency of event I is 0.1 per year and events B and C maybe obtained from the following fault trees (with probabilities assigned to each basic event), determine the frequency of each scenario.

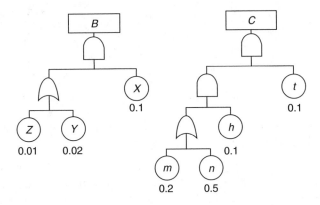

(c) If consequence of each scenario is 100 injuries, what is the total risk of this accident?

3.20 Consider the domestic gas boiler control system shown below. The two gas valves V_1 and V_2 are identical and in series; each valve has two possible failure modes:
(1) Valve stuck open (probability $F_1 = 10^{-4}$).
(2) Major leakage flow through valve (probability $F_2 = 10^{-3}$).

The three switches S_1, S_2, S_3 are identical and in series; each switch has three possible failure modes:

(1) Stuck closed (probability $F_3 = 2 \times 10^{-3}$).
(2) Short circuit (probability $F_4 = 2 \times 10^{-3}$).
(3) Earth fault (probability $F_5 = 10^{-3}$).

If a current flows through the circuit, the solenoids are energized and both valves V_1 and V_2 are opened.

(1) The probability of the pilot flame failing $F_6 = 10^{-1}$.
(2) The probability of a source of ignition existing $F_7 = 1$.
    (a) Draw a fault tree associated with the top event of "hazard of a gas explosion."
    (b) Use the data given to calculate the probability of a gas explosion.

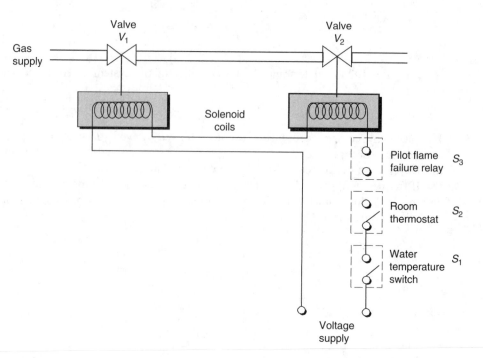

Schematic diagram of the control system for a domestic gas boiler. A system failure corresponds to a gas explosion due to unburnt gas being ignited.

3.21 In a system, three subsystems A, B, and C must work when the system starts. The frequency, I, that the system started is 2 times per month; (i) When the system starts, subsystem A is turned on. If subsystem A properly works, subsystem B would not be needed, after which subsystem C must work. If subsystem C functions properly, the mission is successful, otherwise the mission fails. (ii) If subsystem A fails, then subsystem B can compensate for it. If subsystem B is successful, then subsystem C would be needed. If subsystem C works the mission is not successful, then the mission fails. If subsystem B fails following the failure of A, then the mission fails.

(a) Draw an event tree representing the system above.
(b) If Boolean expressions $A = x + yz$, $B = y + t$, $C = tz$ show the minimal cut sets of subsystems A, B, and C, then find the *minimal* cut sets of each scenario (sequence) of the event tree.
(c) If the probability of the components or the subsystems are as follows, then calculate the frequency (per week) of each scenario: $\Pr(x) = \Pr(y) = 0.1$, $\Pr(z) = \Pr(t) = 0.2$.

3.22 For the system below,

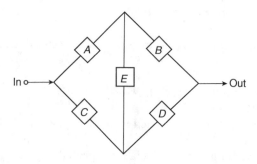

draw a fault tree with the top event "no flow out of the system." Determine the cut sets of the system using the Boolean substitution method. Only consider total failure of the blocks.

3.23 Consider the system below:

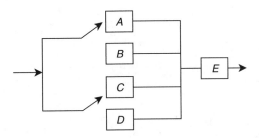

Two standby subsystems (A and B) and (C and D) are placed in parallel. The unit E should be in series with the two standby subsystems. Assume time to failure of A, B, C, D, and E are exponentially distributed and switching is perfect for the standby systems:

  (a) If $\lambda_A = \lambda_B = \lambda_C = \lambda_D = 10^{-3}$ per hour determine the reliability of each standby subsystem (that is reliability (A and B) and reliability (C and D)).
  (b) As a designer, if you are asked to select unit E such that its contribution to the failure of the whole system is less than 10% of the total system failure probability for a mission of $t = 1000$ h, what should be the minimum acceptable failure rate for unit E?

3.24 For the system below,

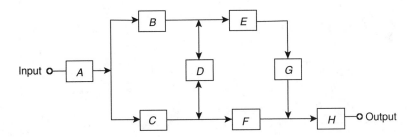

draw a fault tree with the top event (system "output" fails to occur). Write the cut sets using the inspection method. Compare the cut sets to your fault tree to verify them.

3.25 Consider the system below.

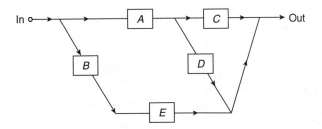

If $\Pr(A) = \Pr(B) = \Pr(C) = \Pr(D) = \Pr(E) = 0.05$ is the failure probability of components of this system for a given mission. Determine system failure probability using a fault tree evaluated with both the minimal cut sets and mutually exclusive sets (truth table).

Table for Exercise 3.25

Component	A	B	C	D	E	F	G	H
Failure rate	0.08	0.02	0.07	0.05	0.01	0.05	0.05	0.05

3.26 Consider the engineering system as shown below:

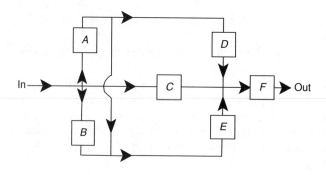

Determine the MTTF of this system if all the components are identical with a failure rate of $\lambda = 0.005$/month.

3.27 Consider the following two fault trees:

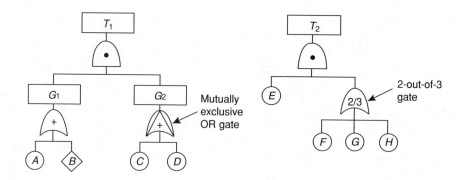

For both trees T_1 and T_2:

    (a) Find the minimal cut sets.
    (b) Find minimal path sets.
    (c) Draw the equivalent success tree.
    (d) Calculate the unreliability function of the two systems based on the constant failure rates.

REFERENCES

1. Stamatelatos, M., et al., *Probabilistic Risk Assessment Procedures Guide for NASA Managers and Practitioners*, Version 1.1, National Aeronautics and Space Administration, Washington DC, 2002.
2. Stamatis, D.H., *Failure Mode and Effect Analysis: FMEA from Theory to Execution*, 2nd edition, ASQ Quality Press, Wisconsin, USA, 2003.
3. Sattison, M.B., et al., Analysis of Core Damage Frequency: Zion, Unit 1 Internal Events, NUREG/CR-4550, 7, Rev. 1, 1990.
4. Modarres, M. and Dezfuli, H.A., Truncation methodology for evaluating large fault trees, *IEEE Transactions on Reliability*, R-33 (4), 325, 1984.
5. Chang, Y.H., Mosleh, A., and Dang, V., Dynamic Probabilistic Risk Assessment: Framework, Tool, and Application, Society for Risk Analysis Annual Meeting, Baltimore, 2003.
6. Dugan, J., Bavuso, S., and Boyd, M., Dynamic fault tree models for fault tolerant computer systems, *IEEE Transactions on Reliability*, 40 (3), 363, 1993.
7. Vesely, W.E., et al., *NASA Fault Tree Handbook with Aerospace Applications*, National Aeronautics and Space Administration (NASA), Washington DC, 2002.
8. Dezfuli, H., Modarres, M., and Meyer, J., Application of REVEAL_W to risk-based configuration control, *Engineering and System Safety Journal*, 44 (3), 243, 1994.
9. Sinnamon, R.M. and Andrews, J.D., Fault tree analysis and binary decision diagrams, in *Annual Reliability and Maintainability Symposium Proceedings*, The International Symposium on Product Quality and Integrity (Cat. No. 96CH35885), New York, 1996, 215.
10. Quantitative Risk Assessment System (QRAS). http://www.enre.umd.edu/srel/research/QRAS.html#one
11. Fleming, K.N., A reliability model for common mode failures in redundant safety systems, in Proceeding of the Sixth Annual Pittsburgh Conference on Modeling and Simulations, Instrument Society of America, Pittsburgh, PA, 1975.
12. Mosleh, A., et al., Procedure for Treating Common Cause Failures in Safety and Reliability Studies, U.S. Nuclear Regulatory Commission, NUREG/CR-4780, Vols. I & II, Washington DC, 1988.
13. Fleming, K.N., Mosleh, A., and Deremer, R.K., A systematic procedure for the incorporation of common cause event, into risk and reliability models, *Nuclear Engineering and Design*, 58, 415–424, 1986.
14. Mosleh, A. and Siu, N.O., A multi-parameter, event-based common-cause failure model, in Proceedings of the Ninth International Conference on Structural Mechanics in Reactor Technology, Lausanne, Switzerland, 1987.
15. Mosleh, A., Common cause failures; an analysis methodology and examples, *Reliability Engineering and System Safety*, 34, 249–292, 1991.
16. Atwood, C.L., Common Cause Failure Rates for Pumps, U.S. Nuclear Regulatory Commission, NUREG/CR-2098, Washington DC, 1983.
17. Rowe, W.D., *An Anatomy of Risk*, John Wiley & Sons, New York, 1977.

4 Performance Assessment: Data and Modeling

4.1 INTRODUCTION

Formal PRA requires probability of failure of hazard barriers. For example, the analyst needs probability of occurrence of failure events (failure of hardware, software, and human) and adverse conditions modeled in the fault trees and event trees of the PRA to estimate the risk values. For this purpose this chapter presents a quick review of the elements of performance assessment of hardware, software, and human elements (e.g., hazard barriers) included in the PRA, followed by techniques used to gather and evaluate performance data.

As discussed in Section 2.2, performance of a hazard barrier is referred to its ability to realize its intended functions at all times. It is composed of measuring its capability, efficiency, and availability. In practice, however, the two core constituents of performance are: *capability* and *availability*. Capability is the ability of the item or hazard barrier (system, structure, or component) to realize its intended function(s) under all possible conditions (normal and abnormal). For example to assure that an emergency core cooling system (a hazard barrier) in a nuclear power plant has the capacity (e.g., required flow) to overcome all challenges (e.g., loss of normal cooling flow) and cool the reactor, the system must have a minimum "flow requirement" or simply have an adequate capability. In a best-estimate approach to measuring capability, the concept of challenge vs. capacity may be used (as discussed in Section 2.2). Since there are uncertainties associated with the measures of capacity (strength, endurance, maximum flow capacity, etc.) and with the measures of challenge (stress, cumulative damage, minimum flow requirements, etc.), such uncertainties must be estimated and characterized. For example, considering challenge and capacity as uncertain random variables, then

$$\text{Capability value} = \text{Pr (capacity} > \text{challenges} \mid \text{all conditions)}$$

Examples of this include:

- Capability value (e.g., probability of maintaining an adequate flow margin) = Pr(emergency cooling flow, natural or forced > flow needed to prevent nuclear reactor fuel or cladding damage | possible loss of normal cooling).
- Capability value (e.g., probability of reactor vessel failure) = Pr(vessel plates and welds fracture toughness > thermally induced stress intensity | possible transients involving high rate of cooling and high vessel pressure).
- Capability value (e.g., probability of support structure failure) = Pr(structural metal yielding point > applied mechanical stresses | possible seismic events).

While performance of certain components, systems, and structures can be expressed by their capability values alone, for most active components and systems that undergo maintenance and experience degradation (such as pumps, motor-operated valves, switches), availability becomes another prime measure of performance. If we had the exact physics and engineering of failure models to estimate the probability of failure of components, systems, and structures, then one could just rely on stress–strength, degradation (or damage)–endurance, and performance–requirement models to calculate the probability of failures. However, these models are limited and not available for all components, systems, and structures. Therefore, time-to-failure (TTF) is considered as a random variable and historical data on time of failures and repairs have been gathered and used to estimate "availability" or its complement "unavailability" as the measure of performance.

Availability is generally the most appropriate performance measure for repairable items (i.e., active maintainable systems and components) because it takes into account both failure data (measured by reliability) and tests and maintenance downtime data (measured by surveillance, preventive, and corrective maintenance). As a quantitative measure, availability is defined as the probability that an item can operate at a specified time, given that it is used under stated conditions in an ideal support environment. If a system operates when in a good condition, availability can be defined as the probability that the hazard barrier is operating at a specified time. Availability at time t can be formally defined as:

$$\text{Availability} = \Pr(\text{hazard barrier is in good operating condition at time } t \mid \text{component or system capable}).$$

In this chapter, we will first discuss the methods of estimating availability (or reliability for nonrepairable barriers), then in Section 4.7 we will discuss the methods used to probabilistically assess capability of hazard barriers.

There are two general types of hazard barriers: active and passive. Active barriers are those that change their states to carry out their prevention, protection, or mitigation functions. Examples are hardware, software, and human which start a device and turn on the power, a fire fighter that responds to fire, and a software routine that calculates the amount of stress applied at a given point in the system. Passive barriers are those that need not change their states and continuously provide the required mitigation, protection, or prevention function. Examples are walls, containment vessels, pipes, natural circulation cooling, and gravity-driven cooling subsystems. Performance analysis of these two types of systems is different. For active systems the performance measure of interest is usually availability, whereas for passive systems it is capability. We will discuss these concepts in more detail.

Passive and active barriers can be further divided to subcategories of those that are: nonrepairable–nonreplaceable (such as software and hardware used in space applications), nonrepairable–replaceable (those involving reactive perfect renewal), repairable (items experiencing reactive and less than perfect renewal), periodically inspected, tested, or maintained with repair (units having proactive, but less than perfect renewal such as in preventive maintenance and condition monitoring).

Before we discuss the concept of reliability, repair and maintenance, and availability, we begin with introducing the concept of renewal process (RP). We then introduce the statistical estimation and methods applicable to repairable hardware items, followed by statistical estimation and methods for nonrepairable–replaceable hardware items. Then, we turn into the methods used to assess performance (i.e., capability) of passive hardware items. In Sections 4.8 and 4.9, performance measures for nonhardware hazard barriers (software and human) will be discussed. Usually reliability is the performance measure of interest to account for software and human errors, because they are nonrepairable. Finally, in cases where limited data are available, performance is estimated through expert elicitation and analysis. Section 4.10 discusses the formal methods used for expert estimation of performance.

4.2   ACTIVE HARDWARE PERFORMANCE ASSESSMENT

Active hardware performance assessment is usually carried out by observing and analyzing historical failures observed. We are also primarily interested in assessing the availability as the prime measure of performance for active hardware components. A second approach to measuring performance of active hardware is to model the performance through some engineering and physics of failure models of performance. This subject will also be discussed later in this chapter. Since availability has two components of failure and repair (and maintenance), we begin with understanding the mathematical underpinnings for modeling failure, repair, and maintenance processes. Therefore, we shall begin with the mathematical concept of renewal.

4.2.1   BASIC RANDOM PROCESSES USED IN PRA

For the situations where the downtime associated with maintenance, repair or replacement actions is negligible, compared with the mean-time-between-failures (MTBFs) of an item, the so-called *point processes* are used as probabilistic models of the failure processes. The point process can be informally defined as a model for randomly distributed events, having negligible duration. The following three-point processes are mainly used as probabilistic failure process models [1]: homogeneous Poisson process (HPP), RP, and nonhomogeneous Poisson process (NHPP). More recently, Kijima and Sumita [2] introduced the concept of generalized renewal process (GRP) to generalize the three-point processes discussed above. These processes will be discussed later in this section. For those situations where the respective downtime is not negligible, compared with MTBF, the so-called, alternating renewal process (ARP) is used.

Usually a point process is related to a single item (e.g., a barrier, subsystem). A *sample path* (*trajectory*) or *realization* of a point process is the successive failure times of the item: $T_1, T_2, \ldots, T_k \ldots$ (see Figure 4.1). We can also use the point process model for studying a group of identical items, if the number of items in the group is constant. We must also remember that the sampling scheme considered is "with instantaneous replacement."

A realization of a point process is expressed in terms of the *counting function*, $N(t)$, which is introduced as the number of failures, which occur during interval $(0, t]$ (see Leemis [1]), i.e., for $t > 0$

$$N(t) = \max(k | T_k \leq t) \tag{4.1}$$

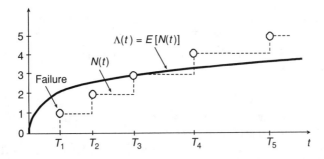

FIGURE 4.1 Geometric interpretation of $f(t)$, $N(t)$, $\Lambda(t)$ for a repairable item.

It is clear that $N(t)$ is a random function. Denote the mean of $N(t)$ by $\Lambda(t)$, i.e., $E[N(t)] = \Lambda(t)$. A realization, $N(t)$, and the respective $\Lambda(t)$ are shown in Figure 4.1. $\Lambda(t)$ and its derivative, $[\Lambda(t)]' = \lambda(t)$, known as the *rate of occurrence of failures* (ROCOF) or the *intensity function*, are the basic characteristics of a point process. Note that notation $\lambda(t)$ can be misleading; it should not be confused with the hazard (failure) rate function for which the same notation is often used. At this point, it is important to make clear that the difference between the failure (hazard) rate function, $h(t)$, and ROCOF, $\lambda(t)$. The hazard (failure) rate function, $h(t)$, is a characteristic of TTF distribution of a replaceable (nonrepairable) hardware item, whereas ROCOF is a characteristic of point process of a repairable item.

The sampling procedures associated with $f(t)$ and $h(t)$ is:

> N items are tested to failure or observed in the field without replacement (so the number of items is time-dependent); or an item is tested or observed to failure with instantaneous replacement by a new one from the same population.

From the standpoint of probabilistic interpretations, $f(t)$ is the unconditional probability density function (pdf), so, $\int_{t_1}^{t_2} f(t)dt$ is the unconditional probability of failure in the interval (t_1, t_2). Meanwhile, the hazard rate (or instantaneous failure rate) function, $h(t)$, is the conditional pdf, and $\int_{t_1}^{t_2} h(t)dt$ is the conditional probability of failure in the interval (t_1, t_2).

Under the sampling procedure for a point process, N items are observed or tested with instantaneous replacement by another one (not necessarily from the same population). The number of items observed or tested is always constant. The respective probabilistic interpretations of ROCOF, $\lambda(t)$, is given by the following equation:

$$\int_{t_1}^{t_2} \lambda(t)dt = E[N(t_2, t_1)]$$

where $E[N(t_1, t_2)]$ is the mean number of failures which occur in the interval (t_1, t_2).

Now, we can summarize the time-dependent performance (reliability or availability) behavior of repairable and nonrepairable items in terms of the hazard rate function and ROCOF [1]. The term *burn-in* is used for a nonrepairable item when its failure (hazard) rate function is decreasing in time, and the term *wear-out* is used when the failure (hazard) rate function is increasing. The life of nonrepairable item is described by the TTF distribution of a single nonnegative random variable.

For repairable items the term *improvement* is used when its ROCOF is decreasing and the term *deterioration* is used when its ROCOF is increasing. The life of repairable items, generally speaking, cannot be described by a distribution of a single nonnegative random variable; in this case such characteristics as time between successive failures are used (the first and the second, the second and the third, and so on).

4.2.2 Homogeneous Poisson Process

HPP is a process having constant ROCOF, λ, and is defined as a point process having the following properties:

(1) $N(0) = 0$.
(2) The process has independent increments (i.e., the numbers of failures observed in nonoverlapping intervals are independent).
(3) The number of failures, observed in any interval of length, t, has the Poisson distribution with mean λt.

The last property of the HPP is not only important for straightforward risk and reliability applications, but also can be used for the hypothesis testing that a random process considered is the HPP. It is obvious that the HPP is stationary. Consider some other useful properties of the HPP.

4.2.2.1   Superposition of the HPPs

As it was mentioned earlier, HPP, RP, and NHPP are used for modeling the failure behavior of a *single* item. In many situations it is important to model the failure pattern of several identical hardware items *simultaneously* (the items must be put in service or on a test at the same time). The superposition of several point processes is the ordered sequence of all failures that occur in any of the individual point process.

The superposition of several HP processes with parameters $\lambda_1, \lambda_2, \ldots, \lambda_k$ is the HPP with $\lambda = \lambda_1 + \lambda_2 + \cdots + \lambda_k$. The well-known example is a series system having multiple units (or barriers) in series.

4.2.2.2   Distribution of Intervals Between Failures

Under the HPP model, the distribution of intervals between successive failures is modeled by the exponential distribution with the constant intensity parameter rate λ. The HPP is the only process for which the failure rate of time-between-failures distribution coincides with its ROCOF.

Assume T_{n_0} is the time from an origin (operation of the item or start of the life test) to the n_0^{th} failure, where n_0 is a fixed (nonrandom) integer. In this notation the time to the first failure is T_1. It is clear that T_{n_0} is the sum of n_0 independent random variables, each exponentially distributed. The random variable, $2\lambda T_{n_0}$, has the chi-squared distribution with $2n_0$ degrees of freedom:

$$2\lambda T_{n_0} = \chi^2_{2n_0} \tag{4.2}$$

Later in this section, we will also be dealing with $\ln(T_{n_0})$. Using relationship (4.2), one can write

$$\ln(T_{n_0}) = \ln(2\lambda) + \ln\left(\chi^2_{2n_0}\right)$$

This expression shows that one has to deal with log chi-squared distribution, for which the following results of Bartlett and Kendall are available [3]. For the large samples the following (asymptotic) normal approximation for the log chi-squared distribution can be used:

$$\begin{aligned}
E(\ln T_{n_0}) &\approx \ln\left(\frac{n_0}{\lambda}\right) - \frac{1}{2n_0 - \frac{1}{3} + \frac{1}{16n_0}} \\
\operatorname{var}(\ln T_{n_0}) &\approx \frac{1}{n_0 - \frac{1}{2} + \frac{1}{10n_0}}
\end{aligned} \tag{4.3}$$

This approximation is used in the following as a basis for a failure trend analysis of hardware items of a hazard barrier.

4.2.3   RENEWAL PROCESS

The RP retains all the properties related to the HPP, except for the last property. In the case of RP the number of failures observed in any interval of length t, generally speaking, does not have to follow the Poisson distribution. Therefore, the time-between-failures distribution of RP can be any continuous distribution. Thus, RP can be considered as a generalization of HPP for the case when the time-between-failures is assumed to have any distribution [1].

The RP-based model is appropriate for the situations where an item is renewed to its original state (as good as new) upon failure. This model is not applicable in the case of a repairable system consisting of several components, if only a failed component is replaced upon failure.

Let $\Lambda(t) = E[N(t)]$, where $N(t)$ is given by (4.1). Function $\Lambda(t)$ is sometimes called the *renewal function*. It can be shown (see Hoyland and Rausand [4]) that $\Lambda(t)$ satisfies the, so-called, *renewal equation*:

$$\Lambda(t) = F(t) + \int_0^t F(t-s)\mathrm{d}\Lambda(s) \tag{4.4}$$

where $F(t)$ is the cumulative density function (cdf) of time-between-failures (t_is). By taking the derivative of both sides of (4.4) with respect to t, one gets the following integral equation for ROCOF, $\lambda(t)$,

$$\lambda(t) = f(t) + \int_0^t f(t-s)\lambda(s)\mathrm{d}s$$

where $f(t)$ is the pdf of $F(t)$. The integral equation obtained can be solved using a Laplace transformation. The solutions for the exponential and gamma distributions can be obtained in closed form. For the Weibull distribution only the recursion procedures are available (Hoyland and Rausand [4]). The possible numerical solutions for other distributions and different types of renewals can be obtained using Monte Carlo simulation. For more information see Kaminskiy and Krivtsov [5].

The statistical estimation of cdf or pdf of time-between-failures distribution on the basis of ROCOF or $\Lambda(t)$ observations is difficult.

For the HPP

$$\frac{\Lambda(t)}{t} = \lambda$$

In general case, the *elementary renewal theorem* states the following asymptotic property of the renewal function

$$\lim_{t\to\infty} \frac{\Lambda(t)}{t} = \frac{1}{\mathrm{MTBF}}$$

Some confidence limits for $\Lambda(t)$ are given in Hoyland and Rausand [4]. Contrary to the HPPs, the superposition of RPs, in general, is not an RP.

Example 4.1
TTF of a repairable hazard barrier in a facility is supposed to follow the Weibull distribution with scale parameter $\alpha = 100$ h and shape parameter $\beta = 1.5$. Assuming that repairs are perfect, i.e., the barrier is renewed to its original state upon failure assess the mean number of repairs during a mission time of 1000 h.

Solution

Use the elementary renewal theorem. The Weibull distribution mean is given by

$$\mathrm{MTBF} = \alpha \cdot \Gamma\left(\frac{\beta + 1}{\beta}\right)$$

where $\Gamma(\,\cdot\,)$ is a gamma function.

So, for the given values of α and β, MTBF $= 90.27$ h. Thus, the mean number of repairs during mission time $t = 1000$ h can be estimated as

$$\Lambda(1000) = \frac{1000}{90.27} = 11.08$$

4.2.3.1 Nonhomogeneous Poisson Process

The definition of the NHPP retains all the properties related to the HPP, except for the last one. In the case of NHPP, ROCOF λ is not constant, and the probability that exactly n failures occur in any interval, (t_1, t_2), has the Poisson distribution with the mean

$$\int_{t_1}^{t_2} \lambda(t)\mathrm{d}t$$

Therefore,

$$\Pr[N(t_2) - N(t_1) = n] = \frac{\left(\int_{t_1}^{t_2} \lambda(t)\mathrm{d}t\right)^n \exp\left(-\int_{t_1}^{t_2} \lambda(t)\mathrm{d}t\right)}{n!} \tag{4.5}$$

for $n = 0, 1, 2, \ldots$. The function

$$\Lambda(t) = \int_0^t \lambda(\tau)\mathrm{d}\tau$$

analogous to the renewal function is often called the *cumulative intensity function* [1], whereas the ROCOF $\lambda(t)$ is called the *intensity function*.

Unlike the HPP or the RP, the NHPP is capable of modeling, improving, and deteriorating systems. If the intensity function (ROCOF) is decreasing, the system is improving, and if the intensity function is increasing, the system is deteriorating. If the intensity function is not changing with time, the process reduces to the HPP.

It should be noted that the NHPP retains the independent increment property, but the time-between-failures are neither exponentially distributed nor identically distributed.

The reliability function for the NHPP can be introduced for a given time interval (t_1, t_2) as the probability of survival over this interval, i.e.,

$$R(t_1, t_2) = \Pr[N(t_2) - N(t_1) = 0] = \frac{\left(\int_{t_1}^{t_2} \lambda(t)\mathrm{d}t\right)^0 \exp\left(-\int_{t_1}^{t_2} \lambda(t)\mathrm{d}t\right)}{0!}$$
$$= \exp\left(-\int_{t_1}^{t_2} \lambda(t)\mathrm{d}t\right) \tag{4.6}$$

It is obvious that in the case of the HPP (where $\lambda = $ const.) this function is reduced to the conditional reliability function for the exponential distribution.

4.2.4 ANALYSIS OF LESS THAN PERFECT REPAIR: GENERALIZED RENEWAL PROCESS

Repairable hazard barriers and systems may end up in one of five possible states after a repair:

1. As good as new.
2. As bad as old.
3. Better than old but worse than new.
4. Better than new.
5. Worse than old.

The RP and the NHPP, discussed earlier, account for the first two states, respectively. However, a practical and accurate approach is needed to address the remaining after repair states. The main reason as to why the last three repair states have not received much attention appears to be the difficulty in developing a mathematically robust and efficient approach to represent them.

Kijima and Sumita [2] have proposed a new probabilistic model to address all after repair states called GRP. Kaminskiy and Krivtsov [5] have extended this GRP approach and have offered a Monte Carlo-type solution for certain application areas. They have also shown that the RP and the NHPP are specific cases of GRP.

Kaminskiy and Krivtsov [5] and Krivtsov [6] describe the details of a Monte Carlo-based approximate solution to the GRP model. The key assumption that makes the Kaminskiy and Krivtsov's Monte Carlo simulation of the GRP possible is that the time to first failure (TTFF) distribution is known and can be estimated from available data. Further, the repair time is assumed negligible so that the failures can be viewed as point processes. Their Monte Carlo solution randomly selects and sets values of the GRP parameters, and makes estimation of the expected number of failures. The expected number of failures is then compared with the actual experience through a formal least-square method. The process is repeated and the expected number of failures with smallest least-square value is selected as the best solution. Considering the complexity of the GRP model and the Monte Carlo simulation, the computation is quite involved and time-consuming.

Kaminskiy and Krivtsov's Monte Carlo approach assumes that a large amount of failure data is available. The availability of such data allows for estimation of the TTFF distribution with a high degree of accuracy. In some cases it would be difficult to obtain the same amount of data, because only limited number of identical equipments is used or they work in different operating and environmental conditions. Further, the population of such equipment is not very large in a given facility. More advanced techniques that do not require establishing TTFF has been proposed by Yanez et al. [7].

A repairable item (hazard barrier or subsystem) is one that undergoes repair and can be restored to an operation by any method other than replacement of the entire system. Figure 4.2 shows a categorization of the stochastic point processes for modeling repairable systems.

The RP assumes that following a repair the system returns to an "as-good-as-new condition," and the NHPP assumes that the system returns to an "as bad as old condition." These are the most commonly used methods for the evaluation of repairable systems. Because of the need to have more accurate analyses and predictions, the GRP can be of great interest to reduce the modeling uncertainty resulting from the above repair assumptions.

4.2.4.1 Generalized Renewal Process

Consider Krivtsov [6] description of the GRP by introducing the concept of virtual age (A_n). Parameter A_n represents the calculated age of the item immediately after the nth repair occurs.

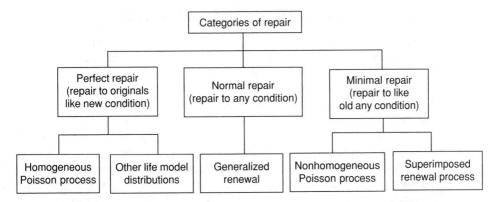

FIGURE 4.2 Categories of stochastic point processes for repairable, Yanez et al. [7].

If $A_n = y$, then the system has a time to the $(n+1)$th failure, X_{n+1}, which is distributed according to the following cumulative distribution function

$$F(X|A_n = y) = \frac{F(X+y) - F(y)}{1 - F(y)} \tag{4.7}$$

where $F(X)$ is the cdf of the TTFF distribution of a new system or component. The summation $S_n = \sum_{i=1}^{n} X_i$, with $S_0 = 0$, is called the real age of the system. It is assumed that the nth repair would only compensate for the damage accumulated during the time between the $(n-1)$th and the nth failure. With this assumption the virtual age of the system after the nth repair is:

$$A_n = A_{n-1} + qX_n = qS_n, \quad n = 1, 2 \ldots \tag{4.8}$$

where q is the *repair effectiveness* parameter (or rejuvenation parameter) and $A_0 = 0$. If the times between the first and second failure, the second and third failure, etc., are considered, they can be expressed as:

$$A_1 = qX_1, \quad A_2 = q(X_1 + X_2), \quad A_n = q(X_1 + X_2 + \cdots + X_n)$$

or simply,

$$A_n = q \sum_{i=1}^{n} X_i$$

According to this model, the result of assuming a value of $q = 0$ leads to an RP (as good as new), whereas the assumption of $q = 1$ leads to a NHPP (as bad as old). The values of q that fall in the interval $0 < q < 1$ represent the after repair state in which the condition of the system is "better than old but worse than new." For the cases where $q > 1$, the system is in a condition of "worse than old." Similarly, cases where $q < 0$ suggests a system restored to a condition of "better than new." Therefore, physically speaking, q can be viewed as an index for representing effectiveness and quality of repair. For $q = 0$, the repair is ideal and the renewal is perfect.

4.3   STATISTICAL DATA ANALYSIS FOR HARDWARE ITEMS

From the discussion in Section 4.2, it is evident that the HPP is the simplest approach for repairable hardware data analysis. For example, in such situations the procedures for estimating parameters of an exponential distribution representing TTF is to use classical interval estimation, and when data are scarce use Bayes' estimation. The main underlying assumption for these procedures, when applied to repairable items, is that ROCOF, λ, is constant and will remain constant over all time intervals of interest. Therefore, the data should be tested for potential increasing or decreasing trends.

The use of the estimators for HPP is justified only after it has been proven that the ROCOF is reasonably constant, i.e., there is no evidence of any increasing or decreasing trend. An increasing trend is not necessarily due to random aging processes. Poor use of equipment, including poor testing, maintenance, and repair work, and out-of-spec (overstressed) operations, can lead to premature aging and be major contributions to increasing trends.

Figure 4.3 depicts three cases of occurrences of failure in a repairable system. The constant ROCOF estimators give the same point and confidence estimates for each of the three situations shown in Figure 4.3, because the number of failures and length of experience are the same for each. Clearly, Case 1 has no trend, while Case 2 shows a decreasing ROCOF, and Case 3 shows an increasing ROCOF. We would therefore expect that, given a fixed time interval in the future, the barrier or item exhibiting Case 3 failures, would be more likely to fail at a time in future than a hazard barrier or item experiencing the other two cases.

This shows the importance of considering trends in ROCOF when predicting performance. Considering Ascher and Feingold [8] and O'Connor [9], the following points should be considered in ROCOF trend analyses:

1. Failure of a barrier or subsystem may be partial, and repair work done on a failed unit may be imperfect. Therefore, the time periods between successive failures are not necessarily independent. This is a major source of trend in ROCOF.
2. Imperfect repairs performed following failures do not renew the item, i.e., the item will not be as good as new following maintenance or repair. The constant ROCOF assumption holds only if the item is assumed to be as good as new.
3. Repairs made by adjusting, lubricating, or otherwise treating barriers or other items that are wearing out provide only a small additional capability for further operation, and do not renew the item. These types of repair may result in a trend of an increasing ROCOF.
4. A hazard barrier may fail more frequently due to aging and wearing out.

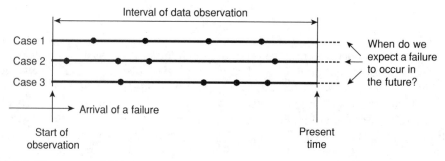

FIGURE 4.3   Three cases of failures occurrence.

In the remainder of this section, we provide a summary of a typical trend analysis process, and discuss the subsequent calculation of performance estimates. Several procedures may be used to check the HPP model assumptions. For example, the goodness-of-fit criteria can be applied to testing the exponential distribution of MTBFs, or the Poisson distribution of the number of failures observed in equal length time intervals.

4.3.1 PARAMETER ESTIMATION FOR THE HPP USING COMPLETE DATA

Suppose that failure process of a hardware item is used as a hazard barrier observed for a predetermined time t_0 during which n failures have been recorded at times $t_1 < t_2 < , \ldots, t_n$, where, obviously, $t_n < t_0$. The process is assumed to follow a HPP.

The corresponding likelihood function can be written as

$$L = \lambda^n \exp\left(-\lambda t_0\right) \tag{4.9}$$

It is clear that, with t_0 fixed, the number of events, n, is a *sufficient statistic* (note that one does not need to know $t_1, t_2, \ldots, t_n$ to construct the likelihood function). Thus, the statistical inference can be based on the Poisson distribution of the number of events. As a point estimate, one usually takes n/t_0, which is the unique unbiased estimate based on the sufficient statistic. Since estimation of the parameter λ of HPP is a special case of NHPP, we will expand on this topic in Section 4.3.2.

4.3.2 PARAMETER ESTIMATION FOR NHPP USING COMPLETE DATA

As it was mentioned above, the NHPP can be used to model improving and deteriorating hazard barrier units: if the intensity function (i.e., the ROCOF) is decreasing, the unit is improving, and if the intensity function is increasing, the unit is deteriorating. Formally, we can test for trend, taking the null hypothesis of no trend, i.e., the events form the HPP and applying a goodness-of-fit test for the exponential distribution of the intervals between successive failures, the Poisson distribution of the number of failures in the time intervals of constant (nonrandom) length. A simple graphical procedure based on this property is to plot the cumulative number of failures versus the cumulative time. Deviation from linearity indicates the presence of a trend.

These tests are not sensitive enough against the NHPP alternatives, so it is better to apply the following methods [3].

Under the NHPP model the intervals between successive events are independently distributed and the probability that, starting from time t_i, the next failure occurs in $(t_{i+1}, t_{i+1} + \Delta t)$ can be approximated by [3]:

$$\lambda(t_{i+1})\Delta t \exp\left(-\int_{t_i}^{t_{i+1}} \lambda(x)dx\right)$$

where the first multiplier is the probability of failure in $(t_{i+1}, t_{i+1} + \Delta t)$, and the second one is the probability of a failure-free operation in the interval (t_i, t_{i+1}).

If the data are the successive failure times, $t_1, t_2, \ldots, t_n$, $(t_1 < t_2 < \cdots < t_n)$ observed in the interval $(0, t_0)$, $t_0 > t_n$, the likelihood function for any $\lambda(t)$ dependence, can be written as

$$\prod_{i=1}^{n} \lambda(t_i) \exp\left(-\int_0^{t_1} \lambda(x)\,dx\right) \exp\left(-\int_{t_1}^{t_2} \lambda(x)\,dx\right) \ldots \exp\left(-\int_{t_{n-1}}^{t_n} \lambda(x)\,dx\right) \exp\left(-\int_{t_n}^{t_0} \lambda(x)\,dx\right)$$

$$= \prod_{i=1}^{n} \lambda(t_i) \exp\left(-\int_0^{t_0} \lambda(x)\,dx\right) \tag{4.10}$$

The corresponding log-likelihood function is given by

$$l = \sum_{i=1}^{n} \ln \lambda(t_i) - \int_0^{t_0} \lambda(x)\,dx \tag{4.11}$$

To avoid complicated notation, consider the case when ROCOF takes the simple log-linear form of

$$\lambda(t) = \exp(\alpha + \beta t) \tag{4.12}$$

Note that the model above is, more general than the linear one, $\lambda(t) = \alpha + \beta t$, which can be considered as a particular case of (4.12), when $\alpha + \beta t \ll 1$.

Plugging (4.12) in (4.10) and (4.11) one gets

$$L_1(\alpha, \beta) = \exp\left[n\alpha + \beta \sum_{i=1}^{n} t_i - \frac{e^{\alpha}(e^{\beta t_0} - 1)}{\beta} \right] \tag{4.13}$$

$$l = \ln[L_1(\alpha, \beta)] = n\alpha + \beta \sum_{i=1}^{n} t_i - \frac{e^{\alpha}(e^{\beta t_0} - 1)}{\beta} \tag{4.14}$$

The conditional likelihood function can also be found by dividing (4.13) by the marginal probability of having observed n failures, which is given by the respective term of the Poisson distribution with mean

$$\int_0^{t_0} \lambda(x)\,dx = \frac{e^{\alpha}(e^{\beta t_0} - 1)}{\beta}$$

The conditional likelihood function is given by Cox and Lewis [3]

$$L_c = \frac{\exp\left(n\alpha + \beta \sum_{i=1}^{n} t_i\right) \exp\left(-\dfrac{e^{\alpha}e^{\beta t_0} - 1}{\beta}\right)}{\dfrac{(e^{\alpha}e^{\beta t_0} - 1)^n}{\beta^n n!} \exp\left(-\dfrac{e^{\alpha}e^{\beta t_0} - 1}{\beta}\right)} = \frac{\beta^n n!}{(e^{\beta t_0} - 1)^n} \exp\left(\beta \sum_{i=1}^{n} t_i\right) \tag{4.15}$$

Because $0 < t_1 < t_2 < \cdots < t_n < t_0$, the conditional likelihood function (4.15) represents the pdf of an ordered sample of size n from the truncated exponential distribution having the pdf

$$f(t) = \frac{\beta}{e^{\beta t_0} - 1} e^{\beta t}, \quad 0 \le t \le t_0, \quad \beta \neq 0 \tag{4.16}$$

Thus, for any β the conditional pdf of the random variable $T = \Sigma t_i$ is the sum of n independent random variables having the pdf (4.16). It is easy to see that for $\beta = 0$, the pdf (4.16) becomes the uniform distribution over $(0, t_0)$. Recent studies have shown that generally the NHPP assumption gives results that are very close to those of GRP. This is an important observation, as NHPP requires far less computational effort.

Example 4.2

In a repairable hardware unit, the following complete eight failures have been observed at: 595, 905, 1100, 1250, 1405, 1595, 1850, and 1995-h. Assume that the time-to-repair is negligible. Test whether these data exhibit any trend in the form of (4.12).

Solution

Taking the derivative of (4.14) with respect to α and β and equating them to zero, results in the following system of equations for maximum likelihood estimates of these parameters

$$\sum_{i=1}^{n} t_i + \frac{n}{\beta} - \frac{nt_n e^{\beta t_n}}{e^{\beta t_n} - 1} = 0$$

$$e^\alpha = \frac{n\beta}{e^{\beta t_n} - 1}$$

For the data given $n = 8$, $t_n = 1995$ h, and $\Sigma t_i = 10{,}695$ h, solving these equations numerically yield estimates of the parameter and, therefore

$$\lambda(t) = \exp(-6.8134 + 0.0011t)$$

Another form of $\lambda(t)$ considered by Bassin [10, 11] and Crow [12] is the Power Law form

$$\lambda(t) = \lambda \beta\, t^{\beta-1} \qquad (4.17)$$

Expression (4.17) has the same form as the failure (hazard) rate of nonrepairable items for the Weibull distribution. Using (4.6), the reliability function of a repairable system having ROCOF (4.17) for an interval $(t, t+t_1)$ can be obtained as follows:

$$R(t, t+t_1) = \exp\left[-\lambda (t+t_1)^\beta + \lambda t^\beta\right] \qquad (4.18)$$

Crow [12] has shown that under the condition of a single system observed to its nth failure, the maximum likelihood estimator (MLE) of parameters of (4.17) can be obtained as:

$$\hat{\beta} = \frac{n}{\sum_{i=1}^{n-1} \ln\left(\frac{t_n}{t_i}\right)} \qquad (4.19)$$

$$\hat{\lambda} = \frac{n}{t_n^{\hat{\beta}}} \qquad (4.20)$$

Where n is the numbered data points, t_i is the arrival time (or demand) and t_n is the observation time.

The $1-\alpha$ confidence limits for inferences on β and λ have been developed and discussed by Bain [13]. Also, the readers can refer to (4.65)–(4.70) for approximate confidence intervals of the parameters. Crow [14] has expanded estimates (4.19) and (4.20) to include situations where data originate from multiunit repairable systems.

Example 4.3

Consider the data shown in Table 4.1, calculate the MLE of β and λ assuming NHPP.

TABLE 4.1
Arrival and Interarrival for the Valve

Date	Interarrival days	Arrival days
April 20, 1980	104	104
September 19, 1980	131	235
October 09, 1985	1597	1832
December 16, 1985	59	1891
December 12, 1985	1895	
July 24, 1987	503	2398
January 22, 1988	157	2555
January 29, 1988	6	2561
June 15, 1988	118	2679
January 01, 1989	173	2852
May 12, 1989	113	2966
July 23, 1989	62	3028
November 17, 1989	101	3129
July 24, 1990	216	3345
November 23, 1990	106	3451
May 04, 1991	140	3591
May 05, 1991	1	3592
August 31, 1991	102	3694
September 04, 1991	3	3697
December 02, 1992	393	4090
March 23, 1993	96	4186
December 16, 1993	232	4418
March 28, 1994	89	4507
June 06, 1994	61	4568
July 19, 1994	37	4605
June 23, 1995	293	4898
July 01, 1995	7	4905
January 08, 1996	165	5070
April 18, 1996	86	5157
August 11, 1996	99	5256

Solution

Using (4.19) and (4.20), we can calculate $\hat{\beta}$ and $\hat{\lambda}$ as 1.59 and 3.71×10^{-5}, respectively. Using $\hat{\beta}$ and $\hat{\lambda}$, the functional form of the demand failure rate can be obtained by using (4.17) as

$$\lambda(t) = 3.71 \times 10^{-5} \times 1.59 t^{0.59},$$

where t represents the time in days.

4.3.2.1 Laplace's Test

Now we are going to use the conditional pdf (4.16) to test the null hypothesis, $H_0: \beta = 0$, against the alternative hypothesis $H_1: \beta \neq 0$. This test is known as the Laplace's test (sometimes it is also called the Centroid test). As mentioned earlier, under the condition of $\beta = 0$, pdf (4.16) is reduced to the uniform distribution over $(0, t_0)$ and $S = \Sigma t_i$ has the distribution of the sum of n independent uniformly distributed random variables. Thus, one can use the distribution of the following statistic:

$$U = \frac{S - \frac{nt_0}{2}}{\sqrt{t_0 \left(\frac{n}{12}\right)}} = \frac{\frac{\sum_{i=1}^{n} t_i}{n} - \frac{t_0}{2}}{t_0 \sqrt{\frac{1}{12n}}} \tag{4.21}$$

which approximately follows the standard normal distribution [3].

If the alternative hypothesis is $H_1: \beta \neq 0$, then the large values of $|U|$ indicate evidence against the null hypothesis. In other words, if U is close to 0, there is no evidence of trend in the data, and the process is assumed to be stationary (i.e., an HPP). If $U < 0$, the trend is decreasing, i.e., the intervals between successive failures (interarrival values) are becoming larger. If $U > 0$, the trend is increasing. For the latter two situations, the process is not stationary (i.e., it is an NHPP).

If the data are failure terminated (Type II censored), statistic (4.21) is replaced by

$$U = \frac{\frac{\sum_{i=1}^{n-1} t_i}{n-1} - \frac{t_n}{2}}{t_n \sqrt{\frac{1}{12(n-1)}}} \tag{4.22}$$

Example 4.4
In a repairable unit, the following six interarrival times-between-failures have been observed: 16, 32, 49, 60, 78, and 182 (in hours). Assume the observation ends at the time when the last failure is observed.

(a) Test whether these data exhibit a trend. If so, estimate the trend model parameters as given in (4.21).
(b) Find the probability that the interarrival time for the seventh failure will be greater than 200 h.

Solution
Use the Laplace's to test the null hypothesis that there is no trend in the data at 10% significance level (the respective acceptance region is $(-1.645, +1.645)$). From (4.22) find

$$U = \frac{\frac{16 + (16+32) + \cdots}{5} - \frac{417}{2}}{417 \sqrt{\frac{1}{12(5)}}} = -1.82$$

Notice that $t_n = 417$. The value of U obtained indicates that the NHPP can be applicable (H_0 is rejected) and the sign of U shows that the trend is decreasing.

Using (4.19) and (4.20), we can find

$$\hat{\beta} = \frac{6}{\ln \frac{417}{16} + \ln \frac{417}{16+32} + \cdots} = 0.712$$

$$\hat{\lambda} = \frac{6}{(417)^{0.712}} = 0.0817 \, h^{-1}$$

Thus, $\hat{\lambda} = 0.058 t^{-0.288}$. From (4.18) with $t_1 = 200$,

Pr(the seventh failure occurring within 200 h) $= 1 - \exp[-\lambda((t_0 + t_1)^\beta + \lambda(t_0)^\beta)] = 0.85$.
The probability that the interarrival time is greater than 200 h is $1 - 0.85 = 0.15$.

4.4 AVAILABILITY AS A MEASURE OF PERFORMANCE FOR REPAIRABLE HARDWARE ITEMS

We define reliability as the probability that a hazard barrier (or its components) will perform its required function without failure over a given time interval (mission). The notion of availability is related to repairable items only. Availability differs from reliability in that it considers renewal in terms of repair and maintenance. As such, availability is the probability that a repairable item will function at time t, as it is supposed to, when called upon to do so. Since it is possible that the item may be under maintenance (e.g., under preventive maintenance and not necessarily failed, but nonfunctional), availability is a broad measure of performance. Conversely, the unavailability of a repairable item, $q(t)$, is defined as the probability that the item is nonfunctional at time t. There are several representations of availability; the most common ones are:

(1) *Instantaneous (point) availability* of a repairable item at time t, $a(t)$, is the probability that the item or hazard barrier is functioning (up) at time t.
(2) *Limiting availability*, a, is defined as the following limit of instantaneous availability, $a(t)$

$$a = \lim_{t \to \infty} a(t) \tag{4.23}$$

(3) *Average availability*, $\bar{a}$ is defined for a fixed time interval, T, as

$$\bar{a} = \frac{1}{T} \int_0^T a(t)\, \mathrm{d}t \tag{4.24}$$

(4) The respective *limiting average availability* is defined as

$$\bar{a}_1 = \lim_{T \to \infty} \frac{1}{T} \int_0^T a(t)\, \mathrm{d}t \tag{4.25}$$

It should be noted that the limiting average availability has narrow applications. We elaborate on these definitions of availability in the remainder of this section.

If a hazard barrier or item is nonrepairable, its availability coincides with its reliability function, $R(t)$, i.e.,

$$a(t) = R(t) = \exp\left[\int_0^t \lambda(\tau)\, \mathrm{d}\tau\right] \tag{4.26}$$

where $\lambda(t)$ is the failure (hazard) rate function. The unavailability, $q(t)$, is obviously related to $a(t)$ as

$$q(t) = 1 - a(t) \tag{4.27}$$

From the modeling point of view, repairable barriers (here hardware items) can be divided into the following two groups:

(1) Repairable hardware items for which failure is immediately detected (revealed faults).
(2) Repairable hardware items for which failure is detected upon inspection (sometimes referred to as periodically inspected or tested items).

4.4.1 INSTANTANEOUS (POINT) AVAILABILITY

For the first group of hardware items, it can be shown (see Modarres et al. [15]) that by using a Markov process representing two state of "down" and "up," the probability that the item remain in each of these states are availability, $a(t)$, and unavailability, $q(t)$, respectively. The following system of ordinary differential equations represents this Markov process:

$$\frac{da(t)}{dt} = -\lambda(t)a(t) + \mu(t)q(t)$$
$$\frac{dq(t)}{dt} = \lambda(t)a(t) - \mu(t)q(t) \tag{4.28}$$

where transition rates between the two states are ROCOF $\lambda(t)$, and the rate of occurrence repair or maintenance, $\mu(t)$. In this case examples of the mathematical forms of $\lambda(t)$ and $\mu(t)$ are those of (4.12) and (4.17).

For the simple case where no trend in the ROCOF and rate of occurrence of repair exist, solving (4.28) yield,

$$a(t) = \frac{\mu}{\lambda + \mu} + \frac{\lambda}{\lambda + \mu} \exp\left[-(\lambda + \mu)t\right]$$
$$q(t) = \frac{\mu}{\lambda + \mu} - \frac{\lambda}{\lambda + \mu} \exp\left[-(\lambda + \mu)t\right] \tag{4.29}$$

Note that in (4.29), $\mu = 1/\tau$, where τ is the average time required to repair or maintain the hardware item. Sometimes τ is referred to as the mean-time-to-repair (MTTR). Clearly, MTBF $= 1/\lambda$ in this case.

For simplicity, the pointwise availability function can be represented in an approximate form. This simplifies availability calculations significantly. For example, for a periodically tested or inspected item, if the maintenance, repair, and test durations are very short compared with the length of operation time, and the maintenance, repair, and test are assumed perfect, one can neglect their contributions to unavailability of the system. This can be shown using Taylor expansion of the unavailability equation (see Lofgren [16]). In this case for each test interval T, the availability and unavailability functions are

$$a(t) \approx 1 - \lambda T$$
$$q(t) \approx \lambda T \tag{4.30}$$

The plot of the unavailability as a function of time, using (4.30), will take a shape similar to that in Figure 4.4. Clearly if the test and repair durations are long they affect the availability and, one must include their contribution to the item's performance.

Vesely et al. [17] have used the approximate pointwise unavailability functions for this purpose. The functions and their plot are shown in Figure 4.5. The average values of the approximate unavailability functions shown in Figure 4.4 and Figure 4.5 are presented in Table 4.2.

It should be noted that, due to imperfect test and repair activities, it is possible that a residual unavailability q_0 would remain following a test and repair. Thus, unlike the unavailability function shown in Figure 4.4, the unavailability function in Figure 4.5 exhibits a residual unavailability q_0 due to these random imperfections. If the operation and maintenance is degrading then q_0 becomes time-dependent. Similarly, f_r, λ, τ_r, and τ_t will be time-dependent.

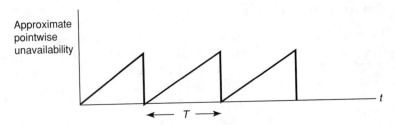

FIGURE 4.4 Approximate pointwise unavailability for a periodically tested item.

The equation representing the pointwise unavailability of periodically tested or inspected items, when there is no trend in ROCOF and rate of occurrence of repair or maintenance is obtained from:

$$\bar{q} = q_0 + \frac{1}{2}\lambda T_0 + f_r \frac{\tau_r}{T} + \frac{\tau_t}{T}$$

(4.31)

$$T = T_0 + \tau_t + f_r \tau_r$$

where τ_t = length of test, preventive maintenance or both, and τ_r = length of repair or corrective maintenance given the test reveals a failure or fault.

4.4.2 LIMITING POINT AVAILABILITY

It is easy to see that some of the pointwise availability equations discussed in Section 4.4.1 has limiting values. For example, (4.29) has the following limiting value:

$$a = \lim_{t \to \infty} a(t) = \frac{\mu}{\lambda + \mu}$$

or its equivalent

$$a = \frac{\text{MTBF}}{\text{MTTR} + \text{MTBF}}$$

(4.32)

Equation (4.32) is also referred to as the asymptotic availability of a repairable system with a constant ROCOF.

TABLE 4.2
Average Availability Functions of Hardware Items

Type of Item/Hazard Barrier	Average Unavailability
Nonrepairable	$\frac{1}{2}\lambda T$
Repairable revealed fault	$\frac{\lambda\tau}{1 + \lambda\tau}$
Repairable periodically tested (assuming no residual unavailability)	$\frac{1}{2}\lambda T_0 + f_r \frac{T_R}{T} + \frac{T_t}{T}$

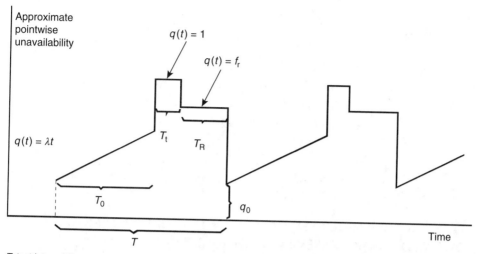

T, test interval; T_R, average repairtime (h); T_t, average test duration (h); f_r, frequency of repair; q_0 residual unavailability.

FIGURE 4.5 Pointwise unavailability for a periodically tested item including test and repair outages.

4.4.3 AVERAGE AVAILABILITY

According to its definition, average availability is a point measure of availability over a period of time T. For noninspected items, T can take on any value (preferably, it should be about the length of the mission, T, which the item is supposed to work). For tested or inspected items, T is normally the interval of time between two successive inspections or tests. If we assume that λt is very small over the period T, then

$$a = \frac{1}{T}\int_0^T (1 - \lambda t)\, dt = 1 - \frac{\lambda T}{2}$$

and

$$q = \frac{\lambda T}{2} \tag{4.33}$$

Vesely et al. [17] have discussed the average unavailability for various types of systems. Table 4.2 shows these functions.

Equations in Table 4.2 can also be applied to periodically tested standby equipment, with λ representing the standby (or demand) ROCOF, and the mission length or operating time replaced by the time between two tests.

Example 4.5

For a periodically inspected and repairable hazard barrier, assume no trends in ROCOF and no residual unavailability. The following data apply: $\lambda = 1.0 \times 10^{-4}$/h, $T_0 = 1$ month, $T_t = 2$ h, $f_r = 0.1$ per inspection, and $\tau_r = 20$ h. Determine the unavailability of the barrier.

Solution

Using (4.31) with $q_0 = 0$,

$$T = T_0 + \tau_t + f_r \tau_r$$

$$\lambda = 0.0001 \text{ h}^{-1}, \quad T_0 = 1 \text{ month} = 720 \text{ h},$$

$$\tau_t = 2 \text{ h}, \quad f_r = 0.1, \quad \tau_r = 20 \text{ h}$$

$$T = 720 + 2 + 0.1(20) = 724$$

$$\bar{q} = \frac{1}{2}(0.0001)720 + 0.1\left(\frac{20}{724}\right) + \frac{2}{724}$$

$$= 0.0365 + 0.00272 + 0.00272 = 0.04194$$

4.5 CLASSICAL PARAMETER ESTIMATION OF DISTRIBUTION MODELS FOR PERFORMANCE ASSESSMENT OF HARDWARE

Performance of nonrepairable (replaceable) items or hazard barriers is usually measured by its reliability (or its probability of failure within a mission of interest). Reliability can be measured through the distribution of TTF (obtained from historical data), or through the stress–strength, damage–endurance, and performance–requirement models. First, we discuss assessment of reliability through the distribution of TTF. Later in Section 4.7 we will examine the physics and engineering-based models of failure. Consider (4.26) for the reliability model of nonrepairable items. This section mainly deals with statistical methods for estimating TTF distribution parameters (from which hazard rate $\lambda(t)$ is estimated), such as parameter λ of the exponential pdf, μ and σ of the normal pdf and lognormal pdf, p of the binomial probability distribution, and α and β of the Weibull pdf. From these TTF distributions, $f(t)$, using the expression $\lambda(t) = f(t)/R(t)$, the expression $\lambda(t)$ in (4.26) can be estimated. The objective is to find a point estimate and a confidence interval for the parameters of these probability densities assuming that we have gathered some data (from test or field observation). It is important to realize why we need to consider confidence intervals in the estimation process. In essence, this need stems from the fact that we have a limited amount of data (e.g., on TTF), and thus we cannot state our point estimation with certainty. Therefore, the confidence interval is a statement about the range within which the actual value of the parameter resides. This interval is greatly influenced by the amount of data available. Of course other factors, such as diversity and accuracy of the data sources and adequacy of the selected model can also influence the state of our uncertainty regarding the estimated parameters. When discussing the goodness-of-fit tests, we are trying to address the uncertainty due to the choice of the probability model form by using the concept of levels of significance. However, uncertainty due to diversity and accuracy of the data sources is a more difficult issue to deal with.

Estimation of parameters of TTF or failure on demand distribution can be based on field data as well as on data obtained from special life (reliability) tests. In life testing, a sample of hardware items from a hypothesized population of such units is placed on test using the environment in which the components are expected to function, and their TTF are recorded. In general, two major types of tests are performed. The first is testing *with replacement* of the failed items, and the second is testing *without replacement*. The test with replacement is sometimes called *monitored*.

Samples of TTF or times-between-failures are seldom *complete* samples. A complete sample is the one in which all items observed have failed during a given observation period, and all the failure times are known (distinct). The likelihood function for a complete sample has been discussed in Appendix A. In the following sections, the likelihood functions for some types of censoring are discussed. Hazard barriers are usually highly reliable so that a complete sample is a rarity, even in accelerated life testing. Thus, as a rule, failure data are incomplete.

When n items are placed on test or observed in the field, whether with replacement or not, it is sometimes necessary, due to the long life of certain components, to terminate the test and perform the reliability analysis based on the observed data up to the time of termination. According to the above discussion, there are two basic types of possible life observation termination. The first type is *time terminated* (which results in the Type I right-censored data), and the second is *failure terminated* (resulting in Type II right-censored data). In the time-terminated life observation, n units are observed and the observation is terminated after a predetermined time elapsed. The number of items failed during the observation time and the corresponding TTF of each component are recorded. In the failure-terminated life observations, n units are monitored and the observation is terminated when a predetermined number of component failures have occurred. The TTF of each failed item including the time of the last failure occurred are recorded.

Type I and Type II life observation or test can be performed with replacement or without replacement. Therefore, four types of life test experiments are possible.

4.5.1 ESTIMATION OF PARAMETERS OF EXPONENTIAL DISTRIBUTION MODEL (BASED ON COMPLETE DATA)

Suppose n nonrepairable (replaceable) hardware items (e.g., hazard barriers) are observed with replacement (i.e., monitored), and the observation is terminated after a specified time t_0. The accumulated (by both failed and unfailed items) time of experience, T (in hours or other time units), is given by:

$$T = nt_0 \tag{4.34}$$

Time, T, is also called the *total time on test or observation*. At each time instant, from the beginning of the observation up to time t_0, exactly n items have been tested or observed in the field. Accordingly, if r failures have been recorded up to time t_0, then assuming the exponential distribution, the maximum likelihood point estimate of the failure rate of the hardware unit, can be estimated from:

$$\hat{\lambda} = \frac{r}{T} \tag{4.35}$$

The corresponding estimate of the mean-time-to-failure (MTTF) is given by

$$\hat{\text{MTTF}} = \frac{T}{r} \tag{4.36}$$

The number of units observed, n', is

$$n' = n + r \tag{4.37}$$

Suppose n items are observed without replacement, and the observation is terminated after a specified time t_0 during which r failures have occurred. The total time of experience, T, for the failed and survived hardware units is

$$T = \sum_{i=1}^{r} t_i + (n - r)t_0 \tag{4.38}$$

where $\sum_{i=1}^{r} t_i$ represents the accumulated time of operation or test of the r failed units (r is random here), and $(n-r)t_0$ is the accumulated time on test of the surviving components. Using

(4.35) and (4.36), the failure rate and MTTF estimates can be obtained, respectively. Since no replacement has taken place, the total number of items tested or observed is $n' = n$. Type I life test without replacement is not as useful and practical as the Type I life test with replacement.

Consider a situation when n units are observed with replacement (i.e., monitored), and an item is replaced with an identical one as soon as it fails (except for the last failure). If the observation is terminated after a time t_r when the rth failure has occurred (that is, r is specified nonrandom), but t_r is random, then the total time of experience, T, associated with failed and unfailed units, is given by

$$T = nt_r \tag{4.39}$$

Note that t_r, unlike t_0, is a variable in this case. If the TTF follows the exponential distribution, λ is estimated as

$$\hat{\lambda} = \frac{r}{T} \tag{4.40}$$

and the respective estimate of MTTF is

$$\hat{\text{MTTF}} = \frac{T}{r} \tag{4.41}$$

The total number of failed units observed, n', is

$$n' = n + r - 1 \tag{4.42}$$

where $(r-1)$ is the total number of failed *and* replaced units. All failed units are replaced except the last one, because the test or observation is terminated when the last hardware unit fails (i.e., the rth failure has been observed).

Finally, consider another situation when n units being observed without replacement, i.e., when a failure occurs, the failed item is not replaced by a new one. The observation is terminated at time t_r when the rth failure has occurred (i.e., r is specified but t_r is random). The total time accumulated of both failed and unfailed items is obtained from

$$T = \sum_{i=1}^{r} t_i + (n - r)t_r \tag{4.43}$$

where $\sum_{i=1}^{r} t_i$ is the accumulated time contribution from the failed items, and $(n-r)t_r$ is the accumulated time contribution from the survived units. Accordingly, the failure rate and MTTF estimates for the exponentially distributed TTF can be obtained using (4.41) and (4.43). It should also be noted that the total number of units in this test is

$$n' = n \tag{4.44}$$

since no components are being replaced.

Example 4.6
Ten replaceable hardware units are observed from when they were new. The observation is terminated at $t_0 = 8500$ h. Eight units fail before 8500 h have elapsed. Determine the

accumulated hours of experience and an estimate of the failure rate and MTTF of the item for the following situation:

- (a) The units are replaced when they fail.
- (b) The units are not replaced when they fail.
- (c) Repeat (a) and (b), assuming the observation is terminated when the eighth unit fails.

The failure times obtained are: 1830, 3180, 4120, 4320, 5530, 6800, 6890, and 7480.

Solution
(a) *Type I test:*

$$T = 10(8500) = 85,000 \text{ unit-hours using (4.34)}.$$
$$\hat{\lambda} = 8/85,000 = 9.4 \times 10^{-5} \, \text{hr}^{-1} \text{ from (4.36), MTTF} = 10,625 \text{ h}.$$

(b) *Type I test:*

$$\sum_{i=1}^{r} t_i = 40,150(n-r)t_0 = (10-8)8500 = 17,000.$$

Thus, $T = 40,150 + 17,000 = 57,150$ unit-hours.

$$\hat{\lambda} = 8/57,150 = 1.4 \times 10^{-4} \, \text{h}^{-1}.$$
$$\text{MTTF} = 7144 \text{ h}.$$

(c) *Type II test:*
Here, t_r is the time-to-the-eighth failure, which is 7480.

$$T = 10(7480) = 74,800 \text{ unit-hours}.$$
$$\hat{\lambda} = 8/74,800 = 1.1 \times 10^{-4} \text{h}^{-1}.$$
$$\text{MTTF} = 9350 \text{h}.$$

$$\sum_{i=1}^{r} t_i = 40,150, (n-r)t_R = (10-8)7480 = 14,960 \text{ unit-hours}.$$

Thus, $T = 40,150 + 14,960 = 55,110$ unit-hours.

$$\hat{\lambda} = 8/55,110 = 1.5 \times 10^{-4} \text{ h}^{-1}.$$
$$\text{MTTF} = 6888 \text{ h}.$$

Note that in reality a life observation with replacement would have yielded different failure times, but in this example we are only interested in the procedure for parameter estimation.

The maximum likelihood for the parameter λ (failure rate, or hazard rate) of the exponential distribution (see Appendix A) yields the point estimator of $\hat{\lambda} = r/T$, where r is the number of failures observed and T is the total observation time. Epstein [18] has shown that if the TTF is exponentially distributed with parameter λ, the quantity $2r\lambda/\hat{\lambda} = 2\lambda T$ has the chi-square distribution with $2r$ degrees of freedom for the Type II censored data (failure-terminated observation). Based on this, one can construct the corresponding confidence intervals. Because uncensored data can be considered as the particular case of the Type II right-censored data (when $r = n$), the same procedure is applicable to complete (uncensored) sample.

Using the distribution of $2r\lambda/\hat{\lambda}$, one can write

$$\Pr\left[\chi^2_{\alpha/2}(2r) \leq \frac{2r\lambda}{\hat{\lambda}} \leq \chi^2_{1-\alpha/2}(2r)\right] = 1 - \alpha \qquad (4.45)$$

By rearranging and using $\hat{\lambda} = r/T$ the two-sided confidence interval for the true value of λ can be obtained as

$$\Pr\left[\frac{\chi^2_{\alpha/2}(2r)}{2T} \leq \lambda \leq \frac{\chi^2_{1-\alpha/2}(2r)}{2T}\right] = 1 - \alpha \qquad (4.46)$$

The corresponding upper confidence limit (the one-sided confidence interval) obviously is

$$\Pr\left[0 \leq \lambda \leq \frac{\chi^2_{1-\alpha/2}(2r)}{2T}\right] = 1 - \alpha \qquad (4.47)$$

Accordingly, confidence intervals for probability of failure of a hardware unit at a time $t = t_0$ can also be obtained in form of one-sided and two-sided confidence intervals from (4.46) and (4.47).

As opposed to Type II censored data, the corresponding exact confidence limits for Type I censored data are not available. The approximate two-sided confidence interval for failure rate, λ, for the Type I data usually are constructed as:

$$\Pr\left[\frac{\chi^2_{\alpha/2}(2r)}{2T} \leq \lambda \leq \frac{\chi^2_{1-\alpha/2}(2r + 2)}{2T}\right] = 1 - \alpha \qquad (4.48)$$

The respective upper confidence limit (a one-sided confidence interval) is given by

$$\Pr\left[0 \leq \lambda \leq \frac{\chi^2_{1-\alpha/2}(2r + 2)}{2T}\right] = 1 - \alpha \qquad (4.49)$$

If no failure is observed during an observation, the formal estimation gives $\hat{\lambda} = 0$, or MTTF $= \infty$. This cannot realistically be true, since we may have a small or limited data. Had the observation been continued, eventually a failure would have been observed. An upper confidence estimate for λ can be obtained for $r = 0$. However, the lower confidence limit cannot be obtained with $r = 0$. It is possible to relax this limitation by conservatively assuming that a failure occurs exactly at the end of the observation period. Then $r = 1$ can be used to evaluate the lower limit for two-sided confidence interval. This conservative modification, although sometimes used to allow a complete statistical analysis, lacks firm statistical basis. Welker and Lipow [19] have shown methods to determine approximate nonzero point estimates in these cases.

Example 4.7
Performance of 25 nonrepairable–replaceable units is observed for 500 h. In this observation, eight failures occur at 75, 115, 192, 258, 312, 389, 410, and 496 h. The failed units are replaced.

Find $\hat{\lambda}$, one-sided and two-sided confidence intervals for λ and MTTF at the 90% confidence level; one-sided and two-sided 90% confidence intervals on the unit's probability of failure at $t_0 = 1000$ h.

Solution

This is a Type I observation. The accumulated time T is

$$T = 25 \times 500 = 12,500 \text{ unit-hours.}$$

The point estimate of failure rate is

$$\hat{\lambda} = 8/12,500 = 6.4 \times 10^{-4} \text{ h}^{-1}.$$

One-sided confidence interval for λ is

$$0 \le \lambda \le \frac{\chi^2 (2 \times 8 + 2)}{2 \times 12,500}$$

From Table B.3,

$$\chi^2_{0.9}(18) = 25.99, \quad 0 \le \lambda \le 1.04 \times 10^{-3} \text{ h}^{-1}$$

Two-sided confidence interval for λ is

$$\frac{\chi^2_{0.05}(2 \times 8)}{2 \times 12,500} \le \lambda \le \frac{\chi^2_{0.95}(2 \times 8 + 2)}{2 \times 12,500}$$

From Table (B.3),

$$\chi^2_{0.05}(16) = 7.96 \quad \text{and} \quad \chi^2_{0.95}(18) = 28.87$$

Thus,

$$3.18 \times 10^{-4} \text{ h}^{-1} \le \lambda \le 1.15 \times 10^{-3} \text{ h}^{-1}$$

One-sided 90% confidence interval for $q(100)$ is

$$0 \le q(100) \le 1 - \exp\left[(-1.04 \times 10^{-3})(100)\right]$$

or

$$0 \le q(100) \le 0.0988$$

Two-sided 90% confidence interval for $q(t)$ is

$$1 - \exp\left[(-3.18 \times 10^{-4})(100)\right] \le q(100) \le 1 - \exp\left[(-1.15 \times 10^{-3})(100)\right]$$

or

$$0.00317 \le q(100) \le 0.1086$$

4.5.2 ESTIMATION OF PARAMETERS OF LOGNORMAL DISTRIBUTION MODEL (BASED ON COMPLETE DATA)

The lognormal distribution is commonly used to represent occurrence of failure of certain events in time that can be varied by one or more orders of magnitude. Because the lognormal distribution has two parameters, parameter estimation poses a more challenging problem than for the exponential distribution. Taking the natural logarithm of data, the analysis is reduced to the case of the normal distribution, so that the point estimates for the two parameters of the lognormal distribution for a complete sample of size n can be obtained from

$$\hat{\mu}_t = \sum_{i=1}^{n} \frac{\ln t_i}{n} \tag{4.50}$$

$$\hat{\sigma}_t^2 = \frac{\sum_{i=1}^{n} (\ln t_i - \bar{\mu})^2}{n-1} \tag{4.51}$$

The confidence interval for μ_t is given by

$$\Pr\left[\hat{\mu}_t - \frac{\hat{\sigma}_t t_{\alpha/2}}{\sqrt{n}} \le \mu_t \le \hat{\mu}_t + \frac{\hat{\sigma}_t t_{\alpha/2}}{\sqrt{n}}\right] = 1 - \alpha \tag{4.52}$$

The respective confidence interval for σ_t^2 is:

$$\Pr\left[\hat{\mu}_t - \frac{\hat{\sigma}_t^2 (n-1)}{\chi_{1-\alpha/2}^2(n-1)} \le \sigma_t^2 \le \hat{\mu}_t + \frac{\hat{\sigma}_t^2 (n-1)}{\chi_{\alpha/2}^2(n-1)}\right] = 1 - \alpha \tag{4.53}$$

In the case of censored data, the corresponding statistical estimation turns out to be much more complicated, readers are referred to Nelson [20] and Lawless [21]. Equations (4.50)–(4.53) are also applicable to estimating normal distribution parameters. Only $\ln(t_i)$ changes to t_i in all equations. Normal distribution is not a very popular model for modeling reliability of nonrepairable items.

4.5.3 ESTIMATION OF PARAMETERS OF WEIBULL DISTRIBUTION MODEL (BASED ON COMPLETE AND CENSORED DATA)

The Weibull distribution can be used for the nonrepairable hardware units exhibiting increasing, decreasing, or constant hazard rate functions. Similar to the lognormal distribution, it is a two-parameter distribution and its estimation, even in the case of complete (uncensored) data, is not a trivial problem.

In the situation when all r units out of n observed fail, at time t_i's the log-likelihood estimates of Weibull distribution is

$$\ln(l) = L = \sum_{i=1}^{r} \ln\left(\frac{\beta}{\alpha}\left(\frac{t_i}{\alpha}\right)^{\beta-1} e^{-(t_i/\alpha)^\beta}\right) - \sum_{j=1}^{n-r}\left(\frac{t_j}{\alpha}\right)^\beta \tag{4.54}$$

using (A.85). The variance of the estimates $\hat{\beta}$ and $\hat{\alpha}$ can be found by the so-called Fisher local information matrix as

$$F = \begin{bmatrix} -\dfrac{\partial^2 L}{\partial \beta^2} & -\dfrac{\partial^2 L}{\partial \beta \partial \alpha} \\ -\dfrac{\partial^2 L}{\partial \beta \partial \alpha} & -\dfrac{\partial^2 L}{\partial \alpha^2} \end{bmatrix} \tag{4.55}$$

For parameter estimation and the terms inside the matrix expressed as

$$\frac{\partial L}{\partial \beta} = \frac{r}{\beta} + \sum_{i=1}^{r} \ln\left(\frac{t_i}{\alpha}\right) - \sum_{i=1}^{r} N_i \left(\frac{t_i}{\alpha}\right)^{\beta} \ln\left(\frac{t_i}{\alpha}\right) - \sum_{j=1}^{n-r} N_j \left(\frac{t_j}{\alpha}\right)^{\beta} \ln\left(\frac{t_j}{\alpha}\right) = 0$$

$$\frac{\partial L}{\partial \alpha} = \frac{-\beta r}{\alpha} + \frac{\beta}{\alpha} \sum_{i=1}^{r} \left(\frac{t_i}{\alpha}\right)^{\beta} + \sum_{j=1}^{n-r} \left(\frac{t_j}{\alpha}\right)^{\beta} = 0$$

$$\frac{\partial^2 L}{\partial \beta^2} = \sum_{i=1}^{r} \left[-\frac{1}{\beta} - \left(\frac{t_i}{\alpha}\right)^{\beta} \ln^2 \left(\frac{t_i}{\alpha}\right)\right] + \sum_{j=1}^{n-r} \left[-\left(\frac{t_i}{\alpha}\right)^{\beta} \ln^2 \left(\frac{t_i}{\alpha}\right)\right]$$

$$\frac{\partial^2 L}{\partial \alpha^2} = \sum_{i=1}^{r} \left[\frac{\beta}{\alpha} - \left(\frac{t_i}{\alpha}\right)^{\beta} \left(\frac{\beta}{\alpha^2}\right)(\beta+1)\right] + \sum_{j=1}^{n-r} \left[-\left(\frac{t_i}{\alpha}\right)^{\beta} \left(\frac{\beta}{\alpha^2}\right)(\beta+1)\right] \qquad (4.56)$$

$$\frac{\partial^2 L}{\partial \beta \partial \alpha} = \sum_{i=1}^{r} \left\{-\frac{1}{\alpha} + \left(\frac{t_i}{\alpha}\right)^{\beta} \left(\frac{1}{\alpha}\right)\left[\beta \ln\left(\frac{t_i}{\alpha}\right) + 1\right]\right\} + \sum_{j=1}^{n-r} \left\{\left(\frac{t_i}{\alpha}\right)^{\beta} \left(\frac{1}{\alpha}\right)\left[\beta \ln\left(\frac{t_i}{\alpha}\right) + 1\right]\right\}$$

Note that $(n-r)$ represents the number of not failed data points at the end of the observation. The variance of the estimate parameters are

$$F^{-1} = \begin{bmatrix} \text{Var}(\hat{\beta}) & \text{Cov}(\hat{\beta}\hat{\alpha}) \\ \text{Cov}(\hat{\beta}\hat{\alpha}) & \text{Var}(\hat{\alpha}) \end{bmatrix} \qquad (4.57)$$

where matrix F^{-1} is the inverse of matrix F shown by (4.55).

Therefore approximate 100% $(1-\alpha)$ confidence intervals for $\hat{\beta}$ and $\hat{\alpha}$ are

$$\hat{\beta} \, exp\left[\frac{-q_{\alpha/2}\sqrt{\text{Var}(\hat{\beta})}}{\hat{\beta}}\right] \leq \beta \leq \hat{\beta} \, exp\left[\frac{q_{\alpha/2}\sqrt{\text{Var}(\hat{\beta})}}{\hat{\beta}}\right]$$

$$\hat{\alpha} \, exp\left[-\frac{q_{\alpha/2}\sqrt{\text{Var}(\hat{\alpha})}}{\hat{\alpha}}\right] \leq \alpha \leq \hat{\alpha} \, exp\left[\frac{q_{\alpha/2}\sqrt{\text{Var}(\hat{\alpha})}}{\hat{\alpha}}\right] \qquad (4.58)$$

where $q_{\alpha/2}$ is the inverse of the standard normal pdf. These confidence intervals are approximate, but approach exact values as the sample size increases. Solution to (4.55)–(4.58) is complex and requires numerical algorithms performed through special computer routines.

Readers are referred to Leemis [1], Bain [13], Nelson [20], Mann et al. [22] for further discussions.

Example 4.8

Ten identical pumps are tested at high-stress conditions for 200 h, and eight of the units failed during the test. The following TTFs are recorded: 56, 89, 95, 113, 135, 178, 188, and 199 h. Estimate the parameters of the Weibull distribution and their 90% confidence interval.

Solution

In this case, eight of the pumps fail, therefore, the maximum likelihood estimates of Weibull parameters are obtained as a solution of (4.56)–(4.58) as $\beta = 2.27$ and $\alpha = 177.08$. The 90% confidence intervals are: $1.53 < \beta < 3.39$ and $143.68 < \alpha < 218.23$.

4.5.4   ESTIMATION OF PARAMETERS OF BINOMIAL DISTRIBUTION MODEL

When the data are in the form of failures of nonrepairable–replaceable items occurring on demand, i.e., x failures observed in n observations, there is a constant probability of failure (or success), and the binomial distribution can be used as an appropriate model. This is often the situation for standby items. For instance, a redundant pump is demanded for operation n times in a given period of test or observation.

The best estimator for p is given by

$$\hat{p} = \frac{x}{n} \tag{4.59}$$

The lower and the upper confidence limits for p can be found, using the, so-called, Clopper–Pearson procedure, see (Nelson [20])

$$p_l = \left\{ 1 + (n - x + 1)x^{-1} F_{1-\alpha/2} \left[2n - 2x + 2; 2x\right] \right\}^{-1} \tag{4.60}$$

$$p_u = \left\{ 1 + (n - x)\left\{(x + 1)F_{1-\alpha/2}[2x + 2; 2n - 2x]\right\}^{-1} \right\}^{-1} \tag{4.61}$$

where $F_{1-\alpha/2}$ (f_1; f_2) is the $(1-\alpha/2)$ quantile (or the $100(1-\alpha/2)$ percentiles) of the F-distribution with f_1 degrees of freedom for the numerator, and f_2 degrees of freedom for the denominator. Table B.5 contains some percentiles of F-distribution. As discussed in Appendix A, the Poisson distribution can be used as an approximation to the binomial distribution when the parameter, p, of the binomial distribution is small and parameter n is large, e.g., $x < n/10$, which means that approximate confidence limits can be constructed according to (4.46) and (4.47) with $r = x$ and $T = n$.

Example 4.9

An emergency cooling system's pump in a nuclear power plant is in a standby mode for cooling purposes when needed. There have been 563 start tests for the pump and only three failures have been observed. No degradation in the pump's physical characteristics or changes in operating environment are observed. Find the 90% confidence interval for the probability of failure per demand.

Solution

Denote $n = 563$, $x = 3$, $\hat{p} = 3/563 = 0.0053$.

$$p_l = [1 + (563 - 3 + 1)/3F_{0.95}(2 \times 563 - 2 \times 3 + 2; 2 \times 3)]^{-1} = 0.0014,$$

where $F_{0.95}$ (1122; 6) = 3.67 from Table B.5.

Similarly,

$$p_u = \{1 + (563 - 3)[(3 + 1)F_{0.95}(2 \times 3 + 2; 2 \times 563 - 2 \times 3)]^{-1}\}^{-1} = 0.0137,$$

therefore,

$$\Pr(0.0014 \le p \le 0.0137) = 90\%.$$

Example 4.10

In a commercial nuclear plant, the performance of the emergency diesel generators has been observed for about 5 years. During this time, there have been 35 real demands with four observed failures. Find the 90% confidence limits and point estimate for the probability of failure per demand. What would the error be if we used (4.48) instead of (4.60) and (4.61) to solve this problem?

Solution

Here, $x = 4$ and $n = 35$. Using (4.59),

$$\hat{p} = \frac{4}{35} = 0.114$$

To find lower and upper limits, use (4.60) and (4.61). Thus,

$$p_l = [1 + (35 - 4 + 1)/4F_{0.95}(2 \times 35 - 2 \times 4 + 2; 2 \times 4)]^{-1} = 0.04$$
$$p_u = \{1 + (35 - 4)[(4 + 1)F_{0.95}(2 \times 4 + 2;\ 2 \times 35 - 2 \times 4)]^{-1}\}^{-1} = 0.243$$

If we used (4.48),

$$p_l = \frac{\chi^2_{0.05}(8)}{2 \times 35} = \frac{2.733}{70} = 0.039$$
$$p_u = \frac{\chi^2_{0.95}(10)}{2 \times 35} = \frac{18.31}{70} = 0.262$$

The error due to this approximation is

$$\text{Lower limit error} = \frac{|0.04 - 0.039|}{0.04} \times 100 = 2.5\%$$
$$\text{Upper limit error} = \frac{|0.243 - 0.262|}{0.243} \times 100 = 7.8\%$$

Note that this is not a negligible error, and (4.48) should not be used, since $x > n/10$.

4.5.5 ESTIMATION OF PARAMETERS OF EXPONENTIAL, LOGNORMAL, AND WEIBULL MODELS WHEN DEALING WITH INCOMPLETE (CENSORED) DATA

In the case of estimating parameters of the Weibull distribution we discussed a special form using censored data. In this section we will discuss an expanded version of this approach to other Weibull and other distributions. The method of determining the parameters discussed in Sections 4.5.1 through 4.5.4 is no longer applicable, and special methods will need to be employed to handle the data. Consider an item (e.g., a hazard barrier) for our observations occurring at specific times t_i at which we make observation and see N_i failures, and specific times t_j at which we make observation and see N_j cases of no failure. The maximum likelihood function is

$$l = \prod_{i=1}^{F} f(t_i; \theta_M) \cdot \prod_{j=1}^{S} [1 - F(t_j; \theta_M)] \qquad (4.62)$$

where θ_M are the unknown parameters estimated from the observed failures at t_i and the observed suspensions (censored data) at t_j. $f(\cdot)$ is the pdf of TTF and $F(\cdot)$ is the cdf of the TTF or time-between-failure. The log-likelihood function is

$$\ln(l) = L = \sum_{i=1}^{F} N_i \ln(f(t_i; \theta_M)) - \sum_{j=1}^{S} N_j F(t_j; \theta_M) \tag{4.63}$$

where F is the number of individual or groups of t_i (failure observation time) points, N_i, number of failures in the ith data group (or individual), θ_M, parameters to be found, t_i, the ith individual or group of failure times (observation times) data, S, number of individual or groups of t_j suspension (censored) data points, N_j, number of suspensions (censored) in the jth individual or group of suspension data points, t_j, the jth individual or group of suspension time (censored time) data.

The parameters θ_M can be estimated by substituting the distributions that represent the data in (4.63), and by maximizing this equation. For example, for the exponential distribution,

$$L = \sum_{i=1}^{F} N_i \ln\left(\lambda e^{-\lambda t_i}\right) - \sum_{j=1}^{S} N_j e^{-\lambda t_j} \tag{4.64}$$

Then the parameter λ is estimated by solving

$$\frac{\partial L}{\partial \lambda} = \sum_{i=1}^{F} N_i \left(\frac{1}{\lambda} - t_i\right) - \sum_{j=1}^{S} N_j t_j = 0 \tag{4.65}$$

Weibull distribution,

$$L = \sum_{i=1}^{F} N_i \ln\left(\frac{\beta}{\alpha} \left(\frac{t_i}{\alpha}\right)^{\beta-1} e^{-(t_i/\alpha)^\beta}\right) - \sum_{j=1}^{S} N_j \left(\frac{t_j}{\alpha}\right)^{\beta} \tag{4.66}$$

This is similar but more general than (4.54)
Then the parameters β on α are estimated by solving

$$\frac{\partial L}{\partial \beta} = \frac{1}{\beta} \sum_{i=1}^{F} N_i + \sum_{i=1}^{F} N_i \ln\left(\frac{t_i}{\alpha}\right) - \sum_{i=1}^{F} N_i \left(\frac{t_i}{\alpha}\right)^{\beta} \ln\left(\frac{t_i}{\alpha}\right) - \sum_{j=1}^{S} N_j \left(\frac{t_j}{\alpha}\right)^{\beta} \ln\left(\frac{t_j}{\alpha}\right) = 0 \tag{4.67}$$

$$\frac{\partial L}{\partial \alpha} = \frac{-\beta}{\alpha} \sum_{i=1}^{F} N_i + \frac{\beta}{\alpha} \sum_{i=1}^{F} N_i \left(\frac{t_i}{\alpha}\right)^{\beta} + \sum_{j=1}^{S} N_j \left(\frac{t_j}{\alpha}\right)^{\beta} = 0 \tag{4.68}$$

Solution to (4.67) and (4.68) require numerical routines or Monte Carlo-based solutions.
   Lognormal distribution,

$$L = \sum_{i=1}^{F} N_i \ln\left(\frac{1}{\sigma_t} \phi\left(\frac{\ln(t_i - \mu')}{\sigma_t}\right)\right) + \sum_{j=1}^{S} N_j \ln\left(1 - \Phi\left(\frac{\ln(t_j - \mu')}{\sigma_t}\right)\right) \tag{4.69}$$

Then the parameters μ' an σ_t are estimated by solving

$$\frac{\partial L}{\partial \mu} = \frac{1}{\sigma_t^2} \sum_{i=1}^{F} N_i (\ln(t_i) - \mu') + \frac{1}{\sigma_t} \sum_{j=1}^{S} N_j \frac{\phi\left(\frac{\ln(t_j) - \mu'}{\sigma_t}\right)}{1 - F\left(\frac{\ln(t_j) - \mu'}{\sigma_t}\right)} = 0 \tag{4.70}$$

$$\frac{\partial L}{\partial \sigma_{T}} = \sum_{i=1}^{F} N_{i}\left(\frac{(\ln(t_{i}) - \mu)^{2}}{\sigma_{t}^{3}} - \frac{1}{\sigma_{t}}\right) - \frac{1}{\sigma_{t}}\sum_{j=1}^{S} N_{j}\frac{\left(\frac{\ln(t_{j}) - \mu}{\sigma_{T}}\right)\phi\left(\frac{\ln(t_{j}) - \mu}{\sigma_{T}}\right)}{1 - \Phi\left(\frac{\ln(t_{j}) - \mu}{\sigma_{t}}\right)} = 0 \qquad (4.71)$$

where

$$\phi(x) = \frac{1}{\sqrt{2\pi}}e^{-1/2(x)^{2}}$$

and

$$\Phi(x) = \frac{1}{\sqrt{2\pi}}\int_{-\infty}^{x} e^{-1/2(t)^{2}}dt$$

Solutions to (4.70) and (4.71) require numerical method such as Monte Carlo simulation.

4.6 BAYESIAN PARAMETERS ESTIMATION OF DISTRIBUTION MODELS FOR PERFORMANCE ASSESSMENT OF HARDWARE

In Section 4.5, we discussed the importance of quantifying point estimate and uncertainties associated with the estimates of various distribution parameters used to estimate performance of hazard barriers. In Section 4.5, we also discussed formal classical statistical methods for estimating reliability and availability parameters of repairable and nonrepairable hardware units. Namely, we discussed the concept of point estimation and confidence intervals of the parameters of interest. It is evident that as more data are available, the more confidence we have. For this reason, the classical statistical approach is sometimes called the frequentist method of treatment. In the framework of Bayesian approach, the parameters of interest are treated as random variables, the true values of which are unknown. Thus, a distribution can be assigned to represent the parameter; the mean (or for some cases the median) of the distribution can be used as an estimate of the parameter of interest. The pdf of a parameter, in Bayesian terms, can then be estimated. In practice, a *prior* pdf is used to represent the relevant prior knowledge, including subjective judgments regarding the characteristics of the parameter and its distribution. When the prior knowledge is combined with other relevant information (often statistics obtained from tests and observations), a *posterior* distribution is obtained, which better represents the parameter of interest. Since the selection of the prior and the determination of the posterior often involve subjective judgments, the Bayesian estimation is sometimes called the subjectivist approach to parameter estimation. Atwood et al. [23] provide a comparison of the frequentist and Bayesian approach as listed in Table 4.3.

The basic concept of the Bayesian estimation is discussed in Appendix A. In essence, the Bayes' theorem can be written in one of the three forms: discrete, continuous, or mixed. Martz and Waller [24] have elaborated on the concept of Bayesian technique and its application to reliability analysis. The discrete form of Bayes' theorem is discussed in Appendix A. The continuous and mixed forms that are the common forms used for parameter estimation in reliability and risk analysis are briefly discussed below.

Let θ be a parameter of interest. It can be a parameter of a TTF distribution or a reliability index, such as MTTF, failure rate, etc. Suppose parameter θ is a continuous random variable, so that the prior and posterior distributions of θ can be represented by continuous pdfs.

TABLE 4.3
Comparison of Frequentist and Bayesian Estimation Techniques used in Risk Assessment (Atwood et al. [23])

	Frequentist	Bayesian
Interpretation of probability	Long-term frequency after many hypothetical repetitions	Measure of uncertainty, quantification of degree of belief
Unknown parameter	Constant, fixed	Constant, but assigned probability distribution, measure current state of belief
Data	Random (before being observed)	Random for intermediate calculations. Fixed (after being observed) for the final conclusion
Typical estimators	MLE, confidence interval	Bayes posterior mean, credible interval
Interpretation of for example 90% interval of a parameter	If many data set are generated, 90% of the resulting confidence intervals will contain the true parameter. We do not know if our interval is one of the unlucky ones	We believe, and would give 9 to 1 odds in a wager, that the parameter is in the interval
Primary uses in risk analysis	Check model assumptions	Incorporate evidence from various sources, as prior distribution
	Provide quick estimates, without work of determining and justifying prior distribution	Propagate uncertainties through fault-tree and event-tree models

Let $h(\theta)$ be a continuous prior pdf of θ, and let $l(\theta|t)$ be the likelihood function based on sample data, t, then the posterior pdf of θ is given by

$$f(\theta|t) = \frac{h(\theta)l(t|\theta)}{\int_{-\infty}^{\infty} h(\theta)l(t|\theta)\mathrm{d}\theta} \qquad (4.72)$$

Relationship (4.72) is the Bayes theorem for a continuous random variable. The Bayesian inference includes the following three stages:

(1) Constructing the likelihood function based on the distribution of interest and type of data available (complete or censored data, grouped data, etc.).
(2) Quantification of the prior information about the parameter of interest in form of a prior distribution.
(3) Estimation of the posterior distribution of the parameter of interest.

The Bayes analog of the classical confidence interval is known as Bayes' probability interval. For constructing Bayes' probability interval, the following obvious relationship based on the posterior distribution is used:

$$\Pr(\theta_l < \theta \le \theta_u) = 1 - \alpha \qquad (4.73)$$

Similar to the classical estimation, the Bayesian estimation procedures can be divided into parametric and nonparametric ones. The following are examples of the parametric Bayesian estimation.

4.6.1 BAYESIAN ESTIMATION OF THE PARAMETER OF EXPONENTIAL DISTRIBUTION MODEL

Consider a test or field observation of n items which results in r distinct times to or between failure $t_{(1)} < t_{(2)} < \cdots < t_{(r)}$ and $n-r$ times to censoring $t_{c1}, t_{c2}, \ldots, t_{c(n-r)}$, so that the total time observed time or test, T, is

$$T = \sum_{i=1}^{r} t_{(i)} + \sum_{i=1}^{n-r} t_{ci} \qquad (4.74)$$

Consider a TTF or time-between-failure which is supposed to have the exponential distribution. The problem is to estimate the parameter λ of the exponential distribution. Suppose a gamma distribution is used as the prior distribution of parameter λ. This distribution is in Appendix A. Rewrite the pdf as a function of λ, which is now considered as a random variable:

$$h(\lambda; \delta, \rho) = \frac{1}{\Gamma(\delta)} \rho^{\delta} \lambda^{\delta-1} \exp(-\rho\lambda), \quad \lambda > 0, \ \rho \geq 0, \ \delta \geq 0 \qquad (4.75)$$

In the Bayesian context, the parameters δ and ρ, as the parameters of prior distribution, are sometimes called hyperparameters. Selection of the hyperparameters is discussed later, but for a time being, suppose that these parameters are known.

For the available data and the exponential TTF distribution, the likelihood function can be written as

$$l(\lambda|t) = \prod_{i=1}^{r} \lambda e^{-\lambda t_{(i)}} \prod_{i=1}^{n-r} e^{-\lambda t_{ci}}$$
$$= \lambda^{r} e^{-\lambda T} \qquad (4.76)$$

where T is the total observation time or test given by (4.74).

Using the prior distribution (4.75), the likelihood function (4.76) and the Bayes' theorem in the form (4.72), one can find the posterior pdf of the parameter λ, as:

$$f(\lambda|T) = \frac{e^{-\lambda(T+\rho)} \lambda^{r+\delta-1}}{\int_{0}^{\infty} e^{-\lambda(T+\rho)} \lambda^{r+\delta-1} d\lambda} \qquad (4.77)$$

Recalling the definition of the gamma function, it is easy to show that the integral in the denominator of (4.77) is

$$\int_{0}^{\infty} \lambda^{r+\delta-1} e^{-\lambda(T+\rho)} d\lambda = \frac{\Gamma(\delta+r)}{(\rho+T)^{\delta+r}} \qquad (4.78)$$

Finally, the posterior pdf of λ can be written as

$$f(\lambda|T) = \frac{(\rho+T)^{\delta+r}}{\Gamma(\delta+r)} \lambda^{r+\delta-1} e^{-\lambda(T+\rho)} \qquad (4.79)$$

Comparing with the prior pdf (4.75), one can conclude that the posterior pdf (4.79) is also a gamma distribution with parameters $\rho' = r + \delta$, and $\lambda' = T + \rho$. Prior distributions which result in posterior distributions of the same family are referred to as *conjugate prior distributions*.

Keeping in mind that the quadratic loss function is used, the point Bayesian estimate of λ is the mean of the posterior gamma distribution with parameters ρ' and λ', so that the point Bayesian estimate, λ_B, is

$$\lambda_B = \frac{\rho'}{\lambda'} = \frac{r+\delta}{T+\rho} \tag{4.80}$$

The corresponding probability intervals can be obtained using (4.73). For example, the $100(1-\alpha)$ level upper one-sided Bayes' probability interval for λ can be obtained from the following equation based on the posterior distribution (4.79). The same upper one-sided probability interval for λ can

$$\Pr(\lambda \leq \lambda_u) = 1 - \alpha \tag{4.81}$$

be expressed in a more convenient form similar to the classical confidence interval, i.e., in terms of the chi-square distribution, as:

$$\Pr\{2\lambda(\rho + T) \leq \chi^2_{1-\alpha}[2(\delta + r)]\} = 1 - \alpha$$

and

$$\lambda_u = \frac{\chi^2_{1-\alpha}[2(\delta + r)]}{2(\rho + T)} \tag{4.82}$$

Note that, contrary to the classical estimation, the number of degrees of freedom, $2(\delta + r)$, for the Bayes' confidence limits is not necessarily an integer. The gamma distribution was chosen as the prior distribution for the purpose of simplicity and performance.

4.6.1.1 Uniform Prior Distribution

This prior distribution has very simple form, so that it is easy to use as an expression of prior information. Consider the prior pdf in the form:

$$h(\lambda; a, b) = \begin{cases} \frac{1}{b-a}, & a < \lambda < b \\ 0, & \text{otherwise} \end{cases} \tag{4.83}$$

Using the likelihood function (4.76), the posterior pdf can be written as

$$f(\lambda|T) = \frac{\lambda^r e^{-\lambda T}}{\int_a^b \lambda^r e^{-\lambda T} d\lambda}, \quad a < \lambda < b \tag{4.84}$$

Example 4.11

An electronic component has the exponential TTF distribution. The uniform prior distribution of λ is given by $a = 1 \times 10^{-6}$ and $b = 5 \times 10^{-6}$/h in (4.83). A life test of the component results in $r = 30$ failures in total time on test of $T = 10^7$ h. Find the point estimate (mean) and 90% two-sided Bayes' probability interval for the constant failure rate parameter λ.

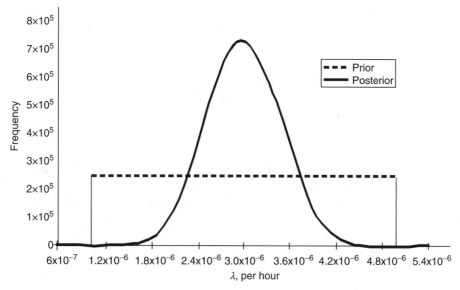

FIGURE 4.6 Prior and posterior distribution of λ.

Solution

Using (4.83) and (4.84), the point estimate $\lambda = 3.1 \times 10^{-6}$/h and the 90% two-sided Bayes' probability interval is $(2.24 \times 10^{-6} < 4.04 \times 10^{-6})$ per hour. Actual prior and posterior distributions of λ have been plotted in Figure 4.6.

4.6.2 BAYESIAN ESTIMATION OF THE PARAMETER OF BINOMIAL DISTRIBUTION MODEL

The binomial distribution plays an important role in measuring performance of many items. Suppose that n identical units have been observed (without replacements) for a specified time, t, and that the observation yields r failures. The number of failures, r, can be considered as a discrete random variable having the binomial distribution with parameters n and $p(t)$, where $p(t)$ is the probability of failure of a single unit during time t. As discussed earlier, $p(t)$, as a function of time, is the TTF cumulative distribution function, as well as $1-p(t)$ is the reliability or survivor function. A straightforward application of the binomial distribution is modeling of a number of failures to start on demand for a redundant unit. The probability of failure in this case might be considered as time-independent. Thus, one should keep in mind two possible applications of the binomial distribution:

(1) The reliability function or cdf for demand to or between failures, and
(2) The binomial distribution itself.

The maximum likelihood estimate of the parameter p is the ratio r/n, which is widely used as classical estimate. To get a Bayesian estimation procedure for the reliability function, let us consider p as the survivor probability in a single Bernoulli trial (so, now the "success" means surviving). If the number of units observed, n, is fixed in advance, the probability distribution of the number, x, of unfailed units during the test or observation (i.e., the number of "successes") is given by the binomial distribution with parameters n and x. The corresponding likelihood function can be written as

$$l(p|x) = cp^x(1-x)^{n-x}$$

where c is a constant which does not depend on the parameter of interest, p. For any continuous prior distribution with pdf $h(p)$ the corresponding posterior pdf can be written as

$$f(p|x) = \frac{p^x(1-p)^{n-x}h(p)}{\int_{-\infty}^{\infty} p^x(1-p)^{n-x}h(p)\mathrm{d}p} \tag{4.85}$$

4.6.2.1   Standard Uniform Prior Distribution

Consider the particular case of uniform distribution, $U(a,b)$, which in the Bayes' context represents "a state of total ignorance." While this seems to have little practical importance, nevertheless, it is interesting from the methodological point of view. For this case one can write

$$h(p) = \begin{cases} 1, & 0 < p \le 1 \\ 0, & \text{otherwise} \end{cases}$$

and

$$f(p|x) = \frac{p^{(x+1)-1}(1-p)^{(n-x+1)-1}}{\int_0^1 p^{(x+1)-1}(1-p)^{(n-x+1)-1}\mathrm{d}p} \tag{4.86}$$

The integral in the denominator can be expressed as

$$\int_0^1 p^{(x+1)-1}(1-p)^{(n-x+1)-1}\mathrm{d}p = \frac{\Gamma(x+1)\Gamma(n-x+1)}{\Gamma(n+2)}$$

So, the posterior cdf can be easily recognized as the pdf of the beta distribution, $f(p; x+1, n-x+1)$, which is discussed in Appendix A. Recalling the expression for the mean value of the beta distribution (A.56), the point Bayes' estimate of p can be written as

$$p_B = \frac{x+1}{n+2} \tag{4.87}$$

Note that the estimate is different as compared with the respective classical estimate (x/n), but when the sample size increases the estimates are getting closer to each other.

Recalling that the cdf of the beta distribution is expressed in terms of the incomplete beta function, the $100(1-\alpha)\%$ two-sided Bayes' probability interval for p can be obtained by solving the following equations:

$$\begin{aligned} \Pr(p < p_l) = I_{p_1}(x+1, n-x+1) = \frac{\alpha}{2} \\ \Pr(p > p_u) = I_{p_u}(x+1, n-x+1) = 1 - \frac{\alpha}{2} \end{aligned} \tag{4.88}$$

where $I_z(a, b)$ is the incomplete beta function,

$$I_z(a,b) = \frac{\Gamma(a+b)}{\Gamma(a)\Gamma(b)} \int_0^z t^{a-1}(1-t)^{b-1}\mathrm{d}t, \quad a>0, b>0, 0<z<1$$

and

$$\int_0^z t^{a-1}(1-t)^{b-1}\,dt \approx z^a \left[\frac{1}{a} + \frac{1-b}{a+1}z + \cdots + \frac{(1-b)\cdots(n-b)}{n!(a-n)}z^n + \cdots \right]$$

It can be mentioned that the probability intervals above are very similar to the corresponding classical confidence intervals.

Example 4.12

Calculate the point estimate and the 95% two-sided Bayesian probability interval for the failure probability of a new items based on observation of 300 items, out of which four have failed. Suppose that for this items no historical information is available. Accordingly, its prior failure probability estimate may be assumed uniformly distributed between 0 and 1.

Solution

Using (4.87), find

$$F = p_B = \frac{4+1}{300+2} = 0.0166$$

Using (4.88), the 95% upper and lower limits are evaluated as 0.0337 and 0.0054, respectively. It is informative to compare the above results with classical ones. The point estimate of the reliability is $F = p = 4/300 = 0.0133$, and the 95% upper and lower limits are 0.0338 and 0.0036, respectively. The higher probability interval based on the Bayesian estimation is because of the nonavailability of prior information.

4.6.2.2 Beta Prior Distribution

The most widely used prior distribution for the parameter, p, of the binomial distribution is the beta distribution. The pdf of the distribution can be written in the following convenient form:

$$h(b; x_0, n_0) = \begin{cases} \frac{\Gamma(n_0)}{\Gamma(x_0)\Gamma(n_0-x_0)} p^{x_0-1} (1-p)^{n_0-x_0-1}, & 0 \le p \le 1 \\ 0 & \text{otherwise} \end{cases} \qquad (4.89)$$

where $n_0 > x_0 \ge 0$.

The pdf provides a great variety of different shapes. It is important to note that the standard uniform distribution is a particular case of the beta distribution. When $x_0 = 1$ and $n_0 = 2$, (4.89) reduces to the standard uniform distribution. Moreover, the beta prior distribution turns out to be a conjugate prior distribution for the estimation of the parameter p of the binomial distribution of interest.

Considering the expression for the mean value of the beta distribution, it is clear that the prior mean is x_0/n_0, so that the parameters of the prior, x_0 and n_0, can be interpreted as a pseudo number of identical units survived (or failed) a pseudo test of no units during pseudo time t. Thus, while selecting the parameters of the prior distribution an expert can express his knowledge in terms of the pseudo test considered (i.e., in terms of x_0 and n_0). On the other hand, an expert can evaluate the prior mean, i.e., the ratio, x_0/n_0, and her/his degree of belief in terms of standard deviation or coefficient of variation of the prior distribution. For example, if the coefficient of variation is used, it can be treated as a measure of uncertainty (relative error) of prior assessment.

Let p_{pr} be the prior mean and k be the coefficient of variation of the prior beta distribution. The corresponding parameters x_0 and n_0 can be found as a solution of the following equation system:

$$p_{pr} = \frac{x_0}{n_0}, \quad n_0 = \frac{1 - p_{pr}}{k^2 p_{pr}} - 1 \tag{4.90}$$

The prior distribution can also be estimated using test or field data collected for analogous products. In this case, the parameters x_0 and n_0 are directly obtained from these test or field data.

Example 4.13
Let the prior mean (point estimate) of the reliability function of a hazard barrier be chosen as $p_{pr} = x_0/n_0 = 0.9$. Select the parameters x_0 and n_0.

Solution
The choice of the parameters x_0 and n_0 can be based on values of the coefficient of variation used as a measure of dispersion (accuracy) of the prior point estimate p_{pr}. Some values of the coefficient of variation and the corresponding values of the parameters x_0 and n_0 for $p_{pr} = x_0/n_0 = 0.9$ are given in the table below.

n_0	x_0	Coefficient of Variation (%)
1	0.9	23.6
9	10	10.0
90	100	3.3
900	1000	1.0

The posterior pdf is

$$f(p|x) = \frac{\Gamma(n + n_0)}{\Gamma(x + x_0)\Gamma(n + n_0 - x - x_0)} p^{(x+x_0)-1}(1 - p)^{(n+n_0-x-x_0)-1} \tag{4.91}$$

which is also a beta distribution pdf. The corresponding posterior mean is given by

$$p_B = \frac{x + x_0}{n + n_0} \tag{4.92}$$

Note that as n approaches infinity, the Bayesian estimate approaches the maximum likelihood estimate, x/n. In other words, the classical inference tends to dominate the Bayes' inference as the amount of data increases.

One should also keep in mind that the prior distribution parameters can also be estimated based on prior data (data collected on similar equipment for example), which is straightforward using the respective sample size, n_0, and the number of failures observed x_0.

It is easy to see that the corresponding $100(1-\alpha)\%$ two-sided Bayesian probability interval for p can be obtained as solutions of the following equations:

$$\Pr(p < p_l) = I_{p_l}(x + x_0, n + n_0 - x - x_0) = \frac{\alpha}{2}$$
$$\Pr(p > p_u) = I_{p_u}(x + x_0, n + n_0 - x - x_0) = 1 - \frac{\alpha}{2} \tag{4.93}$$

Example 4.14

Assume the reliability of a new component at the end of its useful life ($T = 10,000$ h) is 0.75 ± 0.19. A sample of 100 new components has been tested for 10,000 h and 29 failures have been recorded. Given the test results, find the posterior mean and the 90% Bayesian probability interval for the component reliability, if the prior distribution of the component reliability is assumed to be a beta distribution.

Solution

The prior mean is obviously 0.75 and the coefficient of variation is $0.19/0.75 = 0.25$. Using (4.90), the parameters of the prior distribution are evaluated as $x_0 = 3.15$ and $n_0 = 4.19$. Thus, according to (4.92), the posterior point estimate of the new component reliability is $R(10,000) = (3.15 + 71)/(4.19 + 100) = 0.712$. According to (4.93), the 90% lower and upper confidence limits are 0.637 and 0.782, respectively. Figure 4.7 shows the prior and posterior distributions of $1 - p$. See Figure 4.7 for comparing the prior and posterior distribution.

4.6.2.3 Lognormal Prior Distribution

The following example illustrates the case when the prior distribution and the likelihood function do not result in a conjugate posterior distribution, and the posterior distribution obtained cannot be expressed in terms of standard function. This is the case when a numerical integration is required.

Example 4.15

The number of failures to start a diesel generator on demand has a binomial distribution with parameter p. The prior data on the performance of the similar diesel are obtained from field data, and p is assumed to follow the lognormal distribution with known parameters $\mu_y = 0.056$ and $\sigma_y = 0.0421$ (the respective values of μ_t and σ_t are -3.22 and 0.70). A limited test of the diesel generators of interest shows that five failures are observed in 582 demands.

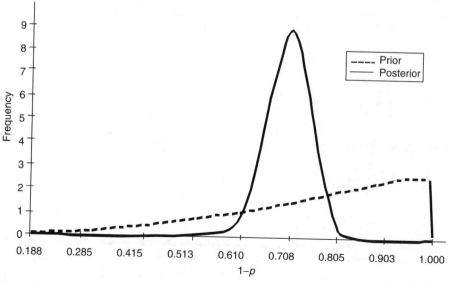

FIGURE 4.7 Prior and posterior distribution of $1 - p$.

Calculate the Bayesian point estimate of p (mean and median) and the 90th percentiles of p. Compare these results with corresponding values for the prior distribution.

Solution
Since we are dealing with a demand failure, a binomial distribution best represents the observed data. The likelihood function is given by

$$\Pr(X|p) = \binom{582}{8} p^8 (1-p)^{574}$$

and the prior pdf is

$$f(p) = \frac{1}{\sigma_t p \sqrt{2\pi}} \exp\left[-\frac{1}{2}\left(\frac{\ln p - \mu_t}{\sigma_t}\right)^\theta\right], \quad p > 0$$

Using the initial data, the posterior pdf becomes

$$f(p|X) = \frac{p^8(1-p)^{574}\exp\left[-\frac{1}{2}\left(\frac{\ln p - 3.22}{0.51}\right)^2\right]}{\int_0^1 p^8(1-p)^{574}\exp\left[-\frac{1}{2}\left(\frac{\ln p - 3.22}{0.51}\right)^2\right]}$$

It is evident that the denominator cannot be expressed in a closed form, so a numerical integration must be applied. A comparison of the prior and posterior distribution of p is illustrated in Figure 4.8. Characteristics of the prior and posterior distributions have been listed in Table 4.4.
The point estimate of the actual data using the classical inference is

$$\hat{p} = \frac{8}{582} = 0.0137$$

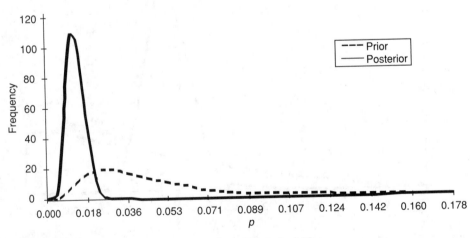

FIGURE 4.8 Prior and posterior distribution of p in Example 4.15.

TABLE 4.4
Comparison of Prior and Posterior Distributions

	Prior	Posterior
Mean	0.0516	0.0130
Median	0.0399	0.0121
Fifth percentile	0.0123	0.0064
95th percentile	0.1293	0.0197

4.6.3   A COMPARISON OF CLASSICAL PARAMETER ESTIMATION AND BAYESIAN PARAMETER ESTIMATION METHODS: ADVANTAGES AND DISADVANTAGES

4.6.3.1   Classical Estimation Methods

Advantages
- The results depend only on the data.
- When the quantity of data is large, classical methods produce good estimates.
- Classical methods are well known and widely used in all areas of science and engineering and thus have historical precedence. Classical estimation is easier to understand and use.

Disadvantages
- A confidence interval cannot be directly interpreted as a probability that the corresponding parameter lies in the interval.
- Relevant information regarding the parameter may exist outside the sample data. While it is possible to model such external information about a parameter using classical estimation, the techniques for doing so are complicated. It is not a straightforward procedure to formally include engineering judgment in the estimation process, and because risk analyses usually deal with rare events, engineering judgment is often the only source of information available about an unknown parameter.
- It is a complicated process to propagate classical confidence intervals through logic tree models to produce corresponding interval estimates on output quantities of interest such as the risk values.
- In most applications of risk analysis, the available data are often a mix of various data sources and types. That is, a host of applicable related data is available. This related data may consist of observed operating experience from similar sources, but for different environmental or operating conditions. Usually the precise differences in conditions are unknown. The result is that the quality of more-or-less relevant data available for use is almost never of the precise form and format required for using classical methods of estimation.
- Classical methods are sensitive to the way in which sample data were generated or collected. In formal risk assessments practice, the precise details of this process are either unknown or unavailable, and classical methods become either difficult or even impossible to use.

4.6.3.2   Bayesian Estimation Methods

Advantages
- Bayesian estimation provides a logical approach to estimation. It can measure uncertainty about parameters using probabilities. To the extent that the information captured

by the prior distribution accurately, Bayesian methods produce good parameter estimates.

- Bayesian methods provide a formal method of explicitly introducing prior information and knowledge into the analysis. This is particularly useful when sample data are scarce, as in the case of rare events. Bayesian estimation permits the use of various types of relevant generic data.
- Bayesian method interprets uncertainty about a parameter using a subjective probability interval. As a direct consequence of this probability can be easily combined with other sources of uncertainty in risk assessment, as will be discussed in Chapter 5.
- Bayesian estimation provides a natural and convenient method for use in updating the state of knowledge about a parameter as future additional sample data become available.
- The reasoning process used in Bayesian estimation is straightforward.
- Bayesian estimation is applicable to a larger class of situations likely to be encountered in risk assessment.

Disadvantages
- A suitable prior distribution must be identified and justified subjectively, which is often a difficult task in practice.
- Bayesian inference may be sensitive to the choice of a prior distribution.
- A risk analyst may find difficulty in convincing the subject matter experts to produce subjective prior distribution.
- A risk analyst using Bayesian estimation is open to the criticism that a self-serving prior distribution has been selected that reflects a biased point of view that may be inappropriate or incorrect.
- Because they are less widely used and new, Bayesian estimation methods sometimes require more effort to understand, obtain, implement, and interpret in practice.

Although there are no universal rules to follow when deciding whether an analyst should use classical or Bayesian estimation methods, there are certain situations in which each has been found to be particularly appropriate. Bayesian estimation methods should be used to determine subjective distributions of input parameters whose uncertainties are required to be propagated through system models using Monte Carlo simulation. Bayesian methods should also be used when generic data are to be combined with facility- or system-specific data to produce a facility- or system-specific distribution of a parameter of interest. On the other hand, classical methods should be used for diagnostic testing of modeling assumptions, such as goodness-of-fit tests of an assumed distribution model.

It is generally believed by most PRA and other risk analysts that, for applications to complex systems and facilities such as nuclear power plants, the advantages of Bayesian methods outweigh the disadvantages; thus, Bayesian estimation methods are widely used in these applications.

When using Bayesian methods, the sensitivity to the choice of the prior distribution should be investigated. When there is strong direct evidence (i.e., when there is a large quantity of observable sample data) both approaches produce similar results.

Empirical Bayes represents another major class of methods of statistical inference that differ markedly in philosophy from Bayesian methods. Empirical Bayes is characterized by the fact that the prior distribution has a relative frequency interpretation in contrast to the degree-of-belief interpretation of Bayesian statistics. For example, if a component belongs to a population of similar components in similar applications, such as a set of similar facilities or systems, then the prior distribution of the component failure rate represents the facility-to-facility variability in the failure rate. The empirical Bayes prior is sometimes referred to as the population variability curve. The prior is empirically determined using observed plant-specific data for a given set of facilities, after which Bayes' theorem may be applied.

4.7 PERFORMANCE OF PASSIVE AND SOME ACTIVE HAZARD BARRIERS THROUGH PHYSICS AND ENGINEERING OF FAILURE MODELING

Many times we are dealing with situations that the performance is better modeled by understanding and modeling the underlying processes and phenomena that apply stress or cause damage and degradation in the unit. This is particularly important in passive system and barrier failures. There are three types of models that can be used to assess reliability, capability, or probability of failure of the various items. These are:

Stress–strength model. This model assumes that the barrier fails if *stress* (mechanical, thermal, etc.) applied to the barrier exceeds the *strength*. The stress represents an aggregate of the operating conditions and challenges as well as external conditions. This failure model may depend on environmental conditions or the occurrence of critical events, rather than the mere passage of time or cycles. Strength is often treated as a random variable representing effect of all conditions affecting (e.g., reducing) the strength, or lack of knowledge about the item's strength (e.g., the item's capability, mechanical strength, and dexterity). Two examples of this model are: (a) a steel bar in tension, compression or bending, and (b) a transistor with a voltage applied across the emitter–collector.

Damage–endurance model. This model is similar to the stress–strength model, but the scenario of interest is that *stress* applied causes damage that accumulates irreversibly, as in corrosion, wear, and fatigue. The aggregate of challenges and external conditions leads to the metric represented as cumulative damage. The cumulative damage may not degrade performance. The item fails when and only when the cumulative damage exceeds the endurance (i.e., the damage accumulates until the endurance of the item is reached such as the crack caused by fatigue grows to reach the critical endurance size dictated by the fracture toughness of the materials). Accumulated damage does not disappear when the stress agents are removed, although sometimes treatments such as annealing are possible. Endurance is often treated as a random variable. Similar to the stress–strength model, endurance is an aggregate measure for effects of challenges and external conditions on the item's capability to withstand cumulative stresses.

Performance–requirement model. When performance characteristics of a unit degrades (e.g., efficiency of a pump or compressor reduces) to a point that the unit does not meet certain operating and functional requirements, the unit may be considered failed. In this model, a distribution of the performance and its degradation over time is estimated and compared to a fixed value or a distribution of minimum requirement. The probability that performance falls below acceptable minimum requirement is considered as probability that unit fails.

The first two modeling approaches have many applications in risk assessment, particularly in assessing passive hazard barriers. A summary of these two methods has been provided in the following sections. The performance–requirement model analytically is very similar to the damage–endurance model.

4.7.1   STRESS–STRENGTH MODELS

In this modeling concept, a failure occurs when the stress applied to an item exceeds its strength. The probability that no failure occurs is equal to the probability that the applied stress is less than the item's strength, i.e.,

$$R = \Pr(S > s), \tag{4.94}$$

where R is the reliability of the item, s is the applied stress, and S is the item's strength.

Examples of stress-related failures include application of extra load to a pressure vessel, such as compressed natural gas container.

Engineers need to ensure that the strength of an item exceeds the applied stress for all possible stress situations. Traditionally, in the deterministic design process, safety factors are used to cover the spectrum of possible applied stresses. This is generally a good engineering principle, but failures occur despite these safety factors. On the other hand, safety factors that are too stringent result in overdesign, high cost, and sometimes poor performance.

If the range of major stresses is known or can be estimated, a probabilistic approach can be used to address the problem. This approach eliminates overdesign, high cost, and failures caused by stresses that are not considered early in the design. If the distribution of S and s can be estimated as $F(S)$ and $g(s)$, then

$$R = \int_0^\infty g(s)\,dg(s)$$
$$F = \int_0^\infty F(S)\left[\int_S^\infty f(s)\,ds\right]dS \tag{4.95}$$

Figure 4.9 shows $F(S)$ and $g(s)$ distributions.

The *safety margin* (SM) is defined as

$$SM = \frac{E(S) - E(s)}{\sqrt{\mathrm{var}(S) + \mathrm{var}(s)}} \tag{4.96}$$

The SM shows the relative difference between the mean values for stress and for strength. The larger the SM, the more reliable the item will be. Use of (4.96) is a more objective way of measuring the safety of items. It also allows for calculation of reliability and probability of failure as compared with the traditional deterministic approach using safety factors. However, good data on the variability of stress and strength are often not easily

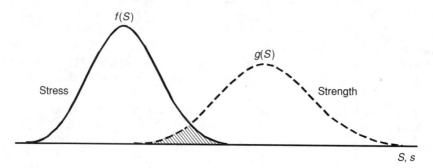

FIGURE 4.9 Stress–strength distributions.

available. In these cases, engineering judgment can be used to obtain the distribution including engineering uncertainty. The section on expert judgment explains methods for doing this in more detail.

The distribution of stress is highly influenced by the way the item is used and the internal- and external-operating environments. The design determines the strength distribution, and the degree of quality control in manufacturing primarily influences the strength variation.

It is easy to show that for a normally distributed S and s,

$$R = \Phi(\mathrm{SM}) \tag{4.97}$$

where $\Phi(\mathrm{SM})$ is the cumulative standard normal distribution with $z = \mathrm{SM}$ (see Table B.1).

Example 4.16

Consider the stress and strength of a beam in a structure represented by the following normal distributions:

$$\mu_S = 420\mathrm{kg/cm}^2 \quad \text{and} \quad \sigma_S = 32\mathrm{kg/cm}^2$$
$$\mu_s = 310\mathrm{kg/cm}^2 \quad \text{and} \quad \sigma_s = 72\mathrm{kg/cm}^2$$

What is the reliability of this structure?

Solution

$$\mathrm{SM} = \frac{420 - 310}{\sqrt{32^2 + 72^2}} = 1.4$$

with $z = 1.4$ and using Table B.1,

$$R = \Phi(1.4) = 0.91$$

Example 4.17

A random variable representing the strength of a nuclear power plant containment building follows a lognormal distribution with the mean strength of 0.905 MPa and standard deviation of 0.144 MPa. Four possible accident scenarios can lead to high-pressure conditions inside the containment that may exceed its strength. The pressures cannot be calculated precisely, but can be represented as another random variable that follows a lognormal distribution.

(a) For a given accident scenario that causes a mean pressure load inside the containment of 0.575 MPa with a standard deviation of 0.117 MPa, calculate the probability that the containment fails.
(b) If the four scenarios are equally probable and each leads to high-pressure conditions inside the containment with the following mean and standard deviations, calculate the probability that the containment fails.

μ_L (MPa)	0.575	0.639	0.706	0.646
σ_L (MPa)	0.117	0.063	0.122	0.061

(c) If the containment strength distribution is divided into the following failure mode contributors with the mean failure pressure and standard deviation indicated, repeat part (a).

Failure Mode	Mean Pressure, μ_S (MPa)	Standard Deviation, σ_S (MPa)
Liner tear around personnel airlock	0.910	1.586×10^{-3}
Basemat shear	0.986	1.586×10^{-3}
Cylinder hoop membrane	1.089	9.653×10^{-4}
Wall–basemat junction shear	1.131	1.586×10^{-3}
Cylinder meridional membrane	1.241	1.034×10^{-3}
Dome membrane	1.806	9.653×10^{-4}
Personnel air lock door buckling	1.241	1.655×10^{-3}

Solution

If S is a lognormally distributed random variable representing strength, and L is a lognormally distributed random variable representing pressure stress (load), then the random variable, $Y = \ln(S) - \ln(L)$, is also normally distributed.

For the lognormal distribution with mean and standard deviation of μ_y and σ_y, the respective mean and standard deviation of the normal distribution, μ_t and σ_t, can be obtained using (A.47) and (A.48). Then:

$$R = \Phi(\text{SM}) = \Phi\left(\frac{\mu_{S_t} - \mu_{L_t}}{\sqrt{\sigma_{S_t}^2 + \sigma_{L_t}^2}}\right)$$

(a)

$$R = \Phi\left(\frac{-0.112 - (-0.574)}{\sqrt{0.158^2 + 0.201^2}}\right) = 0.9642$$

The probability of containment failure:

$$F = 1 - R = 0.0358$$

(b) Because the four scenarios are "equally probable," then the system is equivalent to a series system, such that: $R = R_1 \times R_2 \times R_3 \times R_4$.

$$R_1 = \Phi(\text{SM1}) = \Phi(1.80) = 0.9642$$
$$R_2 = \Phi(\text{SM2}) = \Phi(1.83) = 0.9662$$
$$R_3 = \Phi(SM3) = \Phi(1.07) = 0.8586$$
$$R_4 = \Phi(SM4) = \Phi(1.79) = 0.9631$$

The probability of containment failure:

$$F = 1 - R = 1 - R_1 \times R_2 \times R_3 \times R_4 = 0.2297$$

(c) Because each failure mode may cause a system failure, this case can be treated as a series system. Because we know the mean of the lognormal distribution using the same procedure used in part (a) and (b) we have:

$$R_a = \Phi(SM_a) = \Phi(2.38) = 0.9913$$
$$R_b = \Phi(SM_b) = \Phi(2.78) = 0.9973$$
$$R_c = \Phi(SM_c) = \Phi(3.27) = 0.9995$$
$$R_d = \Phi(SM_d) = \Phi(3.46) = 0.9997$$
$$R_e = \Phi(SM_e) = \Phi(3.92) \approx 1$$
$$R_f = \Phi(SM_f) = \Phi(5.78) \approx 1$$
$$R_g = \Phi(SM_g) = \Phi(3.92) \approx 1$$

The probability of containment failure:

$$F = 1 - R = 1 - R_a \times R_b \times R_c \times R_d \times R_e \times R_f \times R_g = 0.0122.$$

If both the stress and strength distributions are exponential with parameters λ_s and λ_S, the reliability can be estimated as

$$R = \frac{\lambda_s}{\lambda_s + \lambda_S} \qquad (4.98)$$

For more information about stress–strength methods in reliability analysis, the readers are referred to O'Connor [9] and Kapur and Lamberson [25].

4.7.2 DAMAGE–ENDURANCE MODELS

In this modeling technique failure occurs when applied "stress" causes permanent and irreversible damage as a function of time. When the level of damage reaches a level that the item is unable to endure, the hazard barrier fails. In practice, damage and endurance are not precisely known. They may be expressed as pdfs at best. Damage accumulation does not always occur in the same way, and different initial conditions may result in drastically different damage. Endurance limit is not the same as well, because the strength of material varies from unit-to-unit, and endurance is considered as a charac- teristic of material. In any case failure occurs only when accumulated damage exceeds the endurance limit of the material. The interference of variability and uncertainty in accumulated damage and endurance limit and occurrence of failure are illustrated in Figure 4.10. This figure shows how the interference of damage and endurance make a statistical distribution (uncertainty) of TTF, which is the variable of interest in most of the reliability analyses.

Examples of damage include fatigue-caused cracks in passive structures. When the crack size reaches a critical size, the item fails. The critical size is the endurance limit of the item. Also, in nuclear reactors, the pressure vessel (which can be considered as passive hazard containment from the risk assessment point of view) is damaged by irradiation embrittlement, leading to reduction of its endurance over time to a point that it can no longer tolerate certain anticipated or unanticipated transients that cause severe combined thermal and mechanical stresses. This would cause rupture of the pressure vessel. It should be noted that agents that cause damage (activate by mechanisms of failure) act over time and affect both the damage level (e.g., increase it) and the endurance limit (e.g., decrease it). Figure 4.10 depicts the damage–endurance model, and the impact of time.

The problem is to establish a distribution of damage and endurance as a function of time. Clearly, as we are unable to reduce all the uncertainties involved in measuring damage and endurance, they are often represented by pdfs as shown in Figure 4.11. The mathematical concept determining the reliability (and thus performance) of the hazard barrier at a given

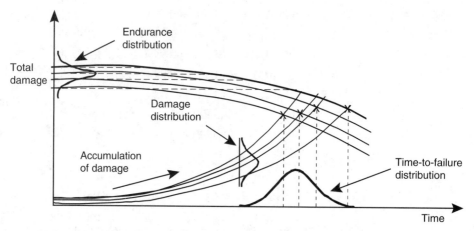

FIGURE 4.10 The damage–endurance modeling model.

time is similar to the strength–strength modeling approach, and (4.94)–(4.97) also apply to damage–endurance model at a given age (time) or cycle of operation. Clearly, the "stress" distribution is now represented by "damage" distribution and "strength" distribution is represented by "endurance" distribution.

To estimate damage and endurance distributions we need to understand the degradation processes leading to damage and reduction of endurance limits, and perform engineering-based degradation analysis. Degradation analysis involves the measurement of degradation as a function of time, relating degradation rate to time and ultimately life and determining the acting failure agent. Many failure mechanisms can be directly linked to degradation agents (agents of failure) and enable the analysts to assess probability of failure of the hazard barrier at any time in the future by extrapolation. Further, tests at elevated stress levels can be performed which speed up the degradation and damage process. The degradation and stress tests can provide the information for assessing the distribution of TTF of the barriers. For more information on this subject the readers are referred to Nelson [26].

In some cases, it is possible to directly measure the degradation over time, as with the wear of a bearing or with the propagation of cracks in structures under random loading (causing fatigue-induced crack growth). In other cases, direct measurement of degradation might not be possible without invasive or destructive measurement techniques that would directly affect the subsequent performance of the item. In such cases, the degradation can be estimated through the measurement of certain performance characteristics, such as using resistance to gauge the degradation of a dielectric material. In both cases, it is necessary to define a level of degradation (endurance limit) at which a failure will occur. With this endurance limit, it is simple to use basic mathematical models to extrapolate and assess failure probabilities

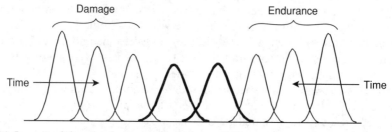

FIGURE 4.11 Impact of time on damage–endurance modeling.

(and thus, the performance measurements) over time to the point where the failure would occur. This is done at different damage level (caused by applied stresses) and, therefore each cycle-to-failure or TTF is also associated with a corresponding stress level. Once the TTF at the corresponding stress levels have been determined, it is merely a matter of analyzing the extrapolated failure times like conventional accelerated TTF data.

Once the level of endurance (or the level of damage beyond which it constitute a failure) is defined, the degradation over time should be measured. The uncertainty in the results is directly related to the number of items and amount of data obtained in the observations at each stress level, as well as in the amount of overstressing with respect to the normal operating conditions. The degradation of these items needs to be measured over time, either continuously or at predetermined intervals. Once this information has been recorded, the next task is to extrapolate the measurements to the defined failure level to estimate the failure time. These models have the following general forms:

$$\text{Linear: } y = ax + b \tag{4.99}$$
$$\text{Exponential: } y = be^{ax} \tag{4.100}$$
$$\text{Power: } y = bx^{a} \tag{4.101}$$
$$\text{Logarithmic: } y = a \ln(x) + b \tag{4.102}$$

where y represents the performance (e.g., item's life or system's output), x represents applied stress, level of damage or level of requirement, and a and b are model parameters to be estimated.

For example, consider developing the damage distribution in a damage–endurance model due to wear of a bearing in rotating machinery as a function of time. As expected from the physics of failure the higher wear values lead to the lower life of the bearing. Note that a main assumption is that the mechanism of failure remains the same.

As illustrated in Figure 4.12, using the engineering of failure model with one wear agent (i.e., $\tau_{\max}$) one can significantly simplify the wear problem and life estimation, instead of

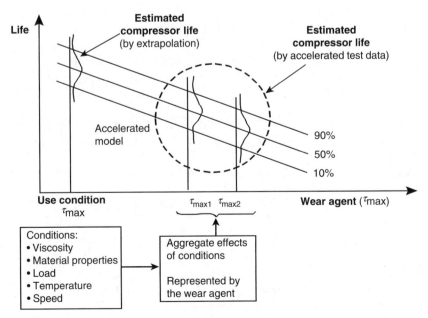

FIGURE 4.12 Wear (damage)–endurance life modeling concept.

dealing with many variables and conditions, as only one value represents the aggregate effects of all others. In this figure, clearly degradation–life model in form of (4.101) is used and plotted in a log–log scale.

Observation of failures in higher stress levels ($\tau_{\text{max}\,1}$ and $\tau_{\text{max}\,2}$) makes the time needed much shorter, and yields better estimate of the probabilistic distribution of life of the item. Typically, lognormal or Weibull distributions best represent the form of these distributions. Life distribution models in high-stress levels are then extrapolated to the normal use level along the life acceleration model. Data from accelerated tests are then used to estimate the parameters of both the life-stress and distribution models. The wear agent is directly related to determining the distribution of damage of the bearing over time.

The degradation analysis using engineering principles show that the maximum shear stress in the vicinity of the surface is an aggregate agent of failure responsible for material removal in the abrasive wear process. The wear rate (and thus the damage) can be expressed as a power relationship similar to (4.101). Having the wear rate, life of the bearing can be estimated by considering an endurance limit beyond which the bearing can be considered failed. This end state can be defined either based on the surface roughness (i.e., RMS of journal and bearing roughness) or a certain amount of material removal on bearing (e.g., the instance that the coating thickness of bearing is removed). Based on the above description the damage of the bearing can be related to the wear agent as,

$$\dot{W} \propto \left[\frac{\tau_{\text{max}}}{\tau_{\text{yp}}}\right]^n$$

$$(1/D) \propto L = \frac{C_0}{\dot{W}}$$

Therefore

$$L = C\left[\frac{\tau_{\text{yp}}}{\tau_{\text{max}}}\right]^n = \frac{K}{[\tau_{\text{max}}]^n} \tag{4.103}$$

where $\dot{W}$ is the wear rate; C_0 the end state wear; C the constant; L, D the life and damage of the bearing in terms of thickness removed (dependent variable); K, n the constants to be estimated from the accelerated test results; τ_{yp} the material shear yield point and τ_{max} the maximum shear stress in the vicinity of the surface (independent variable).

Note that the presence of τ_{yp} (i.e., shear yield point of the bearing material) enables this model to also consider the temperature degradation of the bearing material, which can be critical for coated-type bearings.

Assuming that the life is distributed lognormally, then the joint distribution of life-stress can be obtained from (4.103) as μ is the mean of the natural logarithms of the TTF $= -\ln(K) - n \cdot \ln(\tau_{\text{max}})$.

Therefore,

$$f(t|\mu,\sigma) = \frac{1}{t\sigma\sqrt{2\pi}} \exp\left[-\frac{1}{2}\frac{(\ln t + \ln(K) + n\ln(\tau_{\text{max}}))^2}{\sigma^2}\right] \tag{4.104}$$

Example 4.18

A 2024-T3 Al component was fatigue tested to failure at two different cyclic stress levels of 400 and 450 MPa, and the following data were observed.

	Stress (MPa)	
	400	450
Cycles-to-failure	16	8
	18	9
	20	9
	21	10
	24	11
	27	13
	29	14

Estimate the number of cycles-to-failure at 300 MPa.

Solution

Assuming that fatigue damage mechanism leads to cycles-to-failure that follows a lognormal life distribution, and an inverse power relationship for the stress-life model similar to (4.104) then,

$$f(N|S) = \frac{1}{N\sigma_{N'}\sqrt{2\pi}} \exp\left[-\frac{1}{2}\frac{(\ln N + \ln(K) + n\ln(S))^2}{\sigma_{N'}^2}\right]$$

where N represents cycles-to-failure; $\sigma_{N'}$, the standard deviation of the natural logarithms of the cycles-to-failure, and K, n the power law model parameters.

The parameters are calculated through the MLE approach,

$$K = 2.3E - 18; \ n = 6.3; \text{ and } \sigma_{N'} = 0.2$$

Then the mean life of the conditional joint cycle and stress lognormal life model at the stress level of 300 MPa is calculated from the expected value expression for the lognormal distribution as

$$\bar{T} = \exp\left[-\ln(K) - n\ln(V) + \frac{1}{2}\sigma_{T'}^2\right]$$

$$= \exp\left[-\ln(2.3E - 18) - 6.3\ln(300) + \frac{1}{2}(0.2)^2\right] = 134.2 \text{ cycles}$$

Example 4.19

TTF of a journal bearing is believed to follow a lognormal distribution. Accelerated life tests revealed the median life of the bearing is 2000 h in the first test, where radial force of 750 kg was applied on the bearing, and 1500 h when the radial force was increased to 800 kg in the second test. What is the expected median life for the bearing in normal operation where the radial force is 400 kg?

Solution

Radial force on the bearing is the acceleration variable in these tests. Higher radial force results in higher pressure in the lubricant film and ultimately higher normal stress on the surface. The friction has also a linear relationship with the normal load applied on the bearing. τ_{max}, which is the maximum shear stress in the vicinity of mating surfaces is estimated from:

$$\tau_{max} = ke\sqrt{\left(\frac{\sigma_n}{2}\right)^2 + \tau_f^2}$$

where τ_{max} is the maximum shearing stress; ke the stress concentration factor; σ_n the normal stress on the surface; $\tau_f = \mu\sigma_n$, the friction-generated shear stress, and μ the friction factor.

Since normal stress σ_n is a linear function of bearing radial force, the τ_{max} becomes a linear function of applied radial force and we have:

$$\frac{T_1}{T_2} = \frac{2000}{1500} = \frac{\left(\frac{K}{\tau_{max\,1}^n}\right)}{\left(\frac{K}{\tau_{max\,2}^n}\right)} = \left(\frac{F_2}{F_1}\right)^n = \left(\frac{800}{750}\right)^n \Rightarrow n = 4.457$$

$$\frac{T_{use}}{T_2} = \frac{T_{use}}{1500} = \frac{\left(\frac{K}{\tau_{use}^n}\right)}{\left(\frac{K}{\tau_{max\,2}^n}\right)} = \left(\frac{F_2}{F_{use}}\right)^n = \left(\frac{800}{400}\right)^{4.457} \Rightarrow T_{use} \approx 32,944\,h$$

Example 4.20

An AISI 4130X cylindrical tank is subjected to degradation from corrosion fatigue mechanism. Calculate the crack growth resistance at 30 ksi-in$^{1/2}$, in a wet-gas environment, according to the following data and assume that the crack growth behavior generally follows the Paris equation,

$$da/dN = C(\Delta K)^m$$

where a is the crack length; N the number of cycles; ΔK the stress intensity factor range; and C, m the material constants.

ΔK [ksi-in$^{1/2}$]	40	50	60	70
da/dN [in/cycle]	4.0×10^{-4}	9.8×10^{-3}	9.2×10^{-3}	8.3×10^{-3}

Solution

The resistance of the material to crack growth can be written as the inverse of the crack growth rate,

$$\frac{1}{da/dN} = \frac{1}{C(\Delta K)^m}$$

Then, setting $1/(da/dN) = L(\Delta K)$, where L is the crack resistance,

$$L(\Delta K) = \frac{1}{C(\Delta K)^m}$$

Assuming that corrosion fatigue mechanism follows a lognormal life distribution, and an inverse power law relationship for the stress-life model,

$$f(a', |\Delta K) = \frac{1}{a'\sigma_{a'}\sqrt{2\pi}} \exp\left[-\frac{1}{2}\frac{(\ln(a') + \ln(C) + m\ln(\Delta K))^2}{\sigma_{a'}^2}\right]$$

where a' is the natural logarithm of the crack growth resistance; $\sigma_{a'}$ the standard deviation of the natural logarithms of the crack growth resistance, and C, m the power model parameters.

The paint estimate of parameters are calculated through the maximum likelihood estimator approach as,

$$C = 2.1 \times 10^{11}, \ m = -5.2, \text{ and } \sigma'_a = 0.8$$

then, the mean life of the power-lognormal model at 30 ksi-in$^{1/2}$ is calculated from the expected value expression for a lognormal distribution,

$$\bar{a}' = \exp\left[-\ln(C) - m\ln(\Delta K) + \frac{1}{2}\sigma^2_{a'}\right]$$

$$= \exp\left[-\ln 2.1 \times 10^{11} + 5.2\ln(30) + \frac{1}{2}(0.8)^2\right]$$

$$= 2.9 \times 10^{-4} \text{ in/cycles}$$

4.8   SOFTWARE PERFORMANCE ASSESSMENT

Software plays an important role in complex systems today. It also enters the scenarios leading to exposure of hazards. For example, when used to activate a physical barrier, measure and compute location, and amount of certain hazards or conditions that must be dealt with. Therefore, software can be either a barrier by itself, or supporting another hardware or human barrier. In the business world, everything from supermarkets to the stock exchange are computerized. Software has contributed to some of the disasters and major losses to modern organizations – According to a recent study software errors cost the U.S. economy $59.5 billion annually [27]. For example:

- A software error in the Patriot missile's radar system allowed an Iraqi Scud to penetrate air defense and slam into an American military tents in Saudi Arabia, killing 28 people during the first Gulf war.
- A software error in programs that route calls through the AT&T network was blamed for the 9-h breakdown in 1990 of the long-distance telephone network.
- A software error involving the operation of a switch on the Therac-25, a computer-controlled radiation machine, delivered excessive amount of radiation, killing at least four people.
- A computer error cost American Airlines US$ 50 million in lost ticket sales when its reservation system mistakenly restricted sales of discount tickets.
- A computer error understated the British Retail Price Index by 1% from February 1986 to October 1987, costing the British Government 21 million in compensation to pension and benefit holders, donations to charities, and related administrative cost.
- A software error in Washington's Rainier Teller machines permitted unlimited amount of money to be withdrawn in excess of customer balance. The bank then had to run into the cost of trying to recover the overages.

Software failure risk is the expected economic loss that can result if software fails to operate correctly. The ability to identify and measure software failure risk can be used to manage this risk.

Many techniques have been developed for measuring performance of hardware as discussed in this chapter. However, their extension to software has been problematic for two reasons. First, software faults are design faults, while faults in hardware items are equipment breakage. Second, software systems are more complex than hardware units, so the same performance analysis methods may be impractical to use.

Software has deterministic behavior, whereas hardware behavior is both deterministic and probabilistic. Indeed, once a set of inputs to the software has been selected, and provided that the computer and operating system within which the software will run is error free, the software will either fail or execute correctly. However, our knowledge of the inputs selected, of computer, of the operating system, and of the nature and position of the fault may be uncertain. One may, however, present this uncertainty in terms of probabilities. A software fault is a triggering event that causes software error. A software bug (error in the code) is an example of a fault.

Reliability is the ultimate performance measure of interest in software. Accordingly, we adopt a probabilistic definition for software reliability. Software reliability is the probability that the software product will not fail for a specified time under specified conditions. This probability is a function of the input to and use of the product, as well as a function of the existence of faults in the software. The inputs to the software product will determine whether an existing fault is encountered or not. First, we define software faults and failures.

Faults can be grouped as design faults, operational faults, or transient faults. All software faults are design faults; however, hardware faults may occur in any of the three classes. Faults can also be classified by the source of the fault; software and hardware are two of the possible sources of the fault. Sources of faults are: input data, system state, system topology, humans, environment, and unknown causes. For example, the source of many transient faults is unknown.

Failures in software are classified by mode and scope. A failure mode may be sudden or gradual, partial or complete. All four combinations of these are possible. The scope of failure describes the extent within the system of the effects of the failure. This may range from an internal failure, whose effect is confined to a single small portion of the system, to a pervasive failure, which affects much of the system (see Lawrence [28]).

Software, unlike hardware, is unique in that its only failure modes are the result of design flaws, as opposed to any kind of internal physical mechanisms and external environmental conditions such as aging, for example see McDermid [29]. As a result, traditional reliability techniques, which tend to focus on physical component failures rather than system design faults, have been unable to close the widening gap between the powerful capabilities of modern software systems and the levels of reliability that can be computed for them. The real problem of software reliability is one of managing complexity. There is a natural limitation on the complexity of hardware systems. With the introduction of digital computer systems, however, designers have been able to arbitrarily implement complex designs in software. The result is that the central assumption implicit in traditional reliability theory, that the design is correct and failures are the result of fallible components is no longer valid.

To assess the reliability of software, a software reliability model (SRM) will be needed. In the remainder of the section, details of classes of SRMs, and two of such model are discussed. Also discussed are the models used to assess software life cycle.

4.8.1 SOFTWARE RELIABILITY MODELS

Several SRMs have been developed over the years. These techniques are referred as "analyses" or "models," but there is a distinct difference between the two. An analysis (such as fault tree analysis (FTA)) is carried out by creating a model (the fault tree) of a system, and then using that model to calculate properties of interest, such as reliability.

The standard reliability models such as FTA, event tree analysis (ETA), failure modes and effect analysis (FMEA), and Markov models are adequate for systems whose component remain unchanged for long periods of time. They are less flexible for systems that undergo frequent design changes. If, for example, the failure rate of a component is improved through design or system configuration changes, the reliability model must be re-evaluated. In this model, software is tested for a period of time, during which failures may occur. These failures lead to modification to the design or manufacture of component; the new version then goes back into test. This cycle is continued until design objectives are met.

When these models are applied to software reliability one can group them into two main categories: predictive models and assessment models [30]. Predictive models typically address the reliability of software early in the design cycle. Different elements of a life cycle development of software are discussed later. A predictive model is developed to assess the risks associated with the development of software under a given set of requirements and for specified personnel before the project truly starts. Predictive SRMs are few in number [30], and as such in this section the predictive models are not discussed. Assessment models evaluate present and project future software reliability from failure data gathered when the integration of the software starts.

4.8.1.1 Classification

Most existing SRMs may be grouped into four categories:

(1) Time-between-failure model
(2) Fault-seeding model
(3) Input-domain based model
(4) Failure count model

Each category of models is summarized as follows:

Time-between-failure model. This category includes models that provide an estimate of the times-between-failures in software. Key assumptions of this model are independent time between successive failures, equal probability of exposure of each fault; embedded faults are independent of each other, no new faults introduced during corrective actions.

Specific SRMs that estimate MTBFs are: Jelinski–Moranda model [31], Schick and Wolverton model [32], Littlewood–Verrall's Bayesian model [33], and Goel and Okumoto imperfect debugging model [34].

Fault-seeding model. This category of SRMs includes models that assess the number of faults in the software at time zero via seeding extraneous faults. Key assumptions of this model are: (1) seeded faults are randomly distributed in the software, and (2) indigenous and seeded faults have equal probabilities of being detected.

The specific SRM that falls into this category is Mills Fault-Seeding Model [35]. In this model, an estimate of the number of defects remaining in a program can be obtained by a seeding process that assumes a homogeneous distribution of representative class of defects. The variables in this measure are: the number of seed faults introduced N_S, the number of intentional seed faults found n_s, and the number of faults found n_F that were not intentionally seeded.

Before seeding, a fault analysis is needed to determine the types of faults expected in the code and their relative frequency of occurrence. An independent monitor inserts into the code

N_S faults that are representative of the expected indigenous faults. During testing, both seeded and unseeded faults are identified. The number of seeded and indigenous faults discovered permits an estimate of the number of faults remaining for the fault type considered. The measure cannot be computed unless some seeded faults are found. The maximum likelihood estimate of the unseeded faults is given by

$$\hat{N}_F = n_F N_S / n_S \tag{4.105}$$

Example 4.21
Forty faults of a given type are seeded into a code and, subsequently, 80 faults of that type are uncovered: 32 seeded and 48 unseeded. Estimate remaining faults.

Solution
Using (4.105), $N_F = 60$, and the estimate of faults remaining is

$$N_F(\text{remaining}) = N_F - n_F = 60 - 48 = 12$$

Input-domain based model. This category of SRMs includes models that assess the reliability of software when the test cases are sampled randomly from a well-known operational distribution of software inputs. The reliability estimate is obtained from the number of observed failures during execution. Key assumptions of these models are: (1) input profile distribution is known, (2) random testing is used (inputs are selected randomly), and (3) input domain can be partitioned into equivalence classes.

Specific models of this category are: Nelson's model [36] and Ramamoorthy and Bastani's model [37]. We will further elaborate on Nelson's model.

Nelson's model is typically used for systems with ultrahigh-reliability requirements, such as software used in nuclear power plants that are limited to about 1000 lines of code. The model is applied to the validation phase of the software (acceptance test) to estimate the reliability. Nelson defines the reliability of a software run n times (for n test cases) and which failed n_f times as

$$R = 1 - n_f / n \tag{4.106}$$

where n is the total number of test cases and n_f is the number of failures experienced out of these test cases.

Failure count model. This category of SRMs estimate the number of faults or failures experienced in specific intervals of time. Key assumptions of these models are: (1) test intervals are independent of each other, (2) testing intervals are homogeneously distributed, and (3) number of faults detected during nonoverlapping intervals are independent of each other.

The SRMs that fall into this category are: Shooman's exponential model [38], Goel–Okumoto's NHPP [39], Musa's execution time model [40], Goel's generalized NHPP model [41], Musa–Okumoto's logarithmic Poisson execution time model [42]. We will further elaborate on the Musa model and on the Musa–Okumoto models.

Musa basic execution time model (BETM) model. This model (Musa [40]) assumes that failures occur in form of a NHPP. The unit of failure intensity is the number of failures per central process unit (CPU) time. This relates failure events to the processor time used by the software. In the BETM, the reduction in the failure intensity function remains constant, irrespective of whether any failure is being fixed.

The failure intensity, as a function of number of failures experienced, is obtained from:

$$\lambda(\mu) = \lambda_0 \left(1 - \frac{\mu}{v_0}\right) \tag{4.107}$$

where $\lambda(\mu)$ is the failure intensity (failures per CPU hour), λ_0 is the initial failure intensity at the start of execution, μ is the expected number of failures experienced up to a given point in time, and v_0 is the total number of failures.

The number of failures that need to be fixed to move from present failure intensity, to a target intensity, is given by

$$\Delta\mu = \frac{v_0}{\lambda_0} \left(\lambda_p - \lambda_F\right) \tag{4.108}$$

The execution time required to reach this objective is

$$\Delta\tau = \frac{v_0}{\lambda_0} \ln\left(\frac{\lambda_p}{\lambda_F}\right) \tag{4.109}$$

In these equations, v_0 and λ_0 can be estimated in different ways, see Musa [40].

Musa–Okumoto logarithmic Poisson time model (LPETM) [42]. According to the LPETM, the failure intensity is given by

$$\lambda(\mu) = \lambda_0 \exp(-\theta\mu) \tag{4.110}$$

where θ is the failure intensity decay parameter and λ, μ, λ_0 are the same as in the BETM. This model assumes that repair of the first failure has the greatest impact in reducing failure intensity and that the impact of each subsequent repair decreases exponentially.

In the LPETM, no estimate of v_0 is needed. The expected number of failures that must occur to move from a present failure intensity of λ_p to a target intensity of λ_F is

$$\Delta\mu = \frac{1}{\theta} \ln\left(\frac{\lambda_p}{\lambda_F}\right) \tag{4.111}$$

The execution time to reach this objective is given by

$$\Delta\tau = \frac{1}{\theta} \left(\frac{1}{\lambda_F} - \frac{1}{\lambda_p}\right) \tag{4.112}$$

As we have seen, the execution time components of these models are characterized by two parameters. These are listed in Table 4.5.

Example 4.22
Assume that a piece of software will experience 200 failures in its lifetime. Suppose it has now experienced 100 of them. The initial failure intensity was 20 failures per CPU hour. Using BETM and LPETM calculate the current failure intensity (assume failure intensity decay parameter is 0.02 per failure).

TABLE 4.5
Execution Time Parameters

	Model	
Parameter	Basic	Logarithmic Poisson
Initial failure intensity	λ_0	λ_0
Total failures	ν_0	—
Failure intensity decay parameter	—	θ

Solution
For BETM,

$$\lambda(\mu) = \lambda_0 \left(1 - \frac{\mu}{\nu_0}\right) = 20\left(1 - \frac{100}{100}\right) = 10 \text{ failure per CPU hour}$$

For LPETM,

$$\lambda(\mu) = \lambda_0 \ \exp(-\theta\mu) = 20 \ \exp[-(0.02)(100)] = 2.7 \text{ failures per CPU hour}$$

The most common approach to software reliability analysis is testing. Testing is often performed by feeding random inputs into the software and observing the output produced to discover incorrect behavior. Because of the extremely complex nature of today's modern computer systems, however, these techniques often result in the generation of an enormous number of test cases. For example, Petrella et al. [43] discuss Ontario Hydro's validation testing of its Darlington Nuclear Generating Station's new computerized emergency reactor shutdown systems required a minimum of 7000 separate tests to demonstrate 99.99% reliability at 50% confidence.

4.9   HUMAN RELIABILITY ANALYSIS

The most important performance measure of interest in risk assessment is human reliability. Human reliability analysis (HRA) is an important part of a risk analysis. It has long been recognized that human error has a substantial impact on the reliability of complex systems. Accidents at Three Mile Island Nuclear Plant and many of the aircraft crashes show how human error can defeat engineered safeguards as barriers against hazard and play a dominant role in the progression of accidents and exposure of hazards. At least 70% of aviation accidents are caused by human malfunctions, similar figures apply to the shipping, and process industry. The Reactor Safety Study [44] revealed that more than 60% of the potential accidents in the nuclear industry are related to human errors. In general, the immediate human contribution to overall hazard barrier performance is at least as important as that of hardware reliability.

To obtain a precise and accurate measure of system reliability, human error must be taken into account. Analysis of system designs, procedures, and postaccident reports shows that human error can be an immediate accident initiator or can play a dominant role in the progress of undesired events. Without incorporating human error probabilities (HEPs), the results of risk analysis are incomplete and often underestimated.

To estimate HEPs (and, thus, human reliability), one needs to understand human behavior. However, human behavior is very difficult to model. Literature shows that there is not a strong consensus on the best way to capture all human actions and quantify HEPs. The assumptions, mechanisms, and approaches used by any one specific human model cannot be applied to all human activities. Current human models need further advancement, particularly in capturing and quantifying intentional human errors. Limitations and difficulties in current HRA include the following:

(a) Human behavior is a complex subject that cannot be described as a simple hardware in a system. Human performance can be affected by social, environmental, psychological, and physical factors that are difficult to quantify.
(b) Human actions cannot be considered to have binary success and failure states, as in hardware failure. Furthermore, the full range of human interactions has not been fully analyzed by HRA methods.
(c) The most difficult problem with HRA is the lack of appropriate data on human behavior in extreme situations.

Human error may occur in any phase of the design, manufacturing, construction, and operation of a complex system. Design, manufacturing, and construction errors are also the source of many types of errors during system operation. The most notable errors are dependent failures whose occurrence can cause loss of multiple hazard barriers and system redundancy. Normally, quality assurance programs are designed and implemented to minimize the occurrence of these types of human error.

In applications to risk analysis and hazard barrier performance analysis, we are concerned with human reliability during system operation, where human maintains, supervises, and controls complex systems operation. In the remainder of this section, human reliability models are reviewed, and important models are described in some detail. Emphasis is on the basic ideas, advantages, and disadvantages of each model, and their applicability to different situations. Then, we describe the important area of data analysis in HRA. After the links between models and data are reviewed, the problems of human reliability data sources and respective data acquisition are addressed.

4.9.1 HUMAN RELIABILITY ANALYSIS PROCESS

A comprehensive method of evaluating human reliability is the method called systematic human action reliability procedure (SHARP) developed by Hannaman and Spurgin [45]. The SHARP defines seven steps to perform HRA. Each step consists of inputs, activities, rules, and outputs. The inputs are derived from prior steps, reliability studies, and other information sources, such as procedures and accident reports. The rules guide the activities, which are needed to achieve the objectives of each step. The output is the product of the activities performed by analysts. The goals for each step are as follows:

1. *Definition*: Ensure that all different types of human interactions are considered.
2. *Screening*: Select the human interactions that are significant to system reliability.
3. *Qualitative analysis*: Develop a detailed description of important human actions.
4. *Representation*: Select and apply techniques to model human errors in system logic structures, e.g., fault trees, event trees, MLD, or reliability block diagram.
5. *Impact assessment*: Explore the impact of significant human actions identified in the preceding step on the system reliability model.

6. *Quantification*: Apply appropriate data to suitable human models to calculate probabilities for various interactions under consideration.
7. *Documentation*: Include all necessary information for the assessment to be understandable, reproducible, and traceable.

The relationships among these steps are shown in Figure 4.13. These steps in human reliability consideration are described in more detail below.

Step 1: Definition
The objective of Step 1 is to ensure that key human interactions are included in the human reliability assessment. Any human actions with a potentially significant impact on system reliability must be identified at this step to guarantee the completeness of the analysis.
    Human activities can generally be classified in Figure 4.13 as:

Type 1:   Before any challenge to a system, an operator can affect availability, reliability, and safety by restoring safeguard functions during testing and maintenance.
Type 2:   By committing an error, an operator can initiate a challenge to the system causing the system to deviate from its normal operating envelope.
Type 3:   By following procedures during the course of a challenge, an operator can operate redundant systems (or subsystems) and recover the systems to their normal operating envelope.
Type 4:   By executing incorrect recovery plans, an operator can aggravate the situation or fail to terminate the challenge to the systems.
Type 5:   By improvising, an operator can restore initially failed equipment to terminate a challenge.

    As recommended by the SHARP, HRA should use the above classification and investigate the system to reveal possible human interactions. Analysts can use the above-mentioned characteristics for different types of activities. For example, Type 1 interactions generally involve components, whereas Type 3 and Type 4 interactions are mainly operating actions that can be considered at system level. Type 5 interactions are recovery actions that may affect

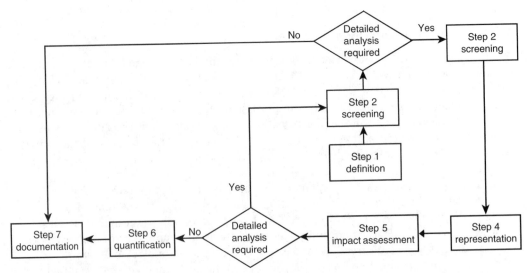

FIGURE 4.13 Systematic human action reliability procedure, Hannaman and Spurgin [45].

both systems and components. Type 2 interactions can generally be avoided by confirming that human-induced errors are included as contributors to the probability of all possible challenges to the system. The output from this step can be used to revise and enrich system reliability models, such as event trees and fault trees, to fully account for human interactions. This output will be used as the input to the next step.

Step 2: Screening

The objective of screening is to reduce the number of human interactions identified in Step 1 to those that might potentially challenge the safety of the system. This step provides the analysts with a chance to concentrate their efforts on the key human interactions. This is generally done in a qualitative manner. The process is judgmental.

Step 3: Qualitative analysis

To incorporate human errors into equipment failure modes, analysts need more information about each key human interaction identified in the previous steps to help in representing and quantifying these human actions. The two goals of qualitative analysis are to:

1. Postulate what operators are likely to think and do, and what kind of actions they might take in a given situation.
2. Postulate how an operator's performance may modify or trigger a challenge to the system.

This process of qualitative analysis may be broken down into four key stages:

1. Information gathering
2. Prediction of operator performance and possible human error modes
3. Validation of predictions
4. Representation of output in a form appropriate for the required function

In summary, the qualitative analysis step requires a thorough understanding of what performance-shaping factors (PSFs) (e.g., task characteristics, experience level, environmental stress, and social-technical factors) affect human performance. Based on this information, analyst can predict the range of plausible human action. The psychological model proposed by Rasmussen [46] is a useful way of conceptualizing the nature of human cognitive activities. The full spectrum of possible human actions following a misdiagnosis is typically very hard to recognize. Computer simulations of performance described by Woods et al. [47] and Amendola et al. [48] offer the potential to assist human reliability analysts in predicting the probability of human errors.

Step 4: Representation

To combine the HRA results with the barrier and logic system analysis models of Chapter 3, human error modes need to be transformed into appropriate representations. Representations are selected to indicate how human actions can affect the operation of a system.

Three basic representations have been used to delineate human interactions: the operator action tree (OAT) described by Wreathall [49], the confusion matrix described by Potash et al. [50], and the HRA event trees described by Swain and Guttman [51]. Figure 4.14 shows an example of OAT.

Step 5: Impact assessment

Some human actions can introduce new impacts on the system response. This step provides an opportunity to evaluate the impact of the newly identified human actions on the system.

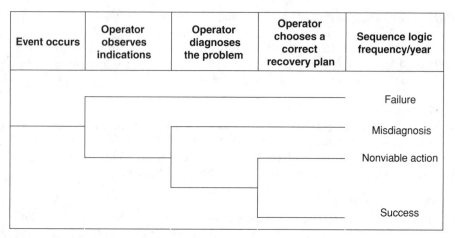

FIGURE 4.14 Operator action tree.

The human interactions represented in Step 4 are examined for their impact on challenges to the system, system reliability, and dependent failures. Screening techniques are applied to assess the importance of the impacts. Important human interactions are found, reviewed, and grouped into suitable categories. If the reexamination of human interactions identifies new human-induced challenges or behavior, the system analysis models (e.g., MLD, fault tree) are reconstructed to incorporate the results.

Step 6: Quantification

The purpose of this step is to assess the probabilities of success and failure for each human activity identified in the previous steps. In this step, analysts apply the most appropriate data or models to produce the final quantitative reliability analysis. Selection of the models should be based on the characteristics of each human interaction.

Guidance for choosing the appropriate data or models to be adopted is provided below.

1. For procedural tasks, the data from Swain and Guttman [51] or equivalent can be applied.
2. For diagnostic tasks under time constrains, time-reliability curves from Hall et al. [52] or the human cognitive reliability (HCR) model from Hannaman and Spurgin [45] can be used.
3. For situations where suitable data are not available, expert opinion approaches, such as paired comparison by Hunns and Daniels [53] and the success likelihood index (SLI) method by Embry et al. [54] can be used.
4. For situations where multiple tasks are involved, the dependence rules discussed by Swain and Guttman [51] can be used to assess the quantitative impact.

Step 7: Documentation

The objective of Step 7 is to produce a traceable description of the process used to develop the quantitative assessments of human interactions. The assumptions, data sources, selected model, and criteria for eliminating unimportant human interactions should be carefully documented. The human impact on the system should be stated clearly.

4.9.2   HUMAN RELIABILITY ANALYSIS MODELS

The HRA models can be classified into the following categories. Representative models in each are also summarized.

1. Simulation methods
   a. Maintenance personnel performance simulation (MAPPS)
   b. Cognitive environment simulation (CES)
2. Expert judgment methods
   a. Paired comparison
   b. Direct numerical estimation (absolute probability judgment)
   c. Success likelihood index methodology (SLIM)
3. Analytical methods
   a. Technique for human error rate prediction (THERP)
   b. Human cognitive reliability correlation (HRC)
   c. Time-reliability correlation (TRC)

We will briefly discuss each of these models. Human error is a complex subject. There is no single model that captures all important human errors and predicts their probabilities. Poucet [55] reports the results of a comparison of the HRA models. He concludes that the methods could yield substantially different results, and presents their suggested use in different contexts.

4.9.2.1   Simulation Methods

These methods primarily rely on computer models that mimic human behavior under different conditions.

Maintenance personnel performance simulation (MAPPS). MAPPS, developed by Siegel et al. [56], is a computerized simulation model that provides human reliability estimation for testing and maintaining tasks. To perform the simulation, analysts must first find out the necessary tasks and subtasks that individuals must perform. Environmental motivational tasks and organizational variables that influence personnel performance reliability are input into the program. Using the Monte Carlo simulation, the model can output the probability of success, time to completion, idle time, human load, and level of stress. The effects of a particular parameter or subtask performance can be investigated by changing the parameter and repeating the simulation.

   The simulation output of task success is based on the difference between the ability of maintenance personnel and the difficulty of the subtask. The model used is

$$Pr(success) = e^y/(1 + e^y), \tag{4.113}$$

where $y > 0$ is the difference between personnel ability and task difficulty.

Cognitive environment simulation (CES). Woods et al. [47] have developed a model based on techniques from artificial intelligence (AI). The model is designed to simulate a limited resources problem solver in a dynamic, uncertain, and complex situation. The main focus is on the formation of intentions, situations, and factors leading to intentional failures, forms of intentional failures, and the consequence of intentional failures.

   Similar to the MAPPS model, the CES model is a simulation approach that mimics the human decision-making process during an emergency condition. But CES is a deterministic approach, which means the program will always obtain the same results if the input is unchanged. The first step in CES is to identify the conditions leading to human intentional failures. CES provides numerous performance-adjusting factors to allow the analysts to test different working conditions. For example, analysts may change the number of people interacting with the system (e.g., the number of operators), the depth or breadth of working knowledge, or the human–machine interface. Human error prone points can be identified by running the CES for different conditions. The human failure probability is evaluated by knowing, *a priori*, the likelihood of occurrence of these error prone points.

In general, CES is not a human rate quantification model. It is primarily a tool to analyze the interaction between problem-solving resources and task demands.

4.9.2.2   Expert Judgment Methods

The primary reason for using expert judgment in HRA is that there often exist little or no relevant or useful human error data. Expert judgment is discussed in more detail in Section 4.10. There are two requirements for selecting experts: (1) they must have substantial expertise; (2) they must be able to accurately translate this expertise into probabilities.

Paired comparison. Paired comparison, described by Hunns and Daniels [53], is a scaling technique based on the idea that judges are better at making simple comparative judgments than making absolute judgments. An interval scaling is used to indicate the relative likelihood of occurrence of each task. Saaty [57] describes this general approach in the context of a decision analysis technique. The method is equally applicable to HRA.

Direct numerical estimation. For the direct numerical estimation method described by Stillwell et al. [58], experts are asked to directly estimate the HEPs and the associated upper/lower bounds for each task. A consistency analysis might be taken to check for agreement among these judgments. Then individual estimations are aggregated by either arithmetic or geometric average.

Success likelihood index methodology (SLIM). The SLIM developed by Embry et al. [54] is a structural method that uses expert opinion to estimate human error rates. The underlying assumption of SLIM is that the success likelihood of tasks for a given situation depends on the combination of effects from a small set of PSFs relevant to a group of tasks under consideration.

In this procedure, the experts are asked to assess the relative importance (weight) of each PSF with regard to its impact on the tasks of interest. An independent assessment is made to the level or the value of the PSFs in each task situation. After identifying and agreeing on the small set of PSFs respective weights and ratings for each PSF are multiplied. These products are then summed to produce the SLI, varying from 0 to 100 after normalization. This value indicates the expert's belief regarding the positive or negative effects of PSFs on task success.

The SLIM approach assumes that the functional relationship between success probability and SLI is exponential, i.e.,

$$\log[\Pr(\text{operator success})] = a(\text{SLI}) + b \qquad (4.114)$$

where a and b are empirically estimated constants. To calibrate a and b, at least two human tasks of known reliability must be used in (6.14), from which constants a and b are calculated.

This technique has been implemented as an interactive computer program. The first module, called multiattribute utility decomposition (MAUD), analyzes a set of tasks to define their relative likelihood of success given the influence of PSFs. The second module, systematic approach to the reliability assessment of humans (SARAH), is then used to calibrate these relative success likelihoods to generate absolute human error probability. The SLIM technique has a good theoretical basis in decision theory. Once the initial database has been established with the SARAH module, evaluations can be performed rapidly. This method does not require extensive decomposition of a task to an elemental level. For situations where no data are available, this approach enables HRA analysts to reasonably estimate human reliability. However, this method makes extensive use of expert judgment, which requires a team of experts to participate in the evaluation process. The resources required to set up the SLIM-MAUD database are generally greater than other techniques.

4.9.2.3   Analytical Methods

These methods generally use a model based on some key parameters that form the value of human reliabilities.

Technique for human error rate prediction (THERP). The oldest and most widely used HRA technique is the THERP analysis developed by Swain and Guttman [51] and reported in the form of a handbook. The THERP approach uses conventional system reliability analysis modified to account for possible human error. Instead of generating equipment system states, THERP produces possible human task activities and the corresponding HEPs. THERP is carried out in the five steps described below.

   1. Define system failures of interest

From the information collected by examining system operation and analyzing system safety, analysts identify possible human interaction points and task characteristics and their impact on the systems. Then screening is performed to determine critical actions that require detailed analysis.

   2. List and analyze related human actions

The next step is to develop a detailed task analysis and human error analysis. The task analysis delineates the necessary task steps and the required human performance. The analyst then determines the errors that could possibly occur.
   The following human error categories are defined by THERP:

   a. Errors of omission (omit a step or the entire task)
   b. Errors of commission, including:
      • Selection error (select the wrong control, choose the wrong procedures)
      • Sequence error (actions carried out in the wrong order)
      • Time error (actions carried out too early/too late)
      • Qualitative error (action is done too little/too much)

At this stage, opportunities for human recovery actions (recovery from an abnormal event or failure) should be identified. Without considering recovery possibilities, overall human reliability might be dramatically underestimated.
   The basic tool used to model tasks and task sequences is the HRA event tree. According to the time sequence or procedure order, the tree is built to represent possible alternative human actions. Therefore, if appropriate error probabilities of each subtask are known and the tree adequately depicts all human action sequences, the overall reliability of this task can be calculated. An example of an HRA event tree is shown in Figure 4.15.

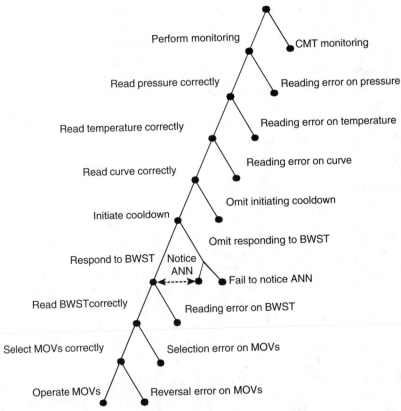

FIGURE 4.15 HRA event tree on operator actions during a small-break loss of coolant in nuclear plants (CMT, computer monitoring; ANN, announciator; BWST, borated water storage tank; MOV, motor-operated valve, Hannaman et al. [59]).

3. Estimate relevant error probabilities

As explained in the previous section, HEPs are required for the failure branches in the HRA event tree. Chapter 20 of Swain and Guttman [51] provides the following information.

- Data tables containing nominal HEPs.
- Performance models explaining how to account for PSFs to modify the nominal error data.
- A simple model for converting independent failure probabilities into conditional failure probabilities.

In addition to the data source of THERP, analysts may use other data sources, such as the data from recorded incidents, trials from simulations, and subjective judgment data, if necessary.

4. Estimate effects of error on system failure events

In the system reliability framework, the human error tasks are incorporated into the system model, such as a fault tree. Hence, the probabilities of undesired events can be evaluated and the contribution of human errors to system reliability or availability can be estimated.

5. Recommend changes to system design and recalculate system reliability

A sensitivity analysis can be performed to identify dominant contributors to system unreliability. System performance can then be improved by reducing the sources of human error or redesigning the safeguard systems.

THERP's approach is very similar to the equipment reliability and performance assessment methods described earlier in this chapter. The integration of HRA and equipment reliability analysis is straightforward using the THERP process. Therefore, it is easily understood by system analysts. Compared with the data for other models, the data for THERP are much more complete and easier to use. The handbook contains guidance for modifying the listed data for different environments. The dependencies among subtasks are formally modeled, although subjective. Conditional probabilities are used to account for this kind of task dependence.

Very detailed THERP analysis can require a large amount of effort. In practice, by reducing the details of the THERP analysis to an appropriate level, the amount of work can be minimized. THERP is not appropriate for evaluating errors involving high-level decisions or diagnostic tasks. In addition, THERP does not model underlying psychological causes of errors. Since it is not an ergonomic tool, this method cannot produce explicit recommendations for design improvement.

HCR correlation. During the development of SHARP, a need was identified to find a model to quantify the reliability of control room personnel responses to abnormal system operations. The HCR correlation, described by Hannaman et al. [59], is essentially a normalized TRC (described below) whose shape is determined by the available time, stress, human–machine interface, etc. Normalization is needed to reduce the number of curves required for a variety of situations. It was found that a set of three curves (skill-, rule-, and knowledge-based ideas, developed by Rasmussen [60] could represent all kinds of human decision behaviors. The application of HCR is straightforward. The HCR correlation curves can be developed for different situations from the results of simulator experiments. Therefore, the validity can be verified continuously. This approach also has the capability of accounting for cognitive and environmental PSFs.

Some of the disadvantages of the HCR correlation are:

- The applicability of the HCR to all kinds of human activities is not verified.
- The relationships of PSFs and nonresponse probabilities are not well addressed.
- This approach does not explicitly address the details of human thinking processes. Thus, information about intentional failures cannot be obtained.

Time-reliability correlation (TRC). Hall et al. [52] concentrate on the diagnosis and decision errors of nuclear power plant operators after the initiation of an accident. They criticize the behavioral approach used by THERP and suggest that a more holistic approach be taken to analyze decision errors.

The major assumption of TRC is that the time available for diagnosis of a system fault is the dominant factor in determining the probability of failure. In other words, the longer people take to think, the more unlikely they are to make mistakes. The available time for decision and diagnosis are delimited by the operator's first awareness of an abnormal situation and the initiation of the selected response. Because no data were available when the TRC was developed, an interim relationship was obtained by consulting psychologists and system analysts. Recent reports confirm that the available time is an important factor in correctly performing cognitive tasks. A typical TRC is shown in Figure 4.16.

Dougherty and Fragola [61] is a good reference for TRC as well as other HRA methods. TRC is very easy and fast to use. However, TRC is still a premature approach. The exact

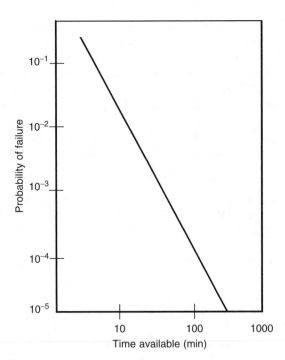

FIGURE 4.16 Time-reliability correlation for operator.

relationship between time and reliability requires more experimental and actual observations. This approach overlooks other important PSFs, such as experience level, task complexity, etc. TRC focuses only on limited aspects of human performance in emergency conditions. The time available is the only variable in this model. Therefore, the estimation of the effect of this factor should be very accurate. However, TRC does not provide guidelines or information on how to reduce human error contributions.

4.9.3 HUMAN RELIABILITY DATA

There is general agreement that a major problem for HRA is the scarcity of data on human performance that can be used to estimate human error rates and performance time. To estimate HEPs, one needs data on the relative frequency of the number of errors and/or the ratio of "near-misses" to total number of attempts. Ideally, this information can be obtained from observing a large number of tasks performed in a given application. However, this is impractical for several reasons. First, error probabilities for many tasks, especially for rare emergency conditions, are very small. Therefore, it is very difficult to observe enough data within a reasonable amount of time to get statistically meaningful results. Second, possible penalties assessed against people who make errors, for example, in a nuclear power plant or in aircraft cockpit, discourages free reporting of all errors. Third, the costs of collecting and analyzing data could be unacceptably high. Moreover, estimation of performance times presents difficulties because data taken from different situations might not be applicable.

Data can be used to support HRA quantification in a variety of ways, e.g., to confirm expert judgment, develop human reliability data, or support development of an HRA model. Currently, available data sources can be divided into the following categories: (1) actual data, (2) simulator data, (3) interpretive information, and (4) expert judgment.

Psychological scaling techniques, such as paired comparisons, direct estimation, SLIM, and other structured expert judgment methods, are typically used to extrapolate error probabilities. In many instances, scarcity of relevant hard data makes expert judgment a very useful data source. This topic is discussed further in the following section.

4.10 EXPERT OPINION IN RISK ASSESSMENT AND PERFORMANCE ASSESSMENT

Expert judgment provides an essential part of the information used in probabilistic risk assessments and performance assessments. Although such judgments here always were used in safety assessments (usually under the name of engineering judgment), a need to trace and defend the information has led to the development of formal methods. This section provides background information on formal techniques for obtaining, evaluating, and processing expert judgment. The reader should gain an understanding of the following:

* Circumstances requiring the use of formal expert judgment
* Formal probability elicitation processes
* Alternative approaches to organizing experts
* Psychological biases affecting expert judgment
* Criteria for evaluating assessed probabilities
* Methods for combining judgments

The use of expert opinions is often desired in reliability analysis, and in many cases it is unavoidable. One reason for using experts is the lack of a statistically significant amount of empirical data necessary to estimate new parameters. Another reason for using experts is to assess the likelihood of a one-time event, such as the chance of rain tomorrow.

However, the need for expert judgment that requires extensive knowledge and experience in the subject field is not limited to one-time events. For example, suppose we are interested in using a new and highly capable microcircuit device currently under development by a manufacturer and expected to be available for use soon. The situation requires an immediate decision on whether or not to design an electronic box around this new microcircuit device. Reliability is a critical decision criterion for the use of this device. Although reliability data on the new device are not available, reliability data on other types of devices employing similar technology are accessible.

Therefore, reliability assessment of the new device requires both knowledge and expertise in similar technology, and can be achieved through the use of expert opinion.

Some specific examples of expert use are the Reactor Safety Study [44]; IEEE Standard 500 [62]; and "Severe Accident Risk: An Assessment for Five U.S. Nuclear Power Plants" [63], where expert opinion was used to estimate the probability of structures and components failure and other rare events. The Electric Power Research Institute Study [64] has relied on expert opinion to assess seismic hazard rates. Other applications include weather forecasting. For example, Clemens and Winkler [65] discuss the use of expert opinion by meteorologists. Another example is the use of expert opinion in assessing human error rates discussed by Swain and Guttman [51].

The use of expert opinion in decision making is a two-step process: elicitation and analysis of expert opinion. The method of elicitation may take the form of individual interviews, interactive group sessions, or the Delphi approach discussed by Dalkey and Helmer [66]. The relative effectiveness of different elicitation methods has been addressed extensively in the literature. Techniques for improving the accuracy of expert estimates include calibration, improvement in questionnaire design, motivation techniques, and other methods, although clearly no technique can be applied to all situations. The analysis portion of expert use involves combining expert opinions to produce an aggregate estimate that can be used by reliability analysts. Again, various aggregation techniques for pooling expert opinions exist, but of particular interest are those adopting the form of mathematical models. The usefulness of each model depends on both the reasonableness of the assumptions (implicit and explicit) carried by the model as it mimics the real-world situation, and the ease of implementation from the user's perspective. The term "expert" generally refers to any source of information that provides an estimate and includes human experts, measuring instruments, and models.

Once the need for expert opinion is determined and the opinion is elicited, the next step is to establish the method of opinion analysis and application. This is a decision task for the analysts, who may simply decide that the single best estimate of the value of interest is the estimate provided by the arithmetic average of all estimates, or an aggregate from a nonlinear pooling method, or some other opinions. Two methods of aggregating expert opinion are discussed in more detail, the geometric averaging technique and the Bayesian technique.

4.10.1 Geometric Averaging Technique

Suppose n experts are asked to make an estimate of the failure rate of an item. The estimates can be pooled using the geometric averaging technique. For example, if λ_i is the estimate of the ith expert, then an estimate of the failure rate is obtained from

$$\hat{\lambda} = \left(\prod_{i=1}^{n}\lambda_i\right)^{1/n} \tag{4.115}$$

This was the primary method of estimating failure rates in IEEE Standard 500 [62]. The IEEE Standard 500 contains rate data for electronic, electrical, and sensing components. The reported values were synthesized primarily from the opinions of some 200 experts (using a form of the Delphi procedure). Each expert reported "low," "recommended," and "high" values for each failure rate under normal conditions, and a "maximum" value that would be applicable under all conditions (including abnormal conditions). The estimates were pooled using (4.115). For example, for maximum values,

$$\hat{\lambda}_{max} = \left(\prod_{i=1}^{n}\lambda_{max_i}\right)^{1/n}$$

As discussed by Mosleh and Apostolakis [67], the use of geometric averaging implies that (1) all the experts are equally competent, (2) the experts do not have any systematic biases, (3) experts are independent, and (4) the preceding three assumptions are valid regardless of which value the experts are estimating, e.g., high, low, or recommended.

The estimates can be represented in the form of a distribution. Apostolakis et al. [68] suggest the use of a lognormal distribution for this purpose. In this approach, the "recommended" value is taken as the median of the distribution, and the error factor (EF) is defined as

$$EF = \left(\frac{\hat{\lambda}_{0.95}}{\hat{\lambda}_{0.05}}\right)^{1/2} \tag{4.116}$$

4.10.2 BAYESIAN APPROACH

As discussed by Mosleh and Apostolakis [67], the challenge of basing estimates on the expert opinion is to maintain coherence throughout the process of formulating a single best estimate based on the experts' actual estimates and their credibilities. Coherence is a notion of internal consistency within a person's state of belief. In the subjectivist school of thought, a probability is defined as a measure of personal uncertainty. This definition assumes that a coherent person will provide his or her probabilistic judgments in compliance with the axioms of probability theory.

An analyst often desires a modeling tool that can aid him or her in formulating a single best estimate from expert opinions in a coherent manner. Informal methods such as simple averaging will not guarantee this coherence. Bayes' theorem, however, provides a framework to model expert belief, and ensures coherence of the analysts in arriving at a new degree of belief in light of expert opinion. According to the general form of the model given by Mosleh and Apostolakis [67], the state-of-knowledge distribution of a failure rate λ, after receiving an expert estimate $\hat{\lambda}$, can be obtained by using Bayes' theorem in the following form:

$$\prod (\lambda|\hat{\lambda}) = \frac{1}{k} L(\hat{\lambda}|\lambda) \pi_0(\lambda) \qquad (4.117)$$

where $\pi_0(\lambda)$ is the prior distribution of λ; $\Pi(\lambda|\hat{\lambda})$ is the posterior distribution of λ; $L(\hat{\lambda}|\lambda)$ is the likelihood of receiving the estimate, given the true failure rate λ; k is a normalizing factor.

One of the models suggested for the likelihood of observing $\hat{\lambda}$ given λ is based on the lognormal distribution in the following form:

$$L(\hat{\lambda}|\lambda) = \frac{1}{\sqrt{2\pi}\sigma\hat{\lambda}} \exp\left[-\frac{1}{2}\left(\frac{\ln\hat{\lambda} - \ln\lambda - \ln b}{\sigma}\right)^2\right] \qquad (4.118)$$

where b is a bias factor ($b=1$ when no bias is assumed) and σ is the standard deviation of logarithm of $\hat{\lambda}$, given λ. When the analyst believes no bias exists among the experts, she or he can set $b=1$. The quantity σ, therefore, represents the degree of accuracy of the experts' estimate as viewed by the analyst. The work by Kim [69], which includes a Bayesian model for a relative ranking of experts, is an extension of the works by Mosleh and Apostolakis [67].

4.10.3 STATISTICAL EVIDENCE ON THE ACCURACY OF EXPERT ESTIMATES

Among the attempts to verify the accuracy of expert estimates, two types of expert estimates are studied — assessment of single values and assessment of distributions.

Notable among the studies on the accuracy of expert assessments of a single estimate is Snaith's study [70]. In this study, observed and predicted reliability parameters for some 130 pieces of different equipment and systems used in nuclear power plants were evaluated. The predicted values included both direct assessments by experts and the results of analysis. The objective was to determine correlations between the predicted and observed values. Figure 4.17 shows the ratio ($R = \lambda|\hat{\lambda}$) of observed to predicted values plotted against their cumulative frequency. As shown, the majority of the points fall within the dashed boundary lines. Predicted values are within a factor of 2 from the observed values, and 93% are within a factor of 4. The figure also shows that $R=1$ is the median value, indicating that there is no systematic bias in either direction. Finally, the linear nature of the curve shows that R tends to be lognormally distributed, at least within the central region. This study clearly supports the use and accuracy of expert estimation.

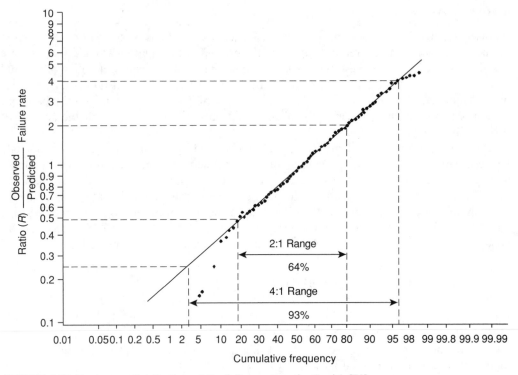

FIGURE 4.17 Frequency distribution of the failure rate ratio, Snaith [70].

Among the studies of expert estimation are the works by cognitive psychologists. For example, Lichtenstein et al. [71] described the results of testing the adequacy of probability assessments and concluded that "the overwhelming evidence from research on uncertain quantities is that people's probability distributions tend to be biased." Commenting on judgmental biases in risk perception, Slovic et al. [72] stated: "A typical task in estimating uncertain quantities like failure rates is to set upper and lower bounds such that there is a 98% chance that the true value fall between them. Experiments with diverse groups of people making different kinds of judgments have shown that, rather than 2% of true values falling outside the 98% confidence bounds, 20% to 50% do so. Thus, people think that they can estimate such values with much greater precision than is actually the case."

Based on the above conclusion Apostolakis [73] has suggested the use of the 20th and 80th percentiles of lognormal distributions instead of the 5th and 95th when using (4.116), to avoid a bias toward low values, overconfidence of experts, or both. When using the Bayesian estimation method based on (4.117), the bias can be accounted for by using larger values of parameter b in (4.118).

Exercises

4.1 Due to the aging process, the failure rate of a nonrepairable item is increasing according to 4.2 $\lambda(t) = \lambda \beta t^{\beta-1}$. Assume that the values of β and λ are estimated as $\hat{\beta} = 1.62$ and $\hat{\lambda} = 1.2 \times 10^{-5}$ h. Determine the probability that the item will fail sometime between 100 and 200 h. Assume an operation beginning immediately after the onset of aging.

4.2 Total test time of a device is 50,000 h. The test is terminated after the first failure. If the pdf of the device TTF is known to be exponentially distributed, what is the probability that the estimated failure rate is not greater than 4.6×10^{-5} (per h).

4.3 The breaking strength X of five specimens of a rope of 1/4 inch diameter are 660, 460, 540, 580, and 550 lbs. Estimate the following:
   a) The mean breaking strength by a 95% confidence level assuming normally distributed strength.
   b) The point estimate of strength value at which only 5% of such specimens would be expected to break if it is assumed to be an unbiased estimate of the true mean, and s^2 is assumed to be the true standard deviation. (Assume $\bar{x}$ is normally distributed.)
   c) The 90% confidence interval of the estimate of the standard deviation.

4.4 Seven pumps have failure times (in months) of 15.1, 10.7, 8.8, 11.3, 12.6, 14.4, and 8.7. (Assume an exponential distribution.)
   a) Find a point estimate of the MTTF.
   b) Estimate the reliability of a pump for $t = 12$ months.
   c) Calculate the 95% two-sided interval of λ.

4.5 A locomotive control system fails 15 times out of the 96 times, it is activated to function. Determine the following:
   a) A point estimate for failure probability of the system.
   b) 95% two-sided confidence intervals for the probability of failure. (Assume that after each failure, the system is repaired and put back in an as-good-as-new state.)

4.6 The following sample of measurements is taken from a study of an industrial process, which is assumed to follow a normal distribution: 8.9, 9.8, 10.8, 10.7, 11.0, 8.0, and 10.8. For this sample, the 95% confidence error on estimating the mean (μ) is 2.2. What sample size should be taken if we want the 99% confidence error to be 1.5, assuming the same sample variance?

4.7 In the reactor safety study, the failure rate of a diesel generator can be described as having a lognormal distribution with the upper and lower 90% bounds of 3×10^{-2} and 3×10^{-4}, respectively. If a given nuclear plant experiences two failures in 8760 h of operation, determine the upper and lower 90% bounds given this plant experience. (Consider the reactor safety study values as prior information.)

4.8 A mechanical life test of 18 circuit breakers of a new design was run to estimate the percentage failed by 10,000 cycle of operation. Breakers were inspected on a schedule, and it is known that failures occurred between certain inspections as shown:

Cycles($\times 1000$)	10–15	15–17.5	17.5–20	20–25	25–30	30+
Number of failures	2	3	1	1	2	9 survived

   a) Assuming Weibull pdf model estimate percentage failing by 10,000 cycles.
   b) Estimate the Weibull parameters.

4.9 The following data were collected by Frank Proschan in 1983. Operating hours to first failure of an engine cooling part in 13 aircrafts are:

Aircraft	1	2	3	4	5	6	7	8	9	10	11	12	13
Hours	194	413	90	74	55	23	97	50	359	50	130	487	102

a) Would these data support an increasing failure rate, decreasing failure rate, or constant failure rate assumption?

b) Based on a graphic nonparametric analysis of these data, confirm the results obtained in part (a).

4.10 A company redesigns one of its compressors and wants to estimate reliability of the new product. Using its past experience, the company believes that the reliability of the new compressor will be higher than 0.5 (for a given mission time). The company's testing of one new compressor showed that the product successfully achieved its mission.

a) Assuming a uniform prior distribution for the above reliability estimate, find the posterior estimate of reliability based on the test data.

b) If the company conducted another test, which resulted in another mission success, what would be the new estimate of the product reliability?

4.11 The following table shows fire incidents during six equal time intervals of 22 chemical plants.

Time Interval	1	2	3	4	5	6
No. of fires	6	8	16	6	11	11

Do you believe the fire incidents are time-dependent? Prove your answer.

4.12 A simplified schematic of the electric power system at a nuclear power plant is shown in the figure below:

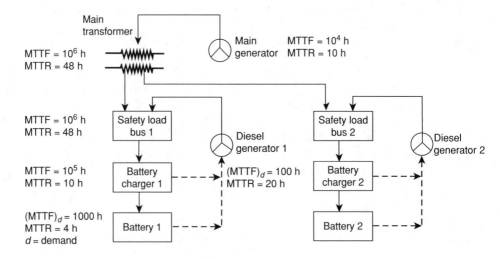

a) Draw a fault tree with the top event "loss of electric power from both safety load buses."

b) Determine the unavailability of each event in the fault tree for 24 h of operation.

c) Determine the top event probability.

Assume the following:

• Either the main generator or one of the two diesel generators is sufficient.

• One battery is required to start the corresponding diesel generator.

• Normally, the main generator is used. If that is lost, one of the diesel generators provides the electric power on demand.

4.13 An operating system is repaired each time it has a failure and is put back into service as soon as possible (monitored system). During the first 10,000 h of service, it fails five times and is out-of-service for repair during the following times:

Out-of-Service (h)
1000–1050
3660–4000
4510–4540
6130–6170
8520–8560

a) Is there a trend in the data?
b) What is the reliability of the system 100 h after the system is put into operation? What is the asymptotic availability assuming no trends in λ and μ?
c) If the system has been operating for 10 h without a failure, what is the probability that it will continue to operate for the next 10 h without a failure?
d) What is the 80% confidence interval for the MTTRs ($\lambda = 1/\mu$)?

4.14 The following cycle-to-failure data have been obtained from a repairable component. The test stopped when the fifth failure occurred.

Repair No.	1	2	3	4	5
Cycle-to-failure (Interarrival of cycles)	5010	6730	4031	3972	4197

a) Is there any significant trend in these data?
b) Determine the ROCOFs.
c) What is the reliability of the component 1000 cycles after the fifth failure is repaired?

4.15 Determine the limiting pointwise unavailability of the system shown below:

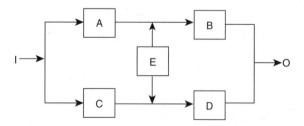

Assume that all components are identical and are repaired immediately after each of them experiences a failure. Rate of occurrence of the failure for each component is $\lambda = 0.001/h$, and MTTR is 15 h.

4.16 We are interested in unavailability of the system shown below:

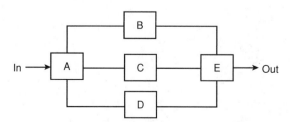

The following information is available:
- A and E are identical components with $\lambda_A = \lambda_E = 1 \times 10^{-5}$/h, $\mu_A = \mu_E = 0.1$/h.
- B, C, and D are identical periodically tested components with $\lambda_B = \lambda_C = \lambda_D = 1 \times 10^{-5}$/h. All test durations are equal ($\tau_t = 1$ h), all frequency of repair per cycle are equal ($f = 0.25$), and all durations of repair are equal ($\tau_r = 15$ h).

Given the above information, calculate unavailability of the system assuming that all components are independent.

4.17 An engine crankshaft is a good example of a high reliability part of a car. Although it is pounded by each cylinder with every piston stroke, that single bar remains intact for a long time. Assume the strength of the shaft is normal with the mean S and standard deviation s, while the load per stroke is L with standard deviation l. Realize that a C cylinder engine hits the shaft at C different places along it, so these events can be considered independent. The problem will be to determine the reliability of the crankshaft.
   a) Express the SM in terms of S, s, L, l, and C.
   b) Estimate the reliability. Assume the motor turns at $X(t)$ revolutions.
   c) Express the total number of reversals. $N(t)$ seen by each piston as a function of time.
   d) If the shaft is subject to fatigue, express the reliability as a function of time. Metals fatigue, generally, following the Manson–Coffin Lae: $S(N) = SN^{(-1/q)}$. Assume, also that the standard deviation, s does not change with N. Also, q is a constant.
   e) Determine the expected life (50% reliability) of the crankshaft turning at a constant rate, R (RPM).

4.18 In response to an RFQ, two vendors have provided a system proposal consisting of subsystem modules A, B, C. Each vendor has provided failure rates and average corrective maintenance time for each module. Determine which vendor system has the best MTTR and which one you would recommend for purchase.

Module	No. in System	Vendor 1		Vendor 2	
		Failure Rate (10^{-4} h)	Mct (min)	Failure Rate (10^{-4}h)	Mct (min)
A	2	45	15	45	20
B	1	90	20	30	15
C	2	30	10	90	10

   a) Describe the advantages of a preventive maintenance program.
   b) Is it worth doing preventive maintenance if the failure rate is constant?

4.19 Use an appropriate Weibull MLE method to determine parameters of the distribution using the following multiple censored data.

Test No.	1	2	3	4	5	6	7	8	9	10
Time-to-failure	309+	386	180	167+	122	229	104+	217+	168	138

"+," censoring.

4.20 The failure time (in hour) of a component is lognormally distributed with parameters $\mu_t = 4$, $\sigma_t = 0.9$.

a) What is the MTTF for this component?

b) When should the component be replaced, if the minimum required reliability for the component is 0.95?

c) What is the value of the hazard function at this time (i.e., the time calculated in Part (b))?

d) What is the mean residual life at this time; if the component has already survived to this time (i.e., the time calculated in Part (b))?

4.21 On surface of a highly stressed motor shaft, there are five crack initiation points observed after the shaft is tested at 6000 rpm for 5 h.

a) Calculate the point estimate of a crack initiation per cycle (per turn of the shaft).

b) Determine the 90% two-sided confidence interval of the probability of a crack initiation per cycle.

c) What is the point estimate of the mean-cycles-to-crack initiation?

d) What is the probability of no crack initiation in a test of 2000 rpm for 10 h.

4.22 The following data represent failure times and repair experiences from the life of a component.

Failure Number	t_0 (age) Months	Repair Duration (h)
1	4	3
2	7	4
3	15.5	7
4	20	3
5	26	7

Assume 30 days/months.

Assuming that the current age of the unit is 30 months.

a) Is there a trend in the rate of occurrence of *failures* and *repairs*?

b) Estimate the instantaneous availability at the present time of 30 months.

4.23 In a periodically tested system with a constant ROCOF of 10^{-4}/h, average constant repair duration of 10 h, average test duration of 5 h, and average constant frequency of repair of 0.05 per test, determine the best (optimum) inspection interval.

4.24 Probability of a massive exposure of a toxic gas is obtained from a Poisson distribution. That is the rate of occurrence of the exposure, A, is constant in time t, and time of occurrence t follows a Poisson distribution.

a) What is the probability that no exposure occurs in time t?

b) If a regulation requires that the probability of a massive exposure be less than 1-in-100 in a 30-year life of a plant, which produces this gas, what is the acceptable rate of occurrence of this exposure? Is this reasonable?

4.25 Fairley defines "Plausible Upper Bound Estimate" for the case of zero occurrence of accidents as

$$(X = 0 | r) = e^{-rt}$$

a) What does this mean?

b) For a small probability 0.01, what is the rate of an accidental airplane crash into a stadium, if no such accident has occurred in 80 years of commercial aviation?

4.26 We are interested to compare reliability of latches from two different suppliers. Life tests were performed to benchmark the performance of the two latch designs. Ten type-I latches from supplier 1 were tested to failures and twelve latches from supplier 2 were tested but the test was suspended after 242×10^3 (cycles). Failure data are as follows:

Supplier 1, Cycles in ($\times10^3$)	Supplier 2, Cycles in ($\times10^3$)
58	124
81	124
91	146
93	179
113	208
171	242S
190	242S
199	242S
258	242S
346	242S
	242S
	242S

S, suspended.

a) Use a Weibull plot approach and determine whether Weibull is a good fit. Plot both curves on the same graph paper.

b) Suppose that you anticipate the latches would normally go through 5000 cycles before the warranty runs out. Which supplier would you use?

4.27 Suppose the following data has been observed in a reparable system:

Failure No.	1	2	3	4	5	6
Time-between-failures (in $\times10^3$ h)	58	50	65	66	48	42

a) Do you observe any trends?

b) Determine the ROCOF.

If the MTTR of this system is 50 h, determine the unavailability of the system assuming that this system s failure is instantly repaired.

4.28 An specifications calls for a power transistor to have a reliability of 0.95 at 2000 h. Five hundred transistors are placed on test for 2000 h with 15 failures observed. The failure units are substituted upon failure.

a) Has this specification been met?

b) What is the chance that that the specification is not met?

4.29 Consider the system below:

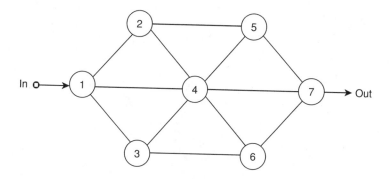

Use any system reliability/availability method to determine the unavailability of this system provided that each unit has a constant ROCOF of 5×10^{-4}/h and a rate of repair of 0.2/h.

4.30 Consider a cut set of a system as following:

$$T = A \times B \times C + A \times B \times D$$

If the total unreliability of $A = B = C = D = 10^{-3}$, and there is a dependency between failure of components A and B with $\beta = 0.05$, but no dependence between A and C, A and D, B and C, B and D, or C and D exists. Determine the unavailability of the system (T).

4.31 A decision has to be made whether to buy two, three, or four diesel generators for the electrical power system for an oil platform. Each generator will normally be working and can supply up to 50% of the total power demand. The reliability of each generator can be specified by a constant failure rate of 0.21 per year. The generators are to be simultaneously tested (the test and repair time is negligible) at 6-monthly intervals.

a) The required system availability must be at least 0.99. How many generators must be bought?

b) Calculate the MTBF for the chosen system.

c) Explain how the average availability of a system with three generators is affected if we assume the probability of generator failures are dependent. Use a β-factor method with $\beta = 0.1$ for this purpose.

4.32 A repairable unit had the following failure characteristic in the past 5 years (we are at the beginning of the 6th year).

Year	Number of Failures that Required Maintenance
1	n
2	$0.9n$
3	$0.88n$
4	$0.72n$
5	$0.88n$

a) Is there any trend?

b) How many failures (in terms of fraction of n) do we expect to see in the 6th year?

4.33 Consider the system below:

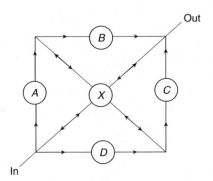

Components A, B, C, D, and X are *identical* and have a Weibull TTF distribution model with shape parameter $\beta = 1.6$.

a) Use the Bayes' theorem to determine the reliability of the system as a function of time.

b) Using the results of part (a), determine the scale parameter if we desire that the total probability of system failure not exceeds 10^{-2} over a 10 year period.

c) Determine the minimal cut sets of this system *by inspection*.

d) Use the minimal cut sets to verify that the probability of failure of this system meets the 10^{-2} over a 10-year period criterion. Explain the reason for any discrepancies.

4.34 For the system represented by the fault tree below, if all the components are periodically tested, determine the *average unavailability* of the system.

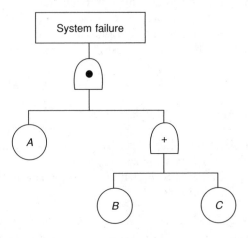

All segments have a test interval of 1-month and average test duration of 1 h.

Components	Failure (rate per hour)	Repair (rate per hour)	Frequency of repair
A	10^{-3}	10^{-1}	0.1
B	10^{-4}	10^{-2}	0.2
C	10^{-4}	10^{-2}	0.3

4.35 Consider the system as represented by the fault tree in Problem (4.34) above. If each component is replaceable and can be represented by an exponential TTF model,

a) Determine the MTTF of the whole system.
b) Estimate MTTFs coefficient variation if the following data applied:

Component	Failure Rate (λ)	Coefficient of Variation
A	10^{-3}	0.2
B	10^{-4}	0.1
C	10^{-4}	0.1

4.36 Based on prior knowledge, the failure probability of a component is believed to be between 0 and 0.2 (uniformly distributed over this range). Suppose that one test is performed and the component properly works as intended. What is the posterior probability of failure of this system?

4.37 The failure time history of a component is shown below:

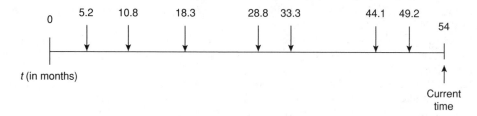

a) Is there practically a trend in ROCOF?
b) What is the conditional probability of no failure (i.e., reliability) in the next 6 months given the component is operating at the current accumulated component age of 54 months?
c) If the observed average downtime for repair is 14 h per repair and repair has no trend, what is the average conditional unavailability (given we are at the age 54 months and the component is operating) in the next 6 months?
d) By assuming no trend in ROCOF, how much error we would make in the unconditional average unavailability?
e) If we decide to adopt a periodically tested policy that reduces the failure rate by a factor of 5 (for example, using a preventive maintenance program) with an average test duration of 5 min, average repair duration of 2 h and average frequency of repair of 0.01, what should we select as our inspection interval (T_0) to maintain the same unavailability value calculated in (c)).

REFERENCES

1. Leemis, L.M., *Reliability: Probabilistic Models and Statistical Methods*, Prentice-Hall, Englewood Cliffs, NJ, 1995.
2. Kijima, M. and Sumita, N., A useful generalization of renewal theory: counting process governed by non-negative Markovian increments, *Journal of Applied Probability*, 23, 71, 1986.
3. Cox, D.R. and Lewis, P.A.W., *The Statistical Analysis of Series of Events*, Methuen and Co., London, 1968.
4. Hoyland, A. and Rausand, M., *System Reliability Theory: Models and Statistical Methods*, John Wiley & Sons, New York, 1994.
5. Kaminskiy, M. and Krivtsov, V., A Monte Carlo approach to repairable system reliability analysis, in *Probabilistic Safety Assessment and Management*, Springer-Verlag, London, 1998, p. 1063.

6. Krivtsov, V., A Monte Carlo Approach to Modeling and Estimation of the Generalized Renewal Process in Repairable System Reliability Analysis, Ph.D. thesis, University of Maryland, 2000.

7. Yanez, M., Joglar, F., and Modarres, M., Mathematical formulation of the generalized renewal process for modeling failures in repairable systems for performance-based applications, *Reliability Engineering and Systems Safety Journal*, 77, 167, 2002.

8. Ascher, H. and Feingold, H., *Repairable System Reliability: Modeling, Inference, Misconceptions and their Causes*, Marcel Dekker, New York, 1984.

9. O'Connor, P., *Practical Reliability Engineering*, 3rd ed., John Wiley & Sons, New York, 1991.

10. Bassin, W.M., Increasing hazard functions and overhaul policy, in *Proceedings 1969 Annual Symposium on Reliability*, IEEE, New York, 1969, p. 173.

11. Bassin, W.M., A Bayesian optimal overhaul interval model for the Weibull restoration process, *Journal of American Statistical Society*, 68, 575, 1973.

12. Crow, L.H., Reliability analysis for complex repairable systems, in *Reliability and Biometry*, Proschan, F. and Serfling, R.J., Eds., SIAM, Philadelphia, 1974.

13. Bain, L.J., *Statistical Analysis of Reliability and Life-Testing Models Theory and Methods*, Marcel Dekker, New York, 1978.

14. Crow, L.H., Evaluating the reliability of repairable systems, in *Annual Reliability and Maintainability Symposium Proceedings*, IEEE, New York, 1990, p. 275.

15. Modarres, M., Kaminskiy, M., and Krivtsov, V., *Reliability Engineering and Risk Analysis*, Marcel Dekker, New York, 1999.

16. Lofgren, E., Probabilistic Risk Assessment Course Documentation, U.S. Nuclear Regulatory Commission, NUREG/CR-4350, Vol. 5, System Reliability and Analysis Techniques, Washington, D.C., 1985.

17. Vesely, W.E. et al., FRANTIC II — A Computer Code for Time-dependent Unavailability Analysis, U.S. Nuclear Regulatory Commission, NUREG/CR-1924, Washington, D.C., 1981.

18. Epstein, B., Estimation from life test data, *Technometrics*, 2, 447, 1960.

19. Welker, E.L. and Lipow, M., Estimating the exponential failure rate from data with no failure events, in *Proceedings of the 1974 Annual Reliability and Maintainability Conference*, IEEE, New York, 1974, p. 420.

20. Nelson, W., *Applied Life Data Analysis*, John Wiley & Sons, New York, 1982.

21. Lawless, J.F., *Statistical Models and Methods for Lifetime Data*, John Wiley & Sons, New York, 1982.

22. Mann, N.R.E., Schafer, R.E., and Singpurwalla, N.D., *Methods for Statistical Analysis of Reliability and Life Data*, John Wiley & Sons, New York, 1974.

23. Atwood C.L. et al., *Handbook of Parameter Estimation for Probabilistic Risk Assessment*, NUREG/CR-6823, Washington, D.C., 2003.

24. Martz, H.F. and Waller, R.A., *Bayesian Reliability Analysis*, John Wiley & Sons, New York, 1982.

25. Kapur, K.C. and Lamberson L.R., *Reliability in Engineering Design*, John Wiley & Sons, New York, 1977.

26. Nelson, W., *Accelerated Testing*, John Wiley & Sons, New York, 1990.

27. National Institute of Standards and Technology, The Economic Impact of Inadequate Infrastructure for software testing. Planning Report 02-3 May 2002.

28. Lawrence, J.D., Software Reliability and Safety in Nuclear Reactor Protection Systems, NUREG/CR-6101, Washington, D.C., 1993.

29. McDermid, J.A., Issues in developing software for safety critical systems, *Reliability Engineering and System Safety*, 32, 1, 1991.

30. Smidts, C., *Software Reliability, The Electronics Handbook*, IEEE Press, New Jersey, 1996.

31. Jelinski, Z. and Moranda, P., Software reliability research, in *Statistical Computer Performance Evaluation*, W. Freiberger, Ed., Academic Press, New York, 1972.

32. Schick, G.J. and Wolverton, R.W., Assessment of software reliability, 11th Annual Meeting German Operation Research Society, DGOR, Hamburg, Germany; also in *Proceedings of the Operation Research*, Physica-Verlag, Wirzberg-Wien, 1973.

33. Littlewood, B. and Verrall, J.K., A Bayesian reliability growth model for computer software, *Journal of Royal Statistical Society Series C-Applied*, 22(3), 332, 1973.

34. Goel, A.L. and Okumoto, K., An analysis of recurrent software errors in a real-time control system, in *Proceedings of the 1978 Annual Conference*, Washington, D.C., 1978, p. 496.

35. Mills, H.D., On the Statistical Validation of Computer Programs, IBM Federal Systems Division, Report 72-6015, Gaithersburg, MD, 1972.

36. Nelson, E., Estimating software reliability from test data, *Microelectronics Reliability*, 17(1), 67, 1978.

37. Ramamoorthy, C.V. and Bastani F.B., Software reliability: status and perspectives, *IEEE Transactions on Software Engineering*, SE-8(4), 354, 1982.

38. Shooman, M.L., Software reliability measurements models, in *Proceedings of the Annual Reliability and Maintainability Symposium*, Washington, D.C., 1975.

39. Goel, A.L. and Okumoto, K., Time-dependent error-detection rate model for software reliability and other performance measures, *IEEE Transactions on Reliability*, 28, 206, 1979.

40. Musa J.D., A theory of software reliability and its application, *IEEE Transactions on Software Engineering*, SE-1(3), 312, 1975.

41. Goel A.L, Software reliability models: assumptions, limitations, and applicability, *IEEE Transactions on Software Engineering*, 11(12), 1411, 1985.

42. Musa, J.D., Iannino, A., and Okumoto, K., *Software Reliability: Measurement, Prediction, Application*, McGraw-Hill, New York, 1987.

43. Petrella, S. et al., Random testing of reactor shutdown system software, in *Proceedings of the International Conference on Probabilistic Safety Assessment and Management*, G. Apostolakis, Ed., Elsevier, New York, 1991.

44. U.S. Nuclear Regulatory Commission, Reactor Safety Study: An Assessment of Accidents in U.S. Commercial Nuclear Power Plants, U.S. Regulatory Commission, WASH-1400, Washington, D.C., 1975.

45. Hannaman, G.W. and Spurgin, A.J., *Systematic Human Action Reliability Procedure (SHARP)*, Electric Power Research Institute, NP-3583, Palo Alto, CA, 1984.

46. Rasmussen, J., Cognitive control and human error mechanisms, in *New Technology and Human Error*, Rasmussen, J., Duncan, K., and LePlate, J., Eds., John Wiley & Sons, New York, 1987, chap. 6.

47. Woods, D.D., Roth, E.M., and Pole, H., Modeling human intention formation for human reliability assessment, *Reliability Engineering and System Safety*, 22,169, 1988.

48. Amendola, A. et al., Modeling operators in accident conditions: advances and perspectives on cognitive model, *International Journal of Man-Machine Studies*, 27, 599, 1987.

49. Wreathall, J., *Operator Action Tree Method*, IEEE Standards Workshops on Human Factors and Nuclear Safety, Myrtle Beach, SC, 1981.

50. Potash, L. et al., Experience in integrating the operator contribution in the PRA of actual operating plants, in *Proceedings of American Nuclear Society, Topical Meeting on Probabilistic Risk Assessment*, New York, 1981.

51. Swain, A.D. and Guttman H.E., Handbook of Human Reliability Analysis with Emphasis on Nuclear Power Applications, U.S. Nuclear Regulatory Commission, NUREG/CR-1278, Washington, D.C., 1983.

52. Hall, R.E., Wreathall, J., and Fragola, J.R., Post Event Human Decision Errors: Operator Action/Time reliability Correlation, U.S. Nuclear Regulatory Commission, NUREG/CR-3010, Washington, D.C., 1982.

53. Hunns, D.M. and Daniels B.K., *The Method of Paired Comparisons*, Third European Reliability Data Bank Seminar, University of Bradford, National Center of System Reliability, United Kingdom, 1980.

54. Embry, D.E. et al., SLIM-MAUD: An Approach to Assessing Human Error Probabilities Using Structured Expert Judgment, U.S. Nuclear Regulatory Commission, NUREG/CR-3518, Washington, D.C., 1984.

55. Poucet, A., Survey of methods used to assess human reliability in the human factors reliability benchmark exercise, *Reliability Engineering and System Safety*, 22(1–4), 257, 1988.

56. Siegel, A.I. et al., Maintenance Personnel Performance Simulation (MAPPS) Model, U.S. Nuclear regulatory Commission, NUREG/CR-3626, Vols I and II, Washington, D.C., 1984.

57. Saaty, T.L., *The Analytic Hierarchy Process*, McGraw-Hill, New York, 1980.

58. Stillwell, W., Seaver, D.A., and Schwartz, J.P., Expert Estimation of Human Error Problems in Nuclear Power Plant Operations: A Review of Probability Assessment and Scaling, U.S. Nuclear Regulatory Commission, NUREG/CR-2255, Washington, D.C., 1982.

59. Hannaman, G.W., Spurgin, A.J., and Lukic, Y.D., *Human Cognitive Reliability Model for PRA Analysis*, NUS Corporation, NUS-4531, San Diego, CA, 1984.

60. Rasmussen, J., Skills, rules and knowledge: signals, signs and symbols and their distinctions in human performance models, *IEEE Transactions on Systems, Man and Cybernetics*, SMC-13(3), 257, 1983.

61. Dougherty E.M. and Fragola J.R., *Human Reliability Analysis: A System Engineering Approach with Nuclear Power Plant Applications*, John Wiley & Sons, New York, 1988.

62. Institute of Electrical and Electronics Engineers, *IEEE Std. 500: Guide to the Collection and Presentation of Electrical, Electronic, Sensing Component and Mechanical Equipment Reliability Data for Nuclear Power Generating Stations*, IEEE Standards, New York, 1984.

63. U.S. Nuclear Regulatory Commission, Severe Accident Risk: An Assessment for Five U.S. Nuclear Power Plants, U.S. Nuclear Regulatory Commission, NUREG-1150, Washington, D.C., 1990.

64. Electric Power Research Institute (EPRI), NP-6395-D, Probabilistic Seismic Hazard Evaluation at Nuclear Plant Sites in the Central and Eastern United States, Resolution of Charleston Issue, April 1989.

65. Clemens, R.J. and Winkler R.L., Unanimity and compromise among probability forecasters, *Management Science*, 36(7), 767, 1990.

66. Dalkey, N. and Helmer O., An experimental application of the Delphi method to the use of experts, *Management Science*, 9(3), 458, 1963.

67. Mosleh, A. and Apostolakis, G., Combining various types of data in estimating failure rates, in *Transaction of the 1983 Winter Meeting of the American Nuclear Society*, San Francisco, CA, 1983.

68. Apostolakis, G., Kaplan, S., Garrick, B.J., and Duphily, R.J., Data specialization plant-specific risk studies, *Nuclear Engineering and Design*, 56, 321–329, 1980.

69. Kim, J.H., A Bayesian Model for Aggregating Expert Opinions, Ph.D. dissertation, Department of Materials and Nuclear Engineering, University of Maryland, College Park, MD, 1991.

70. Snaith, E.R., The Correlation Between the Predicted and Observed Reliabilities of Components, Equipment and Systems, National Center of Systems Reliability, U.K. Atomic Energy Authority, NCSR-R18, 1981.

71. Lichtenstein, S.B., Fischoff, B., and Phillips, L.D., Calibration of probabilities: the state of the art, in *Decision Making and Change in Human Affairs*, Jungermann, H. and deZeeuw G., Eds., Reidel, Dordrecht, The Netherlands, 1982, p. 275.

72. Slovic, P., Fischhoff, B., and Lichtenstein, S., Facts versus fears: understanding perceived risk, in *Judgment Under Uncertainty: Heuristics and Biases*, Kahneman, D., Slovic, P., and Tversky, A., Eds., Cambridge University Press, Cambridge, 1982, p. 463.

73. Apostolakis, G., Data analysis in risk assessment, *Nuclear Engineering and Design*, 71(3), 375, 1982.

5 Uncertainty Analysis

5.1 INTRODUCTION

Uncertainty is a measure of the "goodness" of an estimate, such as performance of a hazard barrier. Without such a measure, it is impossible to judge how closely the estimated value relates or represents the reality, and provides a basis for making decisions related to health and safety risks. Uncertainty arises from lack of or insufficient knowledge about events, system states, processes, and phenomena that factor into the estimation. Epistemologists have spent years trying to understand what it means when we say we know something. Despite this, unfortunately, the literature does not offer a universally acceptable definition of knowledge, but a working definition of knowledge in the context of risk analysis is "...a mixture of experience, values, contextual information, and insight that provides a framework for evaluating and incorporating new experience and information and making rational decisions." Knowledge originates and is applied in the minds of the knower. In human-dominated organizations, it often becomes embedded both in documents or other repositories and in organizational routines, processes, practices, and norms. Clearly knowledge is not a simple element; rather it is mixture of many elements. It is intuitive and therefore hard to capture in words or understand logically, and unlike other things and properties around human beings, knowledge is harder to pin down.

Knowledge derives from information. Knowledge is obtained through the process of changing information into something useful to act upon. For example, knowledge is obtained through comparing information to understand a situation (event, phenomena, etc.), understanding implications of a piece of information, evaluating the relationship of information with other information available and characterizing what others think about the information at hand.

On the other hand, information is the data with meaning attached to it by a variety of means. For example, the data may be analyzed or condensed. Similarly information is derived from raw data by providing some meaning into the data. Finally data are a set of discrete, objective facts about events. Unlike data and information, knowledge contains judgment, therefore it is subjective. Knowledge grows as the bodies holding it interact with the prevailing environments and acquiring more data, and thus more information. As such, knowledge is evolving and growing. When knowledge stops evolving, it turns into dogma. Clearly, as objective data are processed into information and ultimately into knowledge, it becomes highly condensed, but more subjective. Also, in this process a vast amount of objective data can be transformed into a small volume of knowledge.

Engineering systems are often associated with vast amounts of data. Examples of data in risk assessment include historical data on failures of mitigative, protective, and preventive hazard barriers, root causes of the failure events, and actual time of the failures. This data can be transformed into "information" in form of estimates of frequency of certain failure modes and causes, associations between certain environmental or operating conditions and occurrence of certain failure modes, or activation of certain failure mechanisms. Finally, the information can be used to estimate the performance of the system, consequences of system failure, and its risks.

As Bernstein [1] points out, lack of information and knowledge is driven by how much resources are available to us to obtain it. Bernstein states that the information you have is not the information you want. The information you want is not the information you need. The information you need is not the information you can obtain. The information you can obtain costs more than you want to pay. So uncertainty will always remain to be grappled with.

A measure of how much is known about something (an event, process, phenomenon, etc.) and thus a measure of uncertainty is the use of probability. If data are transformed into knowledge without any subjective interpretations, we use the classical interpretation of probability to express our confidence about our knowledge of the proposition we are estimating. On the other hand, in the real world, our transformation of data into knowledge involves subjective interpretations. As such, the so-called subjectivist measures of probability would be the natural way of representing our uncertainty about a thing or proposition. Ayyub [2] describes other nonprobabilistic methods of measuring and representing uncertainty.

In the process of transforming facts (data) into knowledge there is often ambiguity in linguistic expression about knowledge, e.g., the expression "System A provides adequate cooling in emergency situations" is a piece of knowledge which is ambiguous. Ambiguity is a form of uncertainty which can be reduced by improving clarity in explaining our estimations and the nature of our data, information, and knowledge.

It is also important to separate availability of knowledge itself from one's understanding and awareness of such knowledge. For example, the totality of knowledge about global warming in literature is often broader than an expert's knowledge of the same subject. In transforming data into knowledge, one must also be able to characterize knowledge. That is, to explain how much is known, trustworthiness of the known, and knowing how much is not known. It is this knowledge of the knowns and unknowns that can be converted and measured as the "uncertainty" about a thing. The concept discussed above can be captured by recognizing four possible outcomes when transferring data into knowledge. Table 5.1 shows how to characterize and remedy each situation.

Based on Table 5.1, formal uncertainty analysis is the characterization of our state of knowledge about the knowledge of a thing (an event, process, phenomenon, etc.). According to Bedford and Cooke [3], a mathematical representation of uncertainty comprises of three properties: (1) axioms specifying properties of uncertainty; (2) interpretations connecting axioms with observables; (3) measurement procedure describing methods for interpreting axioms. Probability is one of several ways to express uncertainty with its axioms defined by

TABLE 5.1
Categories of Uncertainties and Uncertainty Management Options

Knowledge on the Subject	Awareness of Knowledge	
	Aware	Not Aware
A lot known	No uncertainty (deterministic representation)	Reduce uncertainty by knowledge management
Not a lot unknown	Reduce uncertainty by gathering more knowledge	Use conservative assumptions (e.g., defense-in-depth)

Kolmogorov [4]. Probability has two main interpretations of "frequentist" and "subjectivist," and numerous methods to combine and deal with probabilities.

An adjunct to uncertainty analysis is sensitivity analysis. This type of analysis is frequently performed in risk assessments, especially in PRAs to indicate elements (events, processes, phenomena, etc.) that when changed cause the greatest change in risk results. They are also performed to identify aspects of risk analysis (e.g., simplification or limitations in the scope) to which the analysis results are or are not sensitive. The subject of sensitivity analysis will be further discussed in Chapter 6.

Uncertainty and sensitivity analyses for risk assessment of complex systems are typically iterative. An initial sensitivity analysis is often performed to get an outlook of (i) the behavior of the submodels involved, (ii) strategies for finalizing the risk analysis, and (iii) overview of the most significant parameters and assumptions that affect the uncertainty associated with the output risk estimates. When system behavior is better understood and risk-significant variables are identified, resources can be focused on improving the characterization of the uncertainty in these important input variables.

5.2 TYPES OF UNCERTAINTY

As a matter of practicality and because different types of uncertainties are generally characterized and treated differently, uncertainty may be divided into two types: the *aleatory* and *epistemic*. Aleatory uncertainty is referred to an event that is irreducible within the realm of our ability and resources to gain more knowledge. This type of uncertainty results when an experiment or observation is repeated under identical conditions but with different outcomes. Some authors refer to this type of uncertainty as variability. However, in this book, we consider variability as a form of practically irreducible uncertainty. For example, our inability to assess the exact time of occurrence of a failure of a new light bulb that we just purchased from a hardware store is a representation of an aleatory uncertainty; various light bulbs from the same manufacturer may last at different times. It is possible to characterize this uncertainty by using a Poisson model to estimate the probability that the light bulb fails by a given time (i.e., treating time of failure as a random variable which following a Poisson distribution model), provided that the failure event occurrence rate is constant over time. The probability associated with the estimates of the time that the light bulb fails is an expression of our lack of knowledge (aleatory uncertainty) concerning our inability to express the exact time of the first and final failure.

The epistemic uncertainty, on the other hand, characterizes our state of knowledge about building a model of the world. For example, our uncertainty associated with the choice of the Poisson model to represent the aleatory uncertainty about the exact time of failure of the light bulb itself and the value of the intensity rate (parameter) of the distribution are considered epistemic uncertainties. Such uncertainties may also be represented in form of a pdf.

While aleatory uncertainty may be viewed as being inherent in the physical processes involved, it cannot be practically reduced; however, enlarging the database can be converted to added information about the accuracy of the Poisson probability distribution model and its parameter. This added information reduces the epistemic uncertainty (and thus our knowledge of the failure process), but not the aleatory uncertainty. Consider the Poisson distribution used to represent the probability of occurrence of a failure by time t for the light bulb. The choice of the distribution model itself also involves some modeling uncertainty (epistemic); however, the variability of the failure time t is the aleatory uncertainty. Our epistemic uncertainty about the exact value of the intensity rate in the Poisson distribution may be depicted by another distribution, e.g., a lognormal distribution. In this case, the lognormal distribution model represents the epistemic uncertainty about the building of the Poisson distribution model.

A concern may be raised here that we are also uncertain about the lognormal model and the cycle may continue forever. This is certainly a valid point, but depending on the interpretation of the probability adopted, representation of uncertainties about the parameters of a model is the most practical and meaningful level of uncertainty characterization in risk analyses.

Another way to categorize uncertainties is to divide the uncertainties into two categories of *parameter* and *model*. When we talk about model uncertainty in risk analysis, we are referring to the mathematical or logical models used to represent or estimate properties of events, processes, phenomena that govern operation and behavior of hardware, software, and humans in complex systems.

Hanseth and Monteiro [5] define the model as an abstract representation of reality used to simulate a process, understand a situation, predict an outcome, or analyze a problem. Model uncertainty is a current subject of research and the readers are referred to Droguett and Mosleh [6] for an overview, and a Bayesian approach to model uncertainty analysis.

By parameter uncertainties, we mean possible instances of the model, created due to our inability to uniquely construct the model. For example, in an exponential fire growth model, the parameter of the model (e.g., a constant growth rate) may not be unique and involve epistemic uncertainties. Each value of the constant growth rate parameter leads to an instance of the fire growth model. The parameter is generally a constant in the structure of a mathematical or logical model, subject to the analyst's choice in a specific application or problem, but one that could vary outside of the specific application. For example, in a probability distribution model, a parameter is meant to assume a numeric value which characterizes a specific probability distribution. The mean and variance are typical examples of such parameters that assume distinct values or ranges of values in specific applications. Probability and statistics are usually used to characterize and estimate parameter uncertainties. Since the parameter is a constant in the equation of a model that can be varied to yield a family (many instances) of similar curves relating the independent variables to the dependent variable, some characterize each instance of the family of curves as a separate model, since the shape of the output may change considerably by changing parameters. Most, however, consider the family of the curves as the model, not the specific instances of the family. The latter appears to be the more appropriate definition of the model.

5.2.1 MODEL UNCERTAINTY

Model uncertainty occurs because models are not perfect. Models of physical processes generally have many underlying assumptions and often are not valid for all situations. Sometimes, there are alternative models proposed by different analysts, and it is not known which, if any, of the models is the most appropriate one for the situation at hand (each alternative will have its own strengths and drawbacks). The PRA models, such as the event trees and fault trees, can be constructed in different ways, and those alternative methods of construction can change the resulting models.

As discussed earlier, mathematical and logical model structures are necessarily abstract representations of the phenomena, processes, and events under study. The results and output of such models involve uncertainties. The optimal model is one involving the most simplifications while providing an acceptably low error in the representation of the event, processes, or phenomena of interest. The structural form of the empirical mathematical or logical models in use is often a key source of uncertainty. In addition to the considerable approximations involved in modeling, occasionally competing models may be available. Additionally, the spatial or temporal resolution (e.g., numerical mesh or grid cell size) of many models is also a type of approximation that introduces uncertainty into model results.

Consistent with the Hanseth and Monteiro [5] definition of model, the process of determining the degree to which a model is an accurate representation of the real world is called model uncertainty. There is no "correct model" of anything. Each model only represents or predicts the actual property of interest to some extent. Model uncertainty, in risk assessment applications, is referred to the goodness of the *form* or *structure* of the model used to estimate a quantity of interest associated with an event, process, phenomenon, etc. For example, in estimating the temperature profile due to accidental fire in a room, a fire growth model will be needed. Considering the geometry and characteristics of the burning materials involved, several empirical structural forms of the fire growth model are possible. For example, exponential and second-order polynomial forms are two possible choices, each leading to a different growth profile. Depending on our state of knowledge, information, and data, each of these structural forms can be used to measure the actual temperature at a given location as a function of time with some uncertainty. The model uncertainty is accounted for by superimposing to the model estimation a conditional probability interval within which the real property of interest is expected to lie, given the extent of model extrapolation and values of model parameters used (for example, an instance of the model is created by randomly selecting values from the epistemic uncertainty distribution of the model parameters). Generally, this interval is estimated subjectively considering the state of knowledge about the model accuracy. Also, one can rely on Bayesian estimation to "update" a subjective prior estimate of the probability interval for a property of interest predicted by the model, with actual data and information about this property.

Accounting for model uncertainty is in its infancy and a credible and universally accepted method is yet to emerge. It is possible to recognize two possible situations of model prediction of a property of interest:

1. Single model
2. Multiple models (different structural forms)
   a. Multiple models using the same assumptions, data, and information
   b. Multiple models using different assumptions, but the same data and information
   c. Multiple models using different assumptions, data, and information.

Table 5.2 lists a useful set of criteria that influence characterization, especially quantification of model uncertainty.

The uncertainty characterization approach with the single-model case is to assign a subjective conditional probability interval to the prediction of interest, given the nature of

TABLE 5.2
Criteria on Model Uncertainty

Decision Criteria	Options	
Source of data and information	Multiple sources	Same sources
Assumptions	Different assumptions	Same assumptions
Model form/structure	Multiple-model form/structure	Single-model form/structure
Model extrapolation	Yes	No
Model output	Multiple output	Single output

extrapolation (e.g., animal health data to human extrapolation or significant extrapolation in time or space). This can be done by the subject matter expert(s) either formally (see Section 4.9) or informally. An adjunct approach in characterizing single-model uncertainties is by using a Bayesian updating approach applied to the model outputs. If the epistemic uncertainty associated with a model output is subjectively estimated, *a priori*, in form of a probability density function, it can be updated, if some evidence in form of actual observation from experiments, events, etc. exist (such as the case in Example 5.1). The model is conditional on the validity of its assumptions. The uncertainty associated with the modeling assumptions is usually handled by performing sensitivity studies. These studies are performed to investigate assumptions that suspected are of having a potentially significant impact on the results. Additionally, sensitivity studies are used to assess the sensitivity of model extrapolation results to model input dependencies.

Consider the case of uncertainty characterization when multiple models estimate the same property of interest and when the structural form of the models are different, but the underlying knowledge, information, data, and assumptions in developing the model are the same. In this case, a weighted average of the various model outputs, for exactly the same problem with the same input conditions, can be used to represent an aggregate estimation of the output of interest. Again for each model, *a priori*, a probability density function may be used to represent the epistemic uncertainties of each model output. Then, the weighted averaging method is used to combine the uncertainty from each model. The weights can also be assigned, subjectively by the subject matter experts, or equal weights may be used. A geometric or arithmetic average may be used, as appropriate. Similar to the single-model case, in light of any observed evidence, a Bayesian updating approach may be used to define the averaged uncertainty of the output from multiple-models. The simplest approach is by assuming that all the models available are associated with a probability (weight) that the respective models correctly predict the property of interest. As such, if a set of candidate models are available, one could construct a discrete probability distribution (DPD) (M_i, p_i), where p_i is the degree of belief (in subjectivist terms) in model M_i as being the most appropriate representation of reality (assuming no dependency between the models due to using the same set of data and information to build the models). This has been done for the modeling of seismic hazard, for example, where the result is a DPD on the frequencies of earthquakes. This uncertainty can then be propagated in the same way as the parameter uncertainties. Other methods of combining models are also available. For example, see Mosleh et al. [7].

The uncertainty characterization when the same data, information, and knowledge are used to develop a model, but with different assumptions, is another case of interest. For example, in estimating structural failure caused by fatigue due to random load applied to the structure, one may use a linear elastic fracture mechanic approach to estimate the life of the structure. But, equally one may use an elastic–plastic theory of fatigue to estimate the life. The latter model allows some plasticity at the tip of the crack, but the former assumes no plastic deformation at that location. Thereby, the two model assumptions are fundamentally different in that they refer to different phenomena of crack growth due to fatigues, and both models may satisfy same data and information available. Clearly, the estimate of the two models are considerably different when extrapolated (e.g., in life estimation for a structure), because the two models rely on different underlying theories. In this case the averaging approach is physically meaningless, because for the given problem of interest, one of these models (or even none) may be appropriate. In this case, the characterization of epistemic uncertainty of the models can only be done separately for each model, as in the single-model case and only qualitatively (or semiquantitatively) explain the goodness of each model. It becomes the final decision maker's partiality (e.g., risk manager's decision as to which of the model output he/she prefers to use).

The case of multiple models based on different data, information, and assumptions, in principle, is the same as the multiple models with the same data and information, but different assumptions. Ultimately, multiple models yield different estimates for the output that should not be combined, but the risk manager decides which model to use. The only difference is that it is easier for the risk manager to select the best model output for the problem at hand when the models are based on different data. This is because the efficacy and completeness of the data used to develop the models are easier to judge than the adequacy of model structure and form.

In dealing with multiple-model situations, it is often instructive to understand the impact of a specific assumption on the prediction of the model. The impact of using alternate assumptions or models may be addressed by performing appropriate sensitivity studies, or they may be addressed using qualitative arguments. This may be a part of the model uncertainty evaluation.

Definition and quantification of the uncertainties associated with a model are very complex and cannot always be characterized by a quantitative representation (e.g., probabilistic representation). In fact, most subject matter experts are hesitant and uncomfortable to provide quantitative estimation of uncertainties. Qualitative characterization of uncertainties is clearly an alternative approach. In this case the decision maker needs to consider the uncertainties as described qualitatively to reach a decision. The readers are referred to Morgan and Henrion [8] for more discussion on this topic. Sastry [9] defines different sources of model uncertainties as follows:

Structure/form: Uncertainty arises when there are assumptions for developing a model. In such cases, if the results from alternative models (using other data, information, knowledge) yield the same answer to a problem, then one can be more confident that the results obtained from the model are realistic in the face of uncertainty. If, however, alternative models yield different conclusions, further model evaluation might be required. One evaluation may involve verifying model estimation with the actual data observed or from experiments. Sometimes the uncertainty associated with the risk model assumptions is characterized with sensitivity analysis.

Level of detail: Often, models are simplified for purposes of tractability. An example of this is converting a complex nonlinear model to a simple linear model to trace calculations. Uncertainty in the predictions of simplified models can sometimes be characterized by comparison of their predictions to those of more detailed, inclusive models. Also, certain aspects of a process, phenomena, event or system may not be considered in a model, because the modelers may believe that they are unimportant in comparison to other aspects of the model. Often this aspect of model uncertainty is referred to as completeness uncertainty.

Extrapolation: Models that are validated for one portion of input space may be completely inappropriate for making predictions by extrapolating the model into other regions of the space of interest. For example, a dose–response model based on high-dose, short-term animal tests may involve significant errors when applied to study low-dose, long-term human exposures. Similarly, models that are evaluated only for application in a unique set of conditions may involve enormous uncertainties when they are employed to study significantly different conditions.

Resolution: In the application of mathematical or logical models, selection of a spatial or temporal grid or lattice size often involves uncertainty. On one hand, there is a trade-off between the computation time (hence cost) and prediction accuracy. On the other hand, there is a trade-off between resolution and the validity of the governing equations of the model at such scales. Very often, a coarse grid resolution introduces approximations and uncertainties into model results since certain phenomena, events, or processes may be bypassed or neglected altogether. Sometimes, a finer grid resolution need not necessarily result in more accurate predictions, for example, when a fine-grid produces incorrect results because the governing equation may be insensitive to fine changes.

Boundaries: Any model may have limited boundaries in terms of time, space, number of input variables, and so on. The selection of a model boundary may be a type of simplification. Within the boundary, the model may be an accurate representation, but other overlooked phenomena not included in the model may play key roles. Table 5.3 summarizes the classes of sources of uncertainty which represents Sastry [9] classification with some modifications.

Sometimes an uncertainty called completeness uncertainty is defined as a third type of uncertainty, in addition to the model and parameter uncertainties. This uncertainty accounts for the unknowns related to whether or not all the significant phenomena and relationships have been considered, e.g., whether all important risk scenarios have been identified. Completeness uncertainty, however, is the same as modeling uncertainty, but only occurs at the initial stage in an analysis. In addition to inadequate identification of the physical phenomena, completeness uncertainty can also result from inadequate consideration of human errors, software reliability, or interactions and dependencies among the elements of the process being modeled. As such completeness uncertainty is only a subset of model uncertainty. Completeness uncertainty is rarely treated in a PRA, but should be an important consideration in the future.

5.2.2 PARAMETER UNCERTAINTY

Parameter uncertainties are those associated with the values of the fundamental parameters of the models, such as failure rates in time-to-failure distribution models, event probabilities in

TABLE 5.3
A Classification of the Sources of Uncertainty

Uncertainty in Model Formulation (Structural Uncertainty)	Uncertainty in Model Application (Data/Parametric Uncertainty)
Conceptual Simplifications	Parameter selection Input data Development/selection

Completeness
  Level of details
	Source information
	  Initial and boundary conditions

Mathematical formulation
  Simplifications in mathematical formulation
  Physics-type hypotheses
  Idealizations in formulation
  Independence hypotheses
  Spatial averaging
  Temporal averaging
  Process decoupling
  Lumping of parameters
	Operational model evaluation
	  Uncertainty in model estimates
	  Uncertainty in observations
	  Nonexistence of observations
	
	Response interpretation

Numerical solution
  Discretization
  Numerical algorithm/operator splitting
  Approximations in computer coding	

Representation of results
  Precision
  Bias	

logic models including human and software error probabilities. They are typically character-ized by establishing probability distributions on the parameter values. Parameter uncertainty is generally epistemic in nature and is due to unknowns about correct inputs to the models being used in the analysis. In risk analysis, the parameters of interest may be inputs to either the PRA models themselves, or a variety of physical and process models that influence the PRA models.

Parameter uncertainties can be explicitly represented and propagated through the models used in the risk assessment, and the probability distribution of the relevant metrics can be generated (e.g., distribution of performance measures such as reliability, unavailability, and average downtime for repair). Various measures of central tendency, such as the mean, median, and mode can be evaluated. For example, the distribution can be used to assess the confidence with which some reliability targets are met. The results are also useful to study the contributions from various elements of a model and to see whether or not the tails of the distributions are being determined by uncertainties of a few significant elements of the reliability or risk model. If so, these elements can be identified as candidates for compensatory measures and/or monitoring.

Earlier we discussed methods for quantifying uncertainties of parameter values of distri-bution models for both frequentist and subjectivist (Bayesian) methods. Examples of these parameters are MTBF, mean time to repair, failure rate, rate of occurrence of failure, and probability of failure on demand of a component or a hazard barrier. Uncertainty of the parameters is primarily governed by the amount of test or field data available about failures and repairs of the items. Because of these factors, a parameter does not take a fixed and known value, and has some random variability. Later in this chapter, we will discuss how the parameter uncertainty is propagated in a system to obtain an overall uncertainty about the system failure.

Parameter uncertainties are largely epistemic and involve (a) limited data and information, (b) random errors in measuring and analytic devices (e.g., the imprecision of continuous monitors that measure emissions), (c) systematic biases due to miscalibration of devices, and (d) inaccuracies in the assumptions used to infer the actual quantity of interest from the observed readings of a "surrogate" or "proxy" variable. Other potential sources of uncertainties in estimates of parameters include misclassification, estimation of parameters through a small sample, and estimation of parameters through nonrepresenta-tive samples.

Example 5.1

The crack size in a particular location of an aircraft structure after a fixed number of flight hours was modeled assuming a lognormal distribution model with parameters $\mu_0 = 0.048$ in. and $\sigma_0 = 0.67$. This distribution is considered as our subjective prior estimate. The structure was inspected, and the actual crack sizes measured, taking into account the accuracy of the inspection process, follow a lognormal distribution with parameters $\mu^* = 0.120$ and $\sigma^* = 0.49$. Using Bayesian estimation, calculate the updated model of crack growth distribution.

Solution

The subjective prior distribution is represented by

$$h(a|\mu_0, \sigma_0) = \frac{1}{a\sigma_0\sqrt{2\pi}} \exp\left[-\frac{1}{2}\left(\frac{\ln a - \ln(\mu_0)}{\sigma_0}\right)^2\right]$$

and the likelihood function by

$$\ell\left(a^*/\mu^*, \sigma^*\right) = \frac{1}{a^* \sigma^* \sqrt{2\pi}} \exp\left[-\frac{1}{2}\left(\frac{\operatorname{Ln} a^* - \operatorname{Ln} \mu^*}{\sigma^*}\right)^2\right]$$

A numerical calculation similar to the one discussed in Section 4.6.2 may be used. However, using the Bayes Theorem, it is also possible to show that the posterior distribution is a lognormal posterior in the form of

$$f\left(a|\mu', \sigma'\right) = \frac{1}{a \sigma' \sqrt{2\pi}} \exp\left[-\frac{1}{2}\left(\frac{\ln a - \ln \mu'}{\sigma'}\right)^2\right]$$

in which

$$\mu' = \mu_0^{w_0} \cdot \mu^{*w_1}, \quad \sigma' = \frac{1}{\sqrt{\dfrac{1}{\sigma_0^2} + \dfrac{1}{\sigma^{*2}}}}$$

where

$$w_0 = \frac{\sigma^{*2}}{\sigma_0^2 + \sigma^{*2}} \quad w_1 = \frac{\sigma_0^2}{\sigma_0^2 + \sigma^{*2}} \quad (\text{note that } w_0 + w_1 = 1)$$

By substituting the values in the above equations, $w_0 = 0.348$ and $w_1 = 0.652$

$$\mu' = (0.048)^{0.348}(0.120)^{0.652} = 0.0872 \text{ in.}, \quad \sigma' = \frac{1}{\sqrt{\dfrac{1}{(0.67)^2} + \dfrac{1}{(0.49)^2}}} = 0.3955$$

5.2.2.1 Sensitivity Analysis

An adjunct to uncertainty analysis is sensitivity analysis. This topic will be discussed in more detail in Chapter 6. Sensitivity analyses are also frequently performed in PRAs to indicate elements (events, processes, phenomena, etc.) whose value changes cause the greatest changes in partial or final risk results. They are also performed to identify aspects of risk analysis (e.g., simplification or limitations in the scope) to which the analysis results are or are not sensitive contributors.

In a sensitivity analysis, an input parameter, such as a component failure rate in a fault tree logic model, is changed, and the resulting change in the top event probability is measured. This process is repeated using either different values for the same parameter or changing different parameters by the same amount. An example includes estimating effect of sequentially changing different failure rates in a fault tree model by a factor of 5 or so. Usually the so-called one-at-a-time sensitivity study is used by only changing one parameter at a time. Sometimes two or more parameter values are simultaneously changed to study the interactions among the parameters or measure the impact of a unique family of components (for example, all similar human error probabilities in a risk model).

An important element of sensitivity analysis includes the amount of changes to be applied to each parameter and which parameters or assumptions in the model to change. Sampling-based sensitivity analysis is more effective in searching for analysis of errors than simply running the model for a limited number of cases and then examining the results of these calculations. A sampling-based sensitivity analysis is advisable as part of a model validation and verification effort.

There are several methods for quantification of sensitivity analysis. Depending on the methodology used and according to Saltelli et al. [10], different values for sensitivity of model vs. each parameter may be calculated. Spearman rank, Kendall rank, and Pearson product moment are among the famous sensitivity analysis methods. For example, the Pearson product moment correlation coefficient is the usual linear correlation coefficient (A.76) [11]. The product moment part of the name comes from the way in which it is calculated by summing up the products of the deviations of the scores from the mean.

5.3 MEASURES OF UNCERTAINTY

The most common practice in quantitatively measuring uncertainty in risk assessment is the use of the probability. In this book, we have only used this measure of uncertainty. This also affects the way uncertainty analysis is performed.

The simplest way to represent the probability as a measure of uncertainty is to use mean $\bar{x}$ and variance s^2 of a property associated with an event, process, or phenomena of interest and any epistemic uncertainty associated these estimations. Estimations of $\bar{x}$ and s^2 are themselves subject to some uncertainty. In a classical estimation approach only the confidence intervals of $\bar{x}$ and s^2 can be estimated. However, equally the uncertainty associated with the mean $\bar{x}$ and variance s^2 can be estimated using the Bayesian approach.

Generally, the problem of finding the distribution of a function of random variables is difficult, which is why for the most of reliability and risk assessment applications, the problem is reduced to estimation of mean and variance (or standard deviation) of function of random variables. Such techniques are considered in the following sections. It should be mentioned that the uses of these techniques are, by no mean, limited to risk assessment problems. They are widely used in engineering.

5.4 UNCERTAINTY PROPAGATION METHODS

There are two classes of uncertainty propagation methods available: the method of moments and the probabilistic (mostly Monte Carlo) methods. Further, when (5.1) is very complex to compute (e.g., in Monte Carlo calculations), often one of several possible methods of response surface is used to reduce the computational effort. These methods will be described in the following sections.

5.4.1 METHOD OF MOMENTS

Consider the general case of a model estimation Y (e.g., a variable estimating risk value, consequence system reliability, or unavailability). Based on a model of the risk assessment, risk management or system performance, a general function of uncertain quantities x_i, $i = 1, \ldots, n$ can describe Y, as

$$Y = f(x_1, x_2, \ldots, x_n) \tag{5.1}$$

A simple example would be a system composed of basic nonrepairable (replaceable) elements, each having the exponential time-to-failure distributions. In this case, Y is the MTBF of the system, x_is are the estimates of the MTBF of the basic elements. System risk or performance characteristic, Y, can be the unavailability or probability of occurrence of the top event of a fault tree, risk value, or health effects (response) due to exposure to hazards. The uncertainty of Y as a function of the uncertainty of the basic elements x_i is estimated by the methods of uncertainty propagation.

Using the Taylor's series expansion about the mean of input variable x_i, μ_i, and denoting $(x_1, x_2, \ldots, x_n)$ by X we can write (5.1) in the form of:

$$Y = f(X)$$

$$Y = f(\mu_1, \mu_2, \ldots, \mu_n) + \sum_{i=1}^{n} \left[\frac{\partial f(X)}{\partial X} \right]_{x_i=\mu_i} (x_i - \mu_i) \tag{5.2}$$

$$+ \frac{1}{2!} \sum_{j=1}^{n} \sum_{i=1}^{n} \left[\frac{\partial^2 f(X)}{\partial x_i \, \partial x_j} \right]_{x_i=\mu_i, x_j=\mu_j} (x_i - \mu_i)(x_j - \mu_j) + R$$

where R represents the residual terms.

Taking the expectation of (5.2), one gets

$$E(Y) = f(\mu_1, \mu_2, \ldots, \mu_n) + \sum_{i=1}^{n} \left[\frac{\partial f(X)}{\partial X} \right]_{x_i=\mu_i} E(x_i - \mu_i) \tag{5.3}$$

$$+ \frac{1}{2!} \sum_{j=1}^{n} \sum_{i=1}^{n} \left[\frac{\partial^2 f(X)}{\partial fx_i \, \partial fx_j} \right]_{x_i=\mu_i, x_j=\mu_j} E\left[(x_i - \mu_i)(x_j - \mu_j)\right] + E(R)$$

If the estimates x_i ($i = 1, 2, \ldots, n$) are unbiased with expectations (true values) μ_i, the second term in the above equation may be canceled. Dropping the residual term, $E(R)$, and assuming that the estimates x_i are independent, one gets the following approximation:

$$E(Y) \approx f(\mu_1, \mu_2, \ldots, \mu_n) + \frac{1}{2} \sum_{i=1}^{n} \left[\frac{\partial^2 f(X)}{\partial x_i^2} \right]_{x_i=\mu_i} S^2(x_i) \tag{5.4}$$

where $S(x_i)$ is the standard deviation of parameter x_i.

To get a simple approximation for the variance (as a measure of uncertainty) of the system risk or performance characteristic Y, consider the first two-term approximation for (5.2). Taking the variance and treating the first term as constant and assuming independence between all x_is, one gets

$$\text{Var}\left(\hat{Y}\right) = \text{Var} \sum_{i=1}^{n} \left[\frac{\partial f(X)}{\partial x_i}\right]_{x_i=\mu_i} (x_i - \mu_i)$$

$$= \sum_{i=1}^{n} \left[\frac{\partial f(X)}{\partial x_i}\right]_{x_i=\mu_i}^{2} S^2(x_i) \qquad (5.5)$$

Example 5.2

For the system shown below, the constant failure rate of each component has a mean value of 5×10^{-3} per hour. If the epistemic uncertainty associated with the failure rate parameter can be represented by a random variable which follows a lognormal distribution with a coefficient of variation of 2, calculate the mean and standard deviation of the system performance (unreliability) at $t = 100$ h.

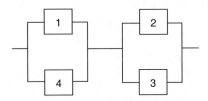

Solution

System performance (unreliability) can be obtained from the following expression:

$$Q = (q_1 \times q_4) + (q_3 \times q_2) - (q_1 \times q_2 \times q_3 \times q_4)$$

since $q_i = 1 - \exp(-\lambda t)$, then

$$Q = \left(1 - e^{-\lambda_1 t}\right)\left(1 - e^{-\lambda_4 t}\right) + \left(1 - e^{-\lambda_3 t}\right)\left(1 - e^{-\lambda_2 t}\right) - \left(1 - e^{-\lambda_1 t}\right)\left(1 - e^{-\lambda_2 t}\right)\left(1 - e^{-\lambda_3 t}\right)$$
$$\times \left(1 - e^{-\lambda_4 t}\right)$$

Note that $\hat{\lambda}_1 = \hat{\lambda}_2 = \hat{\lambda}_3 = \hat{\lambda}_4 = \hat{\lambda} = 5 \times 10^{-3}\,\text{h}^{-1}$. Using (5.4) and neglecting the second term (due to its insignificance):

$$\hat{Q} = \left(1 - e^{-\hat{\lambda}_1 t}\right)\left(1 - e^{-\hat{\lambda}_4 t}\right) + \left(1 - e^{-\hat{\lambda}_3 t}\right)\left(1 - e^{-\hat{\lambda}_2 t}\right) - $$
$$\left(1 - e^{-\hat{\lambda}_1 t}\right)\left(1 - e^{-\hat{\lambda}_2 t}\right)\left(1 - e^{-\hat{\lambda}_3 t}\right)\left(1 - e^{-\hat{\lambda}_4 t}\right)$$
$$\hat{Q} = 2\left(1 - e^{-\hat{\lambda} t}\right)^2 - \left(1 - e^{-\hat{\lambda} t}\right)^4 = 1 - 4e^{-2\hat{\lambda} t} + 4e^{-3\hat{\lambda} t} - e^{-4\hat{\lambda} t}$$
$$\hat{Q}_{100\,\text{h}} = 0.286$$

Calculate the derivatives. For example,

$$\frac{\partial Q}{\partial \lambda_1} = te^{-\lambda_1 t}\left(1 - e^{-\lambda_4 t}\right) - te^{-\lambda_1 t}\left(1 - e^{-\lambda_2 t}\right)\left(1 - e^{-\lambda_3 t}\right)\left(1 - e^{-\lambda_4 t}\right)$$

Repeating for other derivatives of Q with respect to λ_2, λ_3, and λ_4, yields

$$\frac{\partial Q}{\partial \lambda_i} = 4te^{-\lambda_i t}\left[\left(1 - e^{-\lambda t}\right) - \left(1 - e^{-\lambda t}\right)^3\right]$$

and by (5.5)

$$S^2(Q) = 4\,\mathrm{Var}\,(\lambda_i)\left\{te^{-\hat{\lambda}_i t}\left[\left(1 - e^{-\hat{\lambda} t}\right) - \left(1 - e^{-\hat{\lambda} t}\right)^3\right]\right\}^2$$

$$= 4\left(2\hat{\lambda}_i\right)^2\left\{te^{-\hat{\lambda}_i t}\left[\left(1 - e^{-\hat{\lambda} t}\right) - \left(1 - e^{-\hat{\lambda} t}\right)^3\right]\right\}^2$$

Using

$$S\,(\lambda_i) = 2\hat{\lambda} = 2\left(5 \times 10^{-3}\right) = 0.01, \quad \mathrm{Var}\,(\lambda_i) = S^2(\lambda_i) = 10^{-4}, \quad S^2(Q)_{100\,\mathrm{h}} = 4.07 \times 10^{-10}$$

It is now possible to calculate the coefficient of variation for system unreliability as

$$\left.\frac{S(Q)}{\hat{Q}}\right|_{100\,\mathrm{h}} = 7.05 \times 10^{-5}$$

For more detailed consideration of the reliability applications of the method of moments, the reader is referred to Morchland and Weber [12]. Apostolakis and Lee [13] propagate the uncertainty associated with parameters x_i by generating lower order moments, such as the mean and variance for Y, from the lower order moments of the distribution for x_i. A detailed treatment of this method is covered in a comparison study of the uncertainty analysis method by Martz et al. [14]. Atwood et al. [15] is an excellent source for parameter estimation and uncertainty characterization in risk analyses.

Consideration of dependences between x_is is a more complex problem. For a special case where (5.1) is in form of

$$Y = \sum_{i=1}^{n} x_i \tag{5.6}$$

For example, a system comprising of a series of components having the exponential time-to-failure distributions with failure rates x_i, and dependent x_i', the variance of $\hat{Y}$ is given by

$$\mathrm{var}\left[\hat{Y}\right] = \sum_{i=1}^{n} \mathrm{var}\left[x_i\right] + 2\sum_{i=1}^{n-1}\sum_{j=i+1}^{n} \mathrm{cov}\left[x_i, x_j\right] \tag{5.7}$$

In the case where,

$$Y = \prod_{i=1}^{n} x_i \tag{5.8}$$

where x_is are independent (similar to a case of series system composed of components having reliability functions, x_i ($i = 1, 2, \ldots, n$)) estimates of mean and variance of Y are

$$E(Y) = \prod_{i=1}^{n} E(x_i) \tag{5.9}$$

and

$$\mathrm{var}[Y] \approx \left[\sum_{i=1}^{n} \frac{\mathrm{var}(x_i)}{E^2(x_i)} \right] E^2(Y)$$

Dezfuli and Modarres [16] have expanded this approach to efficiently estimate a distribution fit for Y when x_is are highly dependent. The method of moments provides a quick and accurate estimation of lower moments of Y based on the moments of x_i, and the process is simple. However, for highly nonlinear expressions of Y, the use of only low-order moments can lead to significant inaccuracies, and the use of higher moments is complex.

Another classical method is the system reduction (or Maximus) method. The readers are referred to Modarres et al. [17] for a more detailed discussion of this method.

Example 5.3

Consider Example 3.8. Assume vaccination reactions and prevention data in this example are mean values. Calculate uncertainty with the results, assuming lognormal model for representing the associated uncertainty with all input data variables with an error factor (EF) of 10. Only consider the case of General Population and No Vaccination.

Solution:

Since the input events are lognormally distributed, we can calculate the standard deviation of the input variables knowing the mean and the EF from

$$\sigma_i = \mu_i \sqrt{\exp\left(\frac{\ln(\mathrm{EF}_i)}{1.645}\right)^2 - 1}$$

and the coefficient of variation (COV) from

$$\mathrm{COV}_i = \left(\frac{\sigma_i}{\mu_i}\right)$$

By substituting the values of the μ_i for each event in the event tree of Example 3.6, we calculate the uncertainty values. For example, for the Severe Epidemic Season,

$$\sigma_i = 0.1 \sqrt{\exp\left(\frac{\ln 10}{1.645}\right)^2 - 1} = 0.247$$

$$\mathrm{COV}_i^2 = \left(\frac{0.2469}{0.1}\right)^2 = 6.094$$

Using (5.9)

$$\sigma_f^2 \approx \mu_f^2 \left(\sum_{i=1}^{n} \mathrm{COV}_i^2 \right)$$

$$\sigma_R = \sigma_f \text{ (consequence value)}$$

Then for the first scenario, we have

$$\sigma_f = 0.247 \sqrt{6.094 + 6.094} = 0.862$$
$$\sigma_R = 8.62\,(0) = 0$$

The following event tree shows the results for all of the respective scenarios:

Initiating Event	Exposure Potential EP	Type of Flu TF	Fatality Potential FP	Sequence Logic Frequency/year	Risk Value Flu/year
	$\mu_{Ln}=0.6$ $\sigma_{Ln}=1.481$ COV=6.094			$\mu_f=6.0\times10^{-2}$ $\sigma_f=1.3\times10^0$ COV=1.2×10¹	$\mu_R=0$ $\sigma_R=0$
$\mu_{Ln}=0.1$ $\sigma_{Ln}=0.247$ COV=6.094					
		$\mu_{Ln}=0.5$ $\sigma_{Ln}=1.234$ COV=6.094		$\mu_f=2.0\times10^{-2}$ $\sigma_f=1.3\times10^0$ COV=1.8×10¹	$\mu_R=1.0\times10^{-2}$ $\sigma_R=6.4\times10^{-1}$
	$\mu_{Ln}=0.4$ $\sigma_{Ln}=0.987$ COV=6.094				
			$\mu_{Ln}=0.999$ $\sigma_{Ln}=2.466$ COV=6.094	$\mu_f=2.0\times10^{-2}$ $\sigma_f=3.7\times10^0$ COV=2.4×10¹	$\mu_R=2.0\times10^{-2}$ $\sigma_R=3.7\times10^0$
		$\mu_{Ln}=0.5$ $\sigma_{Ln}=1.234$ COV=6.094			
			$\mu_{Ln}=0.001$ $\sigma_{Ln}=0.002$ COV=6.094	$\mu_f=2.0\times10^{-5}$ $\sigma_f=3.7\times10^{-3}$ COV=2.4×10¹	$\mu_R=2.0\times10^{-3}$ $\sigma_R=3.7\times10^{-1}$
					$\mu_R=3.2\times10^{-2}$ $\sigma_R=4.7\times10^0$

Initiating Event	Exposure Potential EP	Type of Flu TF	Fatality Potential FP	Sequence Logic Frequency/year	Risk Value Flu/year
	$\mu_{Ln}=0.85$ $\sigma_{Ln}=2.098$ COV=6.094			$\mu_f=3.4\times10^{-1}$ $\sigma_f=7.2\times10^0$ COV=1.2101	$\mu_R=0$ $\sigma_R=0$
$\mu_{Ln}=0.4$ $\sigma_{Ln}=0.987$ COV=6.094					
		$\mu_{Ln}=0.67$ $\sigma_{Ln}=1.654$ COV=6.094		$\mu_f=4.0\times10^{-2}$ $\sigma_f=2.6\times10^0$ COV=1.8×10¹	$\mu_R=2.0\times10^{-2}$ $\sigma_R=1.29\times10^0$
	$\mu_{Ln}=0.15$ $\sigma_{Ln}=0.370$ COV=6.094				
			$\mu_{Ln}=0.999$ $\sigma_{Ln}=2.466$ COV=6.094	$\mu_f=2.0\times10^{-2}$ $\sigma_f=3.6\times10^0$ COV=2.4×10¹	$\mu_R=1.9\times10^{-2}$ $\sigma_R=3.6\times10^0$
		$\mu_{Ln}=0.33$ $\sigma_{Ln}=0.815$ COV=6.094			
			$\mu_{Ln}=0.001$ $\sigma_{Ln}=0.002$ COV=6.094	$\mu_f=2.0\times10^{-5}$ $\sigma_f=3.6\times10^{-3}$ COV=2.4×10¹	$\mu_R=2.0\times10^{-3}$ $\sigma_R=3.7\times10^{-1}$
					$\mu_R=4.2\times10^{-2}$ $\sigma_R=5.3\times10^0$

Initiating Event	Exposure Potential EP	Type of Flu TF	Fatality Potential FP	Sequence Logic Frequency/year	Risk Value Flu/year
$\mu_{Ln}=0.5$ $\sigma_{Ln}=1.234$ COV $=6.094$	$\mu_{Ln}=0.95$ $\sigma_{Ln}=2.345$ COV $=6.094$			$\mu_f=4.8\text{E}-01$ $\sigma_f=1.0\text{E}+01$ COV $=1.2\times10^1$	$\mu_R=0$ $\sigma_R=0$
	$\mu_{Ln}=0.05$ $\sigma_{Ln}=0.123$ COV $=6.094$	$\mu_{Ln}=0.9$ $\sigma_{Ln}=2.222$ COV $=6.094$		$\mu_f=2.3\times10^{-2}$ $\sigma_f=1.4\times10^0$ COV $=1.8\times10^1$	$\mu_R=1.1\times10^{-2}$ $\sigma_R=7.2\times10^{-1}$
		$\mu_{Ln}=0.1$ $\sigma_{Ln}=0.247$ COV $=6.094$	$\mu_{Ln}=0.999$ $\sigma_{Ln}=2.466$ COV $=6.094$	$\mu_f=2.5\times10^{-3}$ $\sigma_f=4.6\text{E}-01$ COV $=2.4\times10^1$	$\mu_R=2.5\times10^{-3}$ $\sigma_R=4.6\times10^{-1}$
			$\mu_{Ln}=0.001$ $\sigma_{Ln}=0.002$ COV $=6.094$	$\mu_f=2.0\text{E}-05$ $\sigma_f=4.6\text{E}-04$ COV $=2.4\times10^1$	$\mu_R=2.5\times10^{-4}$ $\sigma_R=5.6\times10^{-2}$
					$\mu_R = 1.4\times10^{-2}$ $\sigma_R = 1.2\times10^0$

5.4.2 Probabilistic Methods (Monte Carlo Methods)

In the probabilistic approach to uncertainty propagation, the dependent and independent variables of (5.1) are treated as random variables. Most of the methods to perform the uncertainty propagation are primarily based on some form of Monte Carlo analysis.

In the Monte Carlo simulation, a value is selected (sampled) from pdf of each x_i element in (5.1). This is done using statistical sampling techniques. Each set of data values is used to find one estimate of Y in (5.1). This is repeated a large number of times, such as 10,000 times or more, to obtain many instances of estimate of Y distribution. The distribution of many instances of Y values is then used to obtain the distribution characteristics and a distribution histogram. Figure 5.1 illustrates the Monte Carlo simulation.

Finally, in sampling, the user has the choice of assuming x_i element distributions to be independent or to be coupled in the sampling. In independent sampling, a value is independently sampled from each probability distribution of x_i elements. In coupled sampling, a value sampled for x_i element may be used for multiple elements, or a value sampled for x_i element may impose bias on sampling another element x_j. Generally, the distributions may be coupled if the same data are used to build the distribution of x_i elements, or there are known stochastic dependencies between the elements x_i not explicitly modeled in (5.1). This is the case, for example, if the same generic motor-operated valve failure rate is used for several motor-operated valves in a fault tree in the risk assessment model.

There are four Monte Carlo type techniques for uncertainty propagation and confidence estimation: classical Monte Carlo simulation, Bayes' Monte Carlo method, bootstrap method, and Wilk's tolerance limits.

5.4.2.1 Classical Monte Carlo Simulation

The classical Monte Carlo method is based on classical probabilistic models for characterizing uncertainties of x_i elements which are built based on observed data only. In other words, each element x_i of the system analyzed is associated with a distribution, developed based on real failure data. If we knew exact values of x_i element uncertainty characteristics, we would be able, in principle, to calculate the uncertainty characteristics of Y. Instead of exact uncertainty characteristics of x_i, we often deal with subjective estimates which treat x_i values as random variables. Thus, if there are no data or small data, we have to treat Y as a random variable.

In the framework of the classical Monte Carlo approach, there could be different algorithms for finding uncertainties of Y in (5.1). The following example illustrates the general steps for constructing the confidence limit for Y using this method. These steps are:

1) For each x_i element in (5.1), select an observed data point or a point estimate (e.g., from the maximum likelihood estimate).
2) Calculate the corresponding classical estimate of Y in (5.1).
3) Repeat steps 1–2 for sufficiently large number of times, n (for example, 10,000) to get a large sample of $\hat{Y}$.
4) Using the sample obtained in step 3, and choosing a confidence level $(1-\alpha)$, construct the respective confidence limit for Y.

5.4.2.2 Bayes' Monte Carlo Simulation

The principal and the only difference between the classical Monte Carlo approach and the Bayesian is related to x_i value estimation. Under the Bayes approach, we need to provide prior information for x_i and respective prior distributions for all x_i variables. Then we need to get the corresponding posterior distributions (if any evidence for x_i values is exists). Having these distributions, the same steps as those in the simple random sampling Monte Carlo approach are used.

In simple random sampling, each x_i element is treated as a random variable and characterized by a corresponding pdf. Samples from different regions of the sample space occur in direct relationship to the probability of occurrence of these regions. Each sample is taken independent of other sample elements. Each time, a random value is sampled from a uniform distribution with ranges [0, 1] and used to obtain the corresponding x_i values from the cumulative distribution of x_i.

In simple random sampling, there is no assurance that all points will be sampled. This problem can be addressed by using large number of samples, or other methods of sampling as importance and Latin hypercube sampling (LHS). Importance sampling (IS) operates to ensure the coverage of specified regions in the sample space, but LHS operates to ensure the coverage of the whole regions.

In the absence of prior information about x_i, and if x_is are probabilities of events modeled in a risk assessment, then Martz and Duran [18] recommend using the beta distribution having parameters $\alpha = \beta = 0.5$, as an appropriate prior distribution, which represents a noninformative prior. Note that such noninformative prior should have the mean 0.5 and the coefficient of variation, which is very close to the coefficient of variation of the standard uniform distribution (0, 1). Also, the standard uniform distribution is a particular case of the beta distribution with parameters $\alpha = \beta = 1$.

In certain cases where (5.1) may be a very complex function and requires large amount of computational effort for its evaluation, it is impractical to sample many unbiased samples (say, 10,000 simulation runs of (5.1)). In addition, improvement over the quality of uncertainty and sensitivity analysis results relative to those that could be obtained from simple random sampling of the same size may be needed. To remedy this shortcoming, a number of possible sampling procedures exist, such as the stratified (weighted) sampling and the LHS to reduce the number of samplings necessary, while assuring that the totality of the x_i distributions is sampled.

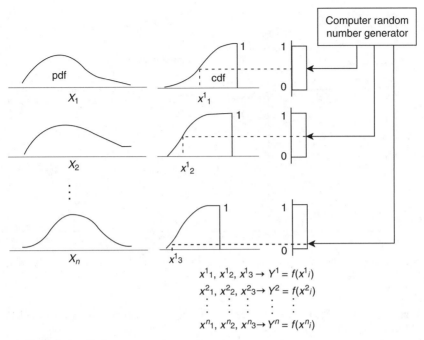

$$x^1_1, x^1_2, x^1_3 \rightarrow Y^1 = f(x^1_i)$$
$$x^2_1, x^2_2, x^2_3 \rightarrow Y^2 = f(x^2_i)$$
$$\vdots \quad \vdots \quad \vdots \qquad \vdots$$
$$x^n_1, x^n_2, x^n_3 \rightarrow Y^n = f(x^n_i)$$

FIGURE 5.1 The Monte Carlo computer procedure.

5.4.2.2.1 Latin hypercube sampling

In the LHS method, the range of probable values for each uncertain x_i model input element is divided into m segments of equal probability. Thus, the whole element x_i space, consisting of n elements, is partitioned into m^n cells, each having equal probability. For example, for the case of four elements and five segments, the x_i space is divided into 5^4 or 625 cells. The next step is to choose m cells from the m^n cells. First, a random sample is generated, and its cell number is calculated. The cell number indicates the segment number that the sample belongs to, with respect to each of the elements. For example, a cell number (2, 5, 2, 3, 1) indicates that the sample lies in the segment 2 with respect to x_1, segment 5 with respect to x_2, segment 2 with respect to x_3, segment 3 with respect to x_4, and segment 1 with respect to x_5. For this sample the corresponding Y is calculated by sampling each x_i from its corresponding m subranges in the sample. At each successive step, a random sample of cells is generated, and is accepted only if it does not agree with any previous sample already evaluated. Typically, 200–400 samples would be sufficient to evaluate uncertainties of Y in (5.1). The advantage of this approach is that the random samples are generated from all the ranges of possible values, thus giving insight into the tails of the probability distributions. The LHS is particularly effective in model verification due to the dense stratification across the range of each sampled variables.

Many large analyses involve a separation of epistemic and aleatory uncertainty. In such analyses, a frequent approach is to use LHS to propagate the effects of epistemic uncertainty, and random or stratified sampling to propagate the effects of aleatory uncertainty. With this approach, the effect of aleatory uncertainty is being calculated conditional on individual cells. Typical analysis outcomes are the so-called Farmer curves, which are complementary cumulative distribution of losses (risk), with each individual distribution arising from aleatory uncertainty and the distributions of complementary cumulative distribution of losses arising

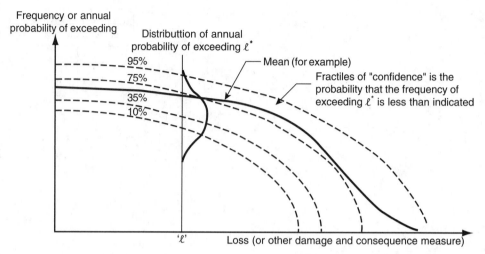

FIGURE 5.2 Display of the effect of epistemic uncertainties on the risk by a family of risk curves. (*Source*: Paté-Cornell [21].)

from epistemic uncertainty. Figure 5.2 depicts this concept. More discussions on LHS may be found in Helton and Davis [19].

5.4.2.2.2 Importance Sampling

In importance sampling IS, the sampling space is divided into a number of nonoverlapping subregions called strata. The number of samples from each stratum depends on the importance of the subregion, but a simple random sampling is performed for each stratum. IS is used to ensure the inclusion of regions with low probability and high consequences. When IS is used, the probability assigned and number of samples from each stratum must be folded back into the analysis before results can be meaningfully presented [19].

It is important to note that the concern over computational cost of carrying out a Monte Carlo analysis may not always justify using alternative sampling techniques such as LHS. In most analyses, the human cost of developing the model, performing the analysis, and documenting and defending the analysis may far exceed the computational cost of the traditional Mote Carlo iterative approach.

5.4.2.2.3 Discrete Probability Distributions

DPD is a probabilistic sampling and propagation technique in which the input distributions are discretized, and a discrete representation of the output distribution is calculated. First, a distribution is determined for each of the input variables (similar to Bayes' Monte Carlo method), then these distributions are divided into discrete intervals. The number of discrete intervals can be different for each input distribution. A value for each interval is chosen subjectively, and the probability that the input variable probability occurs in each interval is calculated. Thus, the distribution of each variable is discretized into m intervals; each interval has a corresponding probability and a corresponding central value. The output variable Y is evaluated $m \times k$ times, where k is the number of independent variables (assuming that pdf of each input variable x_i is divided into m discrete intervals). The result of each evaluation has an associated probability equal to the product of the probabilities of the independent variables. Thus, a DPD can be constructed from these intervals. The DPD technique is a valid method, but becomes quickly impractical for very large problems with many variables. It is sometimes used in limited studies of selected issues, but not in more general risk studies. For more discussions on this topic, see Kaplan [22].

Example 5.5
The top event of a fault tree is represented by the following expression:

$$T = C_1 C_2$$

Assuming that events C_1 and C_2 are normally distributed with parameters $\mu = 0.525$ and $\sigma = 0.183$, and $\mu = 0.600$ and $\sigma = 0.086$, respectively. Calculate the top event pdf using the DPD approach.

Solution
According to the DPD method, we select the discrete intervals for each variable, and calculate the probability for each interval p_i, and midpoint of the interval x_i

$$\{\langle p_1, x_1 \rangle, \langle p_2, x_2 \rangle, \ldots, \langle p_n, x_n \rangle\} = \{\langle p_i, x_i \rangle\} \quad \sum_{i=1}^{n} p_i = 1.0$$

Consider the following subjective discrete intervals:

$$C1 : \{\langle 0.12, 0.30 \rangle \langle 0.34, 0.50 \rangle \langle 0.38, 0.70 \rangle \langle 0.16, 0.90 \rangle\}$$
$$C2 : \{\langle 0.12, 0.50 \rangle \langle 0.38, 0.60 \rangle \langle 0.50, 0.70 \rangle\}$$

Multiplication of DPDs is

$$\{\langle p_i, x_i \rangle\}\{\langle q_j, y_j \rangle\} = \{\langle p_1, q_j, x_i y_j \rangle\} \quad \text{where} \sum_{i=j} p_i q_j = 1.0$$

Then, the top event is

$$T = \{\langle 0.01, 0.15 \rangle \langle 0.04, 0.18 \rangle \langle 0.05, 0.21 \rangle \langle 0.04, 0.25 \rangle \langle 0.13, 0.30 \rangle \langle 0.05, 0.35 \rangle$$
$$\langle 0.17, 0.35 \rangle \langle 0.15, 0.42 \rangle \langle 0.02, 0.45 \rangle \langle 0.19, 0.49 \rangle \langle 0.06, 0.54 \rangle \langle 0.09, 0.63 \rangle\}$$

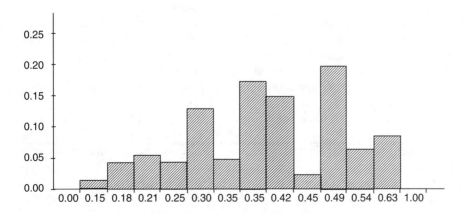

5.4.2.3 The Bootstrap Simulation Method

The bootstrap method introduced by Efron and discussed by Efron and Tibshirani [23] is a Monte Carlo simulation technique in which new samples are generated from the data of an original sample. The method's name, derived from the old saying about pulling yourself up by your own bootstraps, reflects the fact that one available sample gives rise to many others.

Unlike the classical and Bayes' Monte Carlo techniques, the bootstrap method is a universal nonparametric method. To illustrate the basic idea of this method, consider the following simple example, in which the standard error of a median is estimated (Efron and Tibshirani [23]). Consider an original m samples $X_i = x_{i1}, x_{i2}, \ldots, x_{im}$, obtained (e.g., by experiment) from an unknown distribution of x_i. The respective bootstrap sample, x_{i1b}, $x_{i2b}, \ldots$, and $x_{imb} = X_i^b$, is obtained by randomly sampling m times with replacement from the original sample $x_{i1}, x_{i2}, \ldots, x_{im}$. Accordingly, the bootstrap procedure consists of the following steps:

1) Generating a large number, N, of bootstrap samples $X_i^b (i = 1, 2, \ldots, N)$ from all elements x_i (i.e., n_{xN} bootstrap samples),
2) for each element x_i the sample median, $x_{i0.5}$ (from X_i^b) is evaluated and called the bootstrap replication of element x_i.
3) The bootstrap estimate of standard error of the median of interest is calculated as

$$\hat{S}_{x_{0.5}} = \sqrt{\frac{\sum\limits_{i=1}^{N} \left[x_{0.5}\left(X_i^b\right) - \hat{x}_{0.5} \right]^2}{N - 1}} \qquad (5.10)$$

where

$$\hat{x}_{0.5} = \sum\limits_{i=1}^{N} \frac{x_{0.5}\left(X_i^b\right)}{N}$$

Note that no assumption about the distribution of x was introduced.

For some estimation problems, the results obtained using the bootstrap approach coincide with respective known classical ones. This can be illustrated by the following example related to binomial data (Martz and Duran [18]).

Assume that for each element x_i of the system of interest, we have the data collected in the form $\{S_i, N_i\}$, $i = 1, 2, \ldots, n$, as in (5.1), where N_i is the number of units of ith element tested (or observed) during a fixed time interval (the same for all x_i in (5.1)). For illustration of this problem, suppose x_i represents performance of a (barrier) in a risk model and S_i is the respective number of the barriers survived in a reliability test.

The basic steps of the corresponding bootstrap simulation procedure are as follows:

1. For each barrier, obtain the bootstrap estimate of barrier performance. For example reliability, R_i $(i = 1, 2, \ldots, n)$, where n is the number of elements or barriers in the system, generating it from the binomial distribution with parameters N_i and $p = S_i/N_i$. In the case when $S_i = N_i$, i.e., $p = 1$, one needs to smooth the bootstrap, replacing p by $(1 - \varepsilon)$, where $\varepsilon \ll 1$. This procedure is discussed in Efron and Tibshirani [23].
2. Calculate the corresponding classical estimate of the facility or system reliability using (5.1) with R_i $(i = 1, 2, \ldots, n)$ obtained from the results of step 1.
3. Repeat steps 1–2 a sufficiently large number of times, n (for example, 10,000) to get a large sample of R_i.
4. Based on the sample obtained and a chosen confidence level $(1 - \alpha)$, construct the respective lower confidence limit for the system reliability of interest as a sample percentile of level α.

Example 5.6

Consider a fault tree, the top event, T, of which is described by the following expression:

$$T = C_1 + C_2 C_3$$

where C_1, C_2, and C_3 are the cut sets of the system modeled by the fault tree. If the following data are reported for the components representing the respective cut sets, determine a point estimate and 95% confidence interval for the system reliability $\hat{R}_S = 1 - \Pr(T)$ using the bootstrap method

Component	Number of Failures (d)	Number of Trials (N)
C_1	1	1785
C_2	8	492
C_3	4	371

Solution:

The bootstrap estimation can be obtained as follows:

1) Using the failure data for each component, compute the estimate of the binomial probability of failure and treat it as a nonrandom parameter p.
2) Simulate N binomial trials of a component and count the observed number of failures.
3) Obtain a bootstrap replication of p dividing the observed number of failures by number of trials. Once the bootstrap replications are computed for each component, find the estimate of system reliability.
4) Repeat steps 2 and 3 sufficiently large number of times, and use (5.1) to obtain the interval estimates of T.

The procedure and results of the bootstrap solution are summarized in Table 5.4. From the distribution of the system reliability estimates, the 95% confidence bounds can be obtained as the 2.5% and 97.5% sample percentiles.

The system reliability confidence bounds can also be estimated through the use of Clopper–Pearson procedure. The fictitious number of system trials is

$$N_S = \frac{\hat{R}_S \left(1 - \hat{R}_S\right)}{\text{Var}\left(\hat{R}_S\right)} = \frac{0.99926[1 - 0.99926]}{3.45 \times 10^{-7}} = 2143.3$$

Then, the fictitious number of system failures is

$$D_S = N_S(1 - \hat{R}_S) = 2143.3(1 - 0.99926) = 1.6$$

Using the lower confidence limit, $F_l(t)$, at the point t where $S_n(t) = r/n$ ($r = 0, 1, 2, \ldots, n$) is the largest value of p that satisfies $I_p(r, n - r + 1) \le \alpha/2$ and the upper confidence limit, $F_u(t)$, at the same point is the smallest p satisfying the inequality $I_{1-p}(n - r, r + 1) \le \alpha/2$ where $I_t(\alpha, \beta)$ is the incomplete beta function with $n = 2155.5$ and $r = 1.6$, the 95% lower confidence limits for the overall reliability and probability of occurrence of top event T are estimated as

TABLE 5.4
The Bootstrap Solution in Example 5.4

Monte Carlo Run Number	Component	C_1	C_2	C_3	Estimate of System Reliability, R_S
	Number of failures, d	1	8	4	
	Number of trials, N	1785	492	371	
	Binomial probability of failure, $p = d/N$	0.00056	0.01626	0.01078	
1	Observed number of failures in N binomial trials with parameter p, d_i	0	7	4	0.99985
	Bootsrap replication, $p_i^b = d_i/N$	0.00000	0.01423	0.01078	
2	Observed number of failures in N binomial trials with parameter p, d_i	0	7	6	0.99977
	Bootsrap replication, $p_i^b = d_i/N$	0.00000	0.01423	0.01617	
3	Observed number of failures in N binomial trials with parameter p, d_i	1	9	4	0.99924
	Bootsrap replication, $p_i^b = d_i/N$	0.00056	0.01829	0.01078	
$\vdots$	$\vdots$	$\vdots$	$\vdots$	$\vdots$	$\vdots$
10,000	Observed number of failures in N binomial trials with parameter p, d_i	2	10	5	0.99861
	Bootsrap replication, $p_i^b = d_i/N$	0.00112	0.02033	0.01348	
				$E(R_S)$	**0.99926**
				$Var(R_S)$	3.4500×10^{-7}

$$0.99899 \leq \hat{R}_S \leq 0.99944$$
$$0.00056 \leq \hat{T}_S \leq 0.00101$$

Martz and Duran [18] performed some numerical comparisons of bootstrap and Bayes' Monte Carlo methods applied to 20 simple and moderately complex system configurations and simulated binomial data for the system components. They made the following conclusions about the regions of superior performance of the methods:

1) The bootstrap method is recommended for highly reliable and redundant systems.
2) The Bayes' Monte Carlo method is, generally, superior for: (a) moderate to large series systems of reliable components with moderate to large samples of test data, and (b) small series systems, composed of reliable nonrepeated components.

5.4.2.4 Wilks Tolerance Limit

As facilities and systems and their models become more complex and costly to run, the use of tolerance limit uncertainly characterization is gaining popularity. For example, in very complex models containing several uncertain parameters (each represented by a probability or probability density), classical Bayes' and bootstrap Monte Carlo simulation may become impractical. Often in complex computer-based models of (5.1) in which calculation of values requires significant amount of time and effort, the traditional Monte Carlo simulation is not possible. Wilks tolerance limit is used in these cases.

A tolerance interval is a *random* interval (L, U) that contains with probability (or confidence) β at least a fraction γ of the population under study. The probability and fraction β and γ are analyst's selected criteria depending on the confidence desired. The pioneering work in this area is attributed to Wilks [24, 25] and later to Wald [26, 27]. Wilks tolerance limit is an efficient and simple sampling method to reduce sample size from few thousands to around 100 or so. The number of sample size does not depend on the number of uncertain parameters in the model. There are two kinds of tolerance limits:

Nonparametric tolerance limits: Nothing is known about the distribution of the random variable except that it is continuous.
Parametric tolerance limits: The distribution function representing the random variable of interest is known and only some distribution parameters involved are unknown.

The problem in both cases is to calculate a tolerance range (L, U) for a random variable X represented by the observed sample, $x_1, \dots, x_m$, and the corresponding size of the sample.

Consider γ *tolerance limits* L and U for *probability level* β of a limited sample S_1 of size N, the probability β that at least γ proportion of the Xs in another indefinitely large sample S_2 will lie between L and U is obtained from [23–26]

$$p\left(\int_L^U f(x)\,dx \geq \gamma\right) = \beta \qquad (5.11)$$

where $f(x)$ is the probability density function of the random variable X.

Let us consider a complex system's risk model represented in the form of (5.1). Such a model may describe relationship between the output variables (e.g., risk, probability of failure or performance value of a system) as a function of some input variables (e.g., geometry, material properties, etc.). Assume several parametric variables involved in the model. Further, assume that the observed randomness of the output variables is the result of the randomness of the input variables. If we take N samples of each input variable, then we obtain a sample of N output values $\{y_1, \dots, y_N\}$. Note that probability β bears the name confidence level. To be on the conservative side, one should also specify the probability content γ in addition to the confidence level β as large as possible. It should be emphasized that γ is not a probability, although is a nonnegative real number less than 1 [27]. Having fixed β and γ, it becomes possible to determine the number of runs (samples of output) N required to remain consistent with the selected β and γ values.

Let $y_1, \dots, y_N$ be N independent output values of y. Suppose that nothing is known about the pdf of Y except that it is continuous. Arrange the values of $y_1, \dots, y_N$ in an increasing order and denote them by $y(k)$, hence

$$y(1) = \min_{1 \leq k \leq N} y_k, \quad y(N) = \max_{1 \leq k \leq N} y_k$$

and by definition $y(0) = -\infty$, while $y(N+1) = +\infty$. It can be shown [27] that for confidence level β is obtained from

$$\beta = \sum_{j=0}^{k-r-1} \binom{N}{j} \gamma^j (1-\gamma)^{N-1}, \quad 0 \leq r \leq k \leq N, \quad L = y(r), \quad U = y(s) \qquad (5.12)$$

From (5.12) sample sizes N can be estimated. For application of this approach, consider two cases of the tolerance limits: one-sided and two-sided.

TABLE 5.5
Minimum Sample Size (One-Sided)

γ	β		
	0.90	0.95	0.99
0.90	22	45	239
0.95	29	59	299
0.99	44	90	459

One-sided tolerance limits: This is the more common case, for example, measuring tolerance limits of a model output such as a temperature or shear stress at a point on the surface of a structure. In this case, we are interested in assuring that a small sample, for example, estimated temperatures, obtained from the model, and the corresponding upper sample tolerance limit T_U according to (5.12) contains with probability β (say 95%) at least the fraction γ of the temperatures in a fictitious sample containing infinite estimates of such temperatures. Table 5.5 shows values for sample size N based on the values of β and γ. For example, if $\beta = 0.95$, $\gamma = 0.90$; then $N = 45$ samples taken from the model (e.g., by standard Monte Carlo sampling) assure that the highest temperature T_H in this sample represents the 95% upper confidence limit below which 90% of all the possible temperatures lie.

Two-sided tolerance limits: We now consider the two-sided case, which is less common [28]. Table 5.6 shows the Wilks sample size. With γ and β both equal to 95%, we will get $N = 93$ samples. For example, in the 93 samples taken from the model (e.g., by using standard Monte Carlo sampling) we can say that limits $(T_L \ T_H)$ from this sample represent the 95% confidence interval within which 95% of all the possible temperatures lie.

Example 5.6
A manufacturer of steel bars wants to order boxes for shipping their bars. They want to order appropriate length for the boxes, with 90% confidence that at least 95% of the bars do not exceed the box's length. How many samples, N, the manufacturer should select and which one should be used as the measure of the box length?

Solution
From Table 5.5, with $\gamma = 95\%$ and $\beta = 90\%$, the value for N is 29. The manufacturer should order box's length as the x_{29} sampled bar (when samples are ordered).

TABLE 5.6
Minimum Sample Size (Two-Sided)

γ	β			
	0.50	0.90	0.95	0.99
0.50	3	17	34	163
0.80	5	29	59	299
0.90	7	38	77	388
0.95	8	46	93	473
0.99	11	64	130	663

TABLE 5.7
Distribution Characteristics of Some Input Parameters that Influence Pressurized Thermal Shock

Flaw Characterizations	Input Distributions	Lower and Upper Ranges
Flaw size	Uniform	[0, 1.5] in.
C_Dist	Uniform	[0, 3] in.
Aspect ratio	Discrete	2, 6, 10
Temperature	Normal	[M = 382, σ = 192]°F
Heat transfer coefficient	Normal	[M = 0.58, σ = 0.55]
Pressure	Normal	[M = 1176, σ = 947] psi

To compare Wilks tolerance limit with Bayes' Monte Carlo, consider a complex mathematical-based routine [29] (called MD-fracture) used to calculate the probability of a nuclear reactor pressure vessel fracture due to pressurized thermal shock (PTS). Certain transient scenarios can cause a rapid cooling inside the reactor vessel while it is pressurized. For example, a 2.828-in. surge line break in a certain design of nuclear plants may lead to such a condition. Many input variables contribute to the amount of thermal stress and fracture toughness of the vessel. Some of them may involve uncertainties. The temperature, pressure, and heat transfer coefficient are examples of such variables, represented by normal distributions. Also, flaw size, the distance from the flaw inner tip to the interface of reactor vessel (C_Dist), and aspect ratio are unknown and can be represented by random variables with the distributions shown in Table 5.7. To compare the results of vessel fracture due to this scenario using Wilks approach with $\gamma = 95\%$ and $\beta = 95\%$ with the results of the standard 1000 and 2000 trials standard Monte Carlo simulation, three Wilks tolerance limits method runs with 100 samples (assuming $\gamma = 95\%$ and $\beta = 95\%$ with two-sided case as shown in Table 5.6) and two Monte Carlo runs with 1000 and 2000 are performed using the MD-Fracture tool. Results comparing Wilks tolerance limits and simple Monte Carlo sampling are shown in Figure 5.3.

Example 5.7
In Example 5.2, a simple hurricane wind speed model was discussed. Perform an uncertainty analysis based on Wilks tolerance limits and two-sided with $\gamma = 95\%$ and $\beta = 95\%$.

Solution
Same range and distribution is assumed for parameters and with two-sided (95%, 95%) criteria, we design simple sampling with $N = 100$. Results of propagation, and (95%, 95%) tolerance limits are shown in Figure 5.4. A normal distribution is fitted to data for more illustration.

5.4.3 THE RESPONSE SURFACE METHOD

When evaluation of (5.1) is very complex such that many samples in a standard Monte Carlo simulation or even in LHS approach require prohibitive amount of computation time and other resources, then response surface approach is used. Response surface approaches in risk assessment involve developing an approximation to the risk (e.g., the PRA) model. This approximation is used as a surrogate to (5.1) in subsequent uncertainty propagation and sensitivity analyses. In order to develop a response surface, an experimental design process is used to select sets of input parameters for use in the quantification process. Many different experimental design methods are available. It is not necessary that

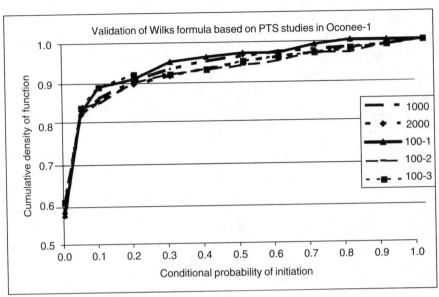

FIGURE 5.3 Comparison between standard Monte Carlo simulation and Wilks tolerance limit method.

a probabilistic approach be used to select the sets of input parameters, although this is often done. Next, the risk model is quantified for each set of selected input parameters. Finally, a response surface approximation is fitted to the results. Often, a least squares technique is used to construct this approximation. Once a response surface approximation has been generated, it can be manipulated in various ways in uncertainty and sensitivity studies. The shape and properties (such as mean and variance) of the output distribution are readily estimated.

The response surface method is similar to the Monte Carlo analysis except that an experimental design is used to select the model input. A variety of possible response surface methods exist, including factorial, central composite, and Placket–Burman. Usually, the method selected depends on many factors, including properties of the model and the type and accuracy of results desired from subsequent uncertainty and sensitivity analysis.

The factorial approach and the central composite are the most common approaches to response surface modeling and we will only discuss these two approaches. Consider (5.1) and let y be the output response variable and x_i the input variables of interest. The general response surface corresponding to (5.1), or response function, can be defined as the Taylor expansion

$$y = a_0 + \sum_i a_i x_i + \sum_{i,j} a_{ij} x_i x_j + \sum_{i,j,k} a_{ijk} x_i x_j x_k + \cdots \tag{5.13}$$

where the first term on the right-hand side of (5.13) is a constant, the second term is the sum of all linear contributions from the independent variables in (5.1), the third term is the sum of all quadratic combinations of the independent variables, the fourth term is the sum of all third-order combinations and so on. The objective of the response surface is to determine constant coefficients a_0, a_i, a_{ij}, a_{ijk}, ... They must be selected so as to adequately capture the behavior of y in (5.1). The basic premise of the response surface is that if enough terms in (5.13) are included then any arbitrary behavior due to changes in the input elements x in (5.1) can be estimated. Equation (5.1) can also be recognized as the Taylor

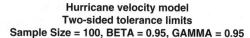
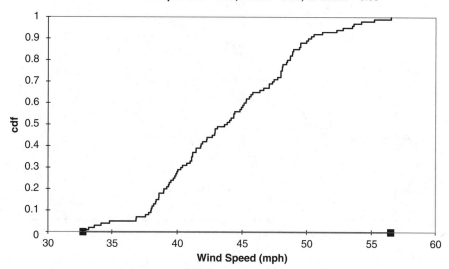

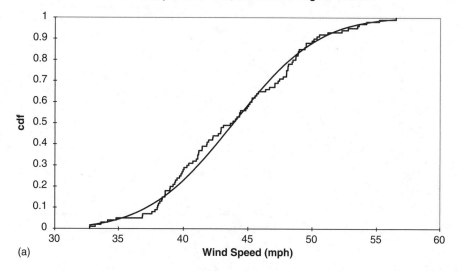

(a)

FIGURE 5.4 Uncertainty distribution for (a) Hurricane wind velocity,

(*continued*)

expansion of y. To control complexity, in practice, only a few first terms on the right-hand side of (5.13) are usually included. For example, for a first-order approximation to y, only the first two terms are retained, and for a second-order approximation, only the first three terms are retained.

Often it is convenient to fit the differences in values of x_i, with the constant term a_0 no longer part of the response equation:

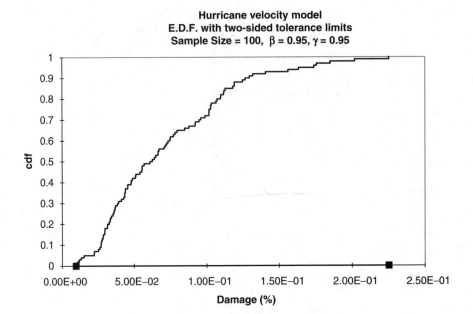

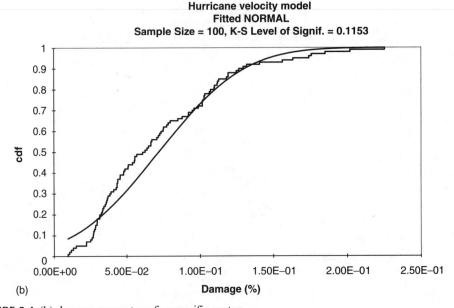

(b)

FIGURE 5.4 (b) damage percentage for specific vertex.

$$\Delta y = \sum_i a_i \Delta x_i + \sum_{i,j} a_{ij} \Delta x_i \Delta x_j + \sum_{i,j,k} a_{ijk} \Delta x_i \Delta x_j \Delta x_k + \cdots \qquad (5.14)$$

where Δy is a difference from a base value due to Δx_is that are also differences from their base values. The base value of y corresponds to the base values of x_i, so that when all $\Delta x_i = 0$, then $\Delta y = 0$.

Sometimes when the values can change by several factors or even by orders of magnitude, the response surface can also be used to fit the logarithm of the output as a function of the logs of the inputs. In this case, in (5.13) or (5.14), $\log y$ or $\log \Delta y$ is shown as a function of

$\log x_i$ or $\log \Delta x_i$, respectively. In (5.14), Δy is then the difference in $\log y$ and Δx_i are the differences in $\log x_i$.

The response surface parameters a_0, a_i, a_{ij}, a_{ijk} can be estimated in different ways. One way is to perform sensitivity studies about a nominal value and use (5.14). For this approach, assume that x_i or Δx_i is changed one at a time and the resulting change in the response variable is determined. Then assume that two input variables are simultaneously changed and the resulting change is determined. The number of input variables simultaneously changed is increased for the orders of interactions to be considered in the response surface.

The parameters in (5.14) can be determined by changing x_i from nominal to an upper bound value x_i^U while all other input variables are kept at their base values. (The superscript U denotes an upper bound value.) Here "upper bound" simply means a higher value. Let y_i^U be the resulting new value from (5.1) for the output response variable corresponding to the change in x_i. Further let

$$\Delta x_i^U = x_i^U - x_i^b$$
$$\Delta y_i^U = y_i^U - y_i^b$$

The quantities Δx_i^U and Δy_i^U are simply the differences in the x_i values and output variable from their base (nominal) values. Then by using (5.14),

$$a_i = \frac{\Delta y_i^U}{\Delta x_i^U} \tag{5.15}$$

Consequently, by changing the variables one at a time from their nominal value to an upper bound, all the parameters may be estimated for the linear terms in the first summation in (5.14) using (5.15). Now assume two input variables x_i and x_j are simultaneously changed to upper bound values with all other input variables kept at their nominal values. Then from (5.14)

$$\Delta y_{ij}^U = a_i \Delta x_i^U + a_j \Delta x_j^U + a_{ij} \Delta x_i^U \Delta x_j^U \tag{5.16}$$

As a_i and a_j have already been estimated, then (5.16) can be used to estimate a_{ij}:

$$a_{ij} = \frac{\Delta y_{ij}^U - a_i \Delta x_i^U - a_j \Delta x_j^U}{\Delta x_i^U \Delta x_j^U} \tag{5.17}$$

The third-order parameters a_{ijk} are estimated similarly by simultaneously changing x_i, x_j, and x_k to upper bound values while keeping the remaining x terms at their nominal values. Therefore,

$$a_{ijk} = \frac{\Delta y_{ijk}^U - a_i \Delta x_i^U - a_j \Delta x_j^U - a_k \Delta x_k^U - a_{ij} \Delta x_i^U \Delta x_j^U - a_{ik} \Delta x_i^U \Delta x_k^U - a_{jk} \Delta x_j^U \Delta x_k^U}{\Delta x_i^U \Delta x_j^U \Delta x_k^U} \tag{5.18}$$

The remaining parameters in (5.14) can be determined in a similar manner. The estimated parameter constants can be further refined by carrying out more sensitivity studies by changing higher numbers of input variables, resulting in adding higher order terms and more refinement to the response surface.

The above equations determine the parameters of the response surface when the input variables are changed to upper bound (i.e., higher values). The same equations apply when the input variables are changed to lower bound values (simply changing the superscript "U" to "L" to denote lower bound). Equations (5.15)–(5.18) apply for the input variables changed to any different value (subjectively selected by the analyst) from their nominal value.

One of the most direct approaches for using the results of both sensitivity studies is to use the total changes to determine the parameters. The values for the parameters in this case are

$$a_i = \frac{\Delta y_i^{\mathrm{U}} + \Delta y_i^{\mathrm{L}}}{\Delta x_i^{\mathrm{U}} + \Delta x_i^{\mathrm{L}}} \tag{5.19}$$

$$a_{ij} = \frac{\hat{\Delta} y_{ij}^{\mathrm{U}} + \hat{\Delta} y_{ij}^{\mathrm{L}}}{\Delta x_i^{\mathrm{U}} \Delta x_j^{\mathrm{U}} + \Delta x_i^{\mathrm{L}} \Delta x_j^{\mathrm{L}}} \tag{5.20}$$

where

$$\hat{\Delta} y_{ij}^{\mathrm{U}} = \Delta y_{ij}^{\mathrm{U}} - a_i^{\mathrm{U}} \Delta x_i^{\mathrm{U}} - a_j^{\mathrm{U}} \Delta x_j^{\mathrm{U}}$$
$$\hat{\Delta} y_{ij}^{\mathrm{L}} = \Delta y_{ij}^{\mathrm{L}} - a_i^{\mathrm{L}} \Delta x_i^{\mathrm{L}} - a_j^{\mathrm{L}} \Delta x_j^{\mathrm{L}}$$

where a_i^{U} is the linear parameter value estimated by using the upper bound sensitivity study from (5.15). The parameter a_i^{L} is the corresponding linear parameter value estimated from the lower bound sensitivity calculation using (5.15) (but with changing the superscript U to L). Similar equations can be obtained for the values for the higher order parameters. Similar equations can also be obtained if combinations of upper and lower bound values are used for the sensitivity studies, instead of using all upper or all lower bound values.

The response surface can be used to specifically estimate the variance of Y from the variances of the input elements (variables). The variance can then be used to describe the uncertainties associated with the output variable y. The response surface can also be used in Monte Carlo simulation discussed in the previous section. The response surface can be used to get a quick, efficient estimate of the output response value for a given set of input values. This saves the time and effort in having to calculate the complex (5.1) for many random set of input values. When the input uncertainties are only roughly known, then the approximation made in using a response surface can provide acceptable results.

The use of the response surface to estimate the variance of the output response is the most basic use of this method in uncertainty evaluations. The determination of the variance does not require any probability distribution assumptions. It only requires estimates of the variances of the input variables, and any dependencies (correlations) considered. The variances for the input variables can be estimated from the two-sigma bounds (or other percentile bounds) on the input variables. Successively more accurate determinations of the variance of the response can be obtained by considering successively higher order interaction terms.

The variance of the response, which uses the variances and possible correlations of the inputs, is estimated from the first-order expression for the response. From (5.14) the variance $\mathrm{Var}(y)$ for y according to (5.7) is estimated from

$$\text{Var}(y) \cong \sum_i a_i^2 \, \text{Var}(x_i) + \sum_{i \neq j} a_i \, a_j \, \text{COV}(x_i, x_j) \tag{5.21}$$

where $\text{Var}(x_i)$ is the variance of x_i. $\text{COV}(x_i, x_j)$ is the covariance between x_i and x_j and is defined in terms of the linear correlation coefficient $\rho(x_i, x_j)$ between x_i and x_j as

$$\text{COV}(x_i, x_j) = \sqrt{\text{Var}(x_i)} \, \sqrt{\text{Var}(x_j)} \, \rho(x_i, x_j) \tag{5.22}$$

If the base value of the response is an approximate midpoint value then approximate upper and lower two-sigma bounds are obtained by adding and subtracting two times the standard deviation to the nominal value. One-sigma bounds are obtained by adding and subtracting the standard deviation to the nominal value. If the evaluations are done on a log scale then the bounds are transformed back to the original scale. Adding and subtracting two-sigma (or one-sigma) to the nominal value on a log scale and then transforming back is equivalent to multiplying and dividing the nominal value by the corresponding EF.

Example 5.8

A hazard barrier's performance in mitigating hazard exposure in a facility is affected by two variables of the process: temperature and reaction time. The nominal mitigating performance (e.g., percent of toxic chemical removal) is estimated as 65.3% at a reaction time of 50 min and temperature of 160 F. We would like to measure the performance using a first-order response surfaces to find the best performance condition (i.e., the corresponding temperature and reaction time).

Because temperature and reaction time have different units, it is better to normalize the parameters before performing the analysis. Suppose variable x_1 is defined for the parameter reaction time, and x_2 is defined for the parameter temperature, each ranging from -1 to 1, and thus the units of the parameters are irrelevant.

To convert the parameters (say, P_i, $i = 1$ and 2) to the response surface variables x_i, the following formula is applied:

$$x_i = 2 \left(\frac{P - P_{\text{midvalue}}}{P_{\text{max}} - P_{\text{min}}} \right)$$

where

$$P_{\text{midvalue}} = \frac{P_{\text{max}} + P_{\text{min}}}{2}$$

The following table shows some tests correlating performance to the upper bound values of the two variables reaction time and temperature, and the corresponding variables extreme x_1 and x_2 values. Note that the nominal values of the variables are arbitrarily set at $x_1 = -1$ and $x_2 = -1$.

	Time (min), P_1	Temperature (F), P_2	Performance (%)	x_1	x_2
1	50	160	65.3	-1	-1
2	60	160	68.2	1	-1
3	50	170	66	-1	1
4	60	170	69.8	1	1

Develop the corresponding response surface and find the best performance of this barrier.

Solution:

The Δx_i values are calculated by changing x_i from nominal to an upper bound value x_i^U while all other input variables are kept are their base or nominal values using the data provided in the table. For x_1 (reaction time), the value of change is calculated by keeping the parameter temperature at its nominal value. Assuming performance is measured by the dependent variable y then

$$\Delta x_1^U = x_1^U(60,160) - x_1^b(50,160) = 1 - (-1) = 2$$
$$\Delta y_1^U = y_1^U - y_1^b = 68.2 - 65.3 = 2.9$$

$$a_1 = \frac{\Delta y_1^U}{\Delta x_1^U} = \frac{2.9}{2} = 1.450$$

For x_2, change the temperature to the upper bound while keeping the reaction time at nominal value

$$\Delta x_2^U = x_2^U(50,170) - x_2^b(50,160) = 1 - (-1) = 2$$
$$\Delta y_2^U = y_2^U - y_2^b = 66 - 65.3 = 0.7$$

$$a_2 = \frac{\Delta y_2^U}{\Delta x_2^U} = \frac{0.7}{2} = 0.350$$

and from (5.16)

$$a_{12} = \frac{\Delta y_{12}^U - a_1\Delta x_1^U - a_2\Delta x_2^U}{\Delta x_1^U \Delta x_2^U} = \frac{(69.8 - 65.3) - 1.45 \cdot 2 - 0.35 \cdot 2}{2 \cdot 2} = 0.225$$

Then from (5.14)

$$\Delta y = 1.450\Delta x_1 + 0.350\Delta x_2 + 0.225\Delta x_1 \Delta x_2$$

The constant a_0 and the response values can be estimated from (5.13)

$$a_0 = 69.8 - [1.450(1) + 0.350(1) + 0.225(1)(1)] = 67.78$$

Finally, we have the general response surface as

$$y = 67.78 + 1.450x_1 + 0.350x_2 + 0.225x_1 x_2$$

Figure 5.5 shows the three-dimensional surface plot for this model. From these results, we can observe that the maximum performance is about 69.5%, obtained at approximately reaction time of 60 min and temperature of 170 F. This response surface may be used for Monte Carlo calculation or other as part of the risk assessment.

If there is a large nonlinearity in (5.1), then a polynomial of higher degree must be used, such as the second-order method [30]. The central composite design (CCD) is a popular response surface method for fitting second-order response surfaces [31]. The CCD has three main properties:

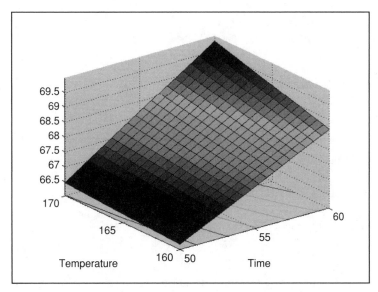

FIGURE 5.5 Three-dimensional surface plot for response surface of Example 5.8.

1. A CCD can be run sequentially. It can be naturally partitioned into two subsets of points; the first subset estimates linear and two-factor interaction effects while the second subset estimates curvature effects. The second subset need not be run when analysis of the data from the first subset points indicates absence of significant curvature effects.
2. CCDs are very efficient in providing information on the effect of independent variable and their overall error in estimating the dependent variable using a minimum number of required runs.
3. CCDs are very flexible. The availability of several varieties of CCDs enables their use under different regions of interest and operability.

There are three methods of CCD: face-centered, circumscribed, and inscribed. To make the right selection, the analyst must first understand the differences between these varieties in terms of the experimental region of interest and region of operability, according to the following definitions:

Region of interest: a geometric region defined by lower and upper limits of variable sets combinations of interest.

Region of operability: a geometric region defined by lower and upper limits of variable of interest set combinations that can be physically and operationally achieved and that will result in realistic value.

A CCD consists of [32]:

1. A complete 2^k factorial design. The factor levels are coded to the usual $-1, +1$. This is known as the factorial portion of the design.
2. n_0 center points ($n_0 \geq 1$).
3. Two axial points on the axis of each design variable at a distance of α from the design center. This portion is known as the axial portion of the design.

The total number of points is $N = 2^k + 2^k + n_0$. The points for a circumscribed CCD method with two variables are represented graphically in Figure 5.6.

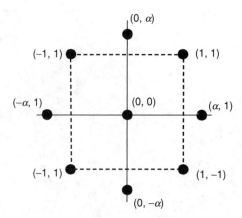

FIGURE 5.6 Points for a circumscribed method with two variables.

The value of α in Figure 5.6 depends on the number of data points (or runs):

$$\alpha = [\text{number of factorial runs}]^{1/4}$$

If the factorial is a full factorial, then

$$\alpha = [2^k]^{1/4}$$

For example if $k = 2$, the scaled value for α relative to ± 1 is

$$\alpha = [2^2]^{1/4} = 1.41$$

The readers are referred to Khuri and Cornell [33] for a detailed discussion on response surface and CCD.

5.5 COMPARISON OF UNCERTAINTY PROPAGATION METHODS

No approach to the propagation and analysis of uncertainty can be ideal for all applications. For example, in a specific case, simple random sampling may be more appropriate than LHS for the estimation of the extreme quantiles of a distribution. Similarly, method of moments may be selected when measuring the effects of small perturbations away from nominal values. However, generally speaking, Monte Carlo analysis with LHS is the most useful approach to the propagation and analysis of uncertainty. Below a simple description of pros and cons of using the propagation method discussed in this book has been presented.

5.5.1 MONTE CARLO METHOD

The advantages of Monte Carlo simulation methods include:

1. Extensive sampling from the ranges of the input elements.
2. Direct use of the model without relying on surrogate models (e.g., using Taylor series expansion or response surface that require additional simplifying assumptions).
3. Extensive modifications and manipulation of the original model are not required.

4. The extensive sampling from the individual variables allows for the identification of nonlinearities, thresholds, and discontinuities.
5. Approach is conceptually simple and easy to use with many applications available in the literature.

The only drawback of this method is the computational cost when dealing with complex forms of (5.1), or when dealing with probabilities very close to zero or one. Depending on the accuracy desired, Wilks tolerance limit and LHS can dramatically reduce sample size with approximately the same quality of output. The LHS method has the advantages of:

1. Increased accuracy in the output distribution compared to the classical Monte Carlo for the same number of observations.
2. Assurance that all parts of the input distributions are sampled, leading to less chance of a sample that is nonrepresentative of the output distribution.
3. Because of the reduced computer costs, LHS is the method of choice for most risk analyses, especially PRA studies.

However LHS has disadvantages including:

1. Time needed by the analyst to set up the problem and run it can become significant in complex problems.
2. In problems involving highly skewed input distributions with long tails, it is difficult to obtain accuracy in the tails (and sometimes the mean value) of the output distribution. This problem is less significant in the classical Monte Carlo estimation.

Because of the reduced computer costs, LHS is the method of choice for most PRA studies.

5.5.2 DIFFERENTIAL ANALYSIS (METHOD OF MOMENTS)

This method is based on developing a Taylor series approximation to the risk model under consideration. The advantages of this method are:

1. The effects of small perturbations away from the base value at which the Taylor series is developing are revealed.
2. Uncertainty and sensitivity analysis based on variance propagation are straightforward once the Taylor series is developed.
3. The approach has been widely studied and applied.

There are three primary drawbacks:

1. The method provides local solutions.
2. It can be difficult to implement and can require large amounts of analyst and/or computer time.
3. It is best for simple models.

5.5.3 RESPONSE SURFACE METHODOLOGY

This method is based on using an experimental design to select model input and then developing a response surface replacement for the original model that is used in subsequent Monte Carlo uncertainty and sensitivity analysis. The advantages include:

1. Complete control over structure of the model input through the experimental design method selected.
2. Good choice for a model whose dependent variables are linear or quadratic function of the independent (input) variables.
3. Uncertainty and sensitivity analysis are straightforward once the response surface model has been developed.

The drawbacks of the response surface approach include:

1. Experimental design process can be tedious.
2. Input variables are often very limited.
3. Sometimes it needs a large number of design points or runs.
4. It is possible to miss thresholds, discontinuities, and nonlinearities in the model.
5. Correlations between input variables can be difficult to capture.
6. Choice of an appropriate response surface approximation to a model is difficult.

5.6 GRAPHIC REPRESENTATION OF UNCERTAINTY

The results of a probabilistic uncertainty analysis should be presented in a clear manner that aids analysts in developing appropriate qualitative insights. Generally, we will discuss three different ways of presenting probability distributions (so their use is not limited by uncertainty analysis): plotting the pdf or the cdf, or displaying selected percentiles, as in a Tukey [33] box plot (sometimes referred to as just a box plot).

Figure 5.7 shows examples. The probability density function shows the relative probabilities of different values of the parameters. One can easily see the areas or ranges where high densities (occurrences) of the random variable occur (e.g., the modes). One can easily judge symmetry and skewness and the general shape of the distribution (e.g., bell-shaped vs. J-shaped). The cdf is best for displaying percentiles (e.g., median) and the respective confidence intervals. It is easily used for both continuous and discrete distributions.

The standard Tukey box shows a horizontal line from the 10th to 90th percentiles, a box between the lower percentiles (e.g., from the 25th to 75th percentiles), and a vertical line at the median, and points at the minimum and maximum observed values. This method clearly shows the important quantities of the random variable.

In cases where confidence limits or probability bounds are estimated, the Tukey box can be used to describe the confidence intervals. Consider a case where the distribution of a variable Y is estimated and described by a pdf. For example, a pdf of time-to-failure (the aleatory model) can be represented by an exponential distribution and the value of λ for this exponential distribution is represented, using the Bayes approach, by a lognormal distribution to capture the epistemic uncertainties. Then, $f(Y|\lambda)$ for various values of λ can be plotted by the families of curves which can show an aggregate effect of both kinds of uncertainty. For example, see Figure 5.8.

In general, a third method can be used by actually displaying the probability densities of λ in a multidimensional form. For example, Figure 5.9 presents such a case for a two-dimensional distribution. In this figure, $f(Y|\lambda)$ is shown for various values of λ.

Example 5.9

An aluminum plate (20204-T4) used for a certain aircraft structure is subjected to cyclic loading between $\sigma_{min} = 165$ and $\sigma_{max} = 420$ MPa. Assuming that the variation in the material properties follows a normal distribution with a 5% coefficient of variation, calculate the life for this component using the Smith, Watson, and Topper (SWT) relationship:

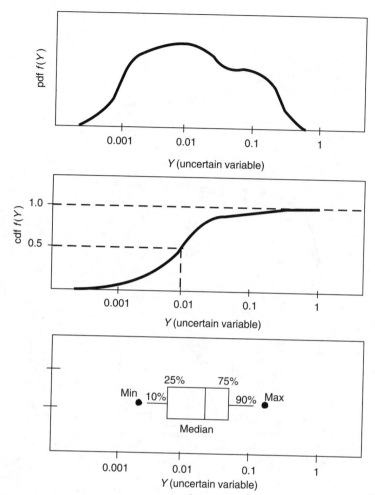

FIGURE 5.7 Three conventional methods of displaying distribution.

$$N_f = \frac{1}{2}\sqrt[b]{\frac{\sigma_{max}}{\sigma_f'}}\sqrt{\frac{1-R}{2}}$$

where N_f is the life, constants $\sigma_f' = 900$ MPa and $b = -0.102$ are material properties for this aluminum alloy, and $R = \sigma_{max}/\sigma_{min}$ is the stress ratio. Show the uncertainty in life of the structure using a box plot.

Solution

By substituting the above values in the SWT relationship and performing a standard Monte Carlo simulation in order to consider the uncertainty of the material properties, we obtain the following results:

median life = 304,865 cycles
standard deviation of life = 300,700

Note that the method of moments can also be used to determine the mean and standard deviation of life.

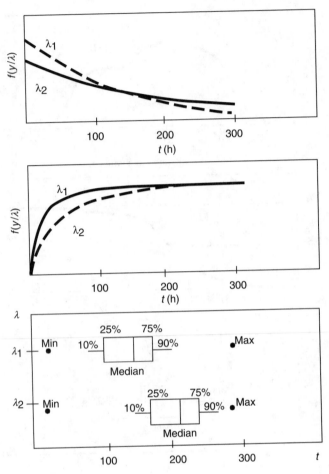

FIGURE 5.8 Representation of uncertainties.

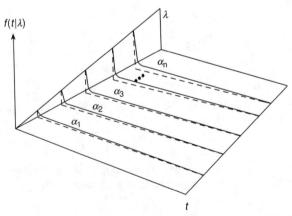

$\alpha_1, \alpha_2, ..., \alpha_n$ probability intervals associated with each exponential distribution

FIGURE 5.9 Two-dimensional uncertainty representation.

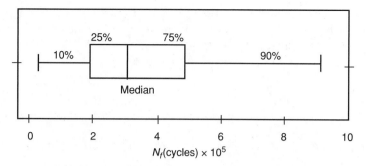

FIGURE 5.10 Uncertainty representation of life of the plate for Example 5.9.

Figure 5.10 shows the box plot with the uncertainty representation.

Exercises

5.1 Consider two resistors in parallel configuration. The mean and standard deviation for the resistance of each are as follows:

$$\mu_{R1} = 25\sigma_{R1} = 0.1\mu_{R1}$$
$$\mu_{R2} = 50\sigma_{R2} = 0.1\mu_{R2}$$

Using one of the statistical uncertainty techniques, obtain:

a) Mean and standard deviation of the equivalent resistor.
b) In what ways the uncertainty associated with the equivalent resistance is different from the individual resistor? Discuss the results.

5.2 The results of a bootstrap evaluation give: $\mu = 1 \times 10^{-4}$ and $\sigma = 1 \times 10^{-3}$. Evaluate the number of pseudofailures F in N trials for an equivalent binomial distribution. Estimate the 95% confidence limits of μ.

5.3 A class of components is temperature sensitive in that they will fail if temperature is raised too high. Uncertainty associated with a component's failure temperature is characterized by a continuous uniform distribution such as shown below:

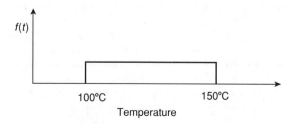

If the temperature for a particular component is uncertain but can be characterized by an exponential distribution with $\lambda = 0.05$ per degree celsius, calculate reliability of this component.

5.4 The risk may be evaluated by using $R = \Sigma f_i C_i$ where f_i is the frequency of scenario i and C_i is the consequence of this scenario. For a given scenario if the mean frequency is 1×10^{-8}/year standard derivation of 1×10^{-8}/year and mean consequence is 1000 injuries with standard derivation of 100 injuries, determine the mean risk of scenario and its associated standard deviation.

REFERENCES

1. Bernstein, P.L., *Against the Gods: The Remarkable Story of Risk*, John Wiley & Sons, New York, 1996.
2. Ayyub, B.M., *Elicitation of Expert Opinions for Uncertainty and Risks*, CRC Press, Boca Raton, FL, 2002.
3. Bedford, T. and Cooke, R., *Probabilistic Risk Analysis: Foundations and Methods*, Cambridge University Press, Cambridge, 2001.
4. Kolmogorov, A.N., *Grundbegriffen der Wahrscheinlichkeitsrechnung*, Ergebnisse der Mathematik und ihrer Grenzebiete, Springer, 1933.
5. Hanseth, O. and Monteiro, E., Modelling and the representation of reality: some implications of philosophy on practical systems development, *Scandinavian Journal of Information Systems*, 6(1), 25, 1994.
6. Droguett, E. and Mosleh, A., Methodology for the treatment of model uncertainty, Proceedings of the International Conference on Probabilistic Safety Assessment and Management — PSAM 5, Osaka, Japan, 2000.
7. Mosleh, A., Siu, N., Model uncertainty: its characterization and quantification, Center for Reliability Engineering, University of Maryland International Workshop Series on Advanced Topics in Reliability and Risk Analysis, Annapolis, Maryland, October 20–22, 1995.
8. Morgan, M.G. and Henrion, M., *Uncertainty: A Guide to Dealing with Uncertainty in Quantitative Risk and Policy Analysis*, Cambridge Press, Cambridge, 1990.
9. Sastry, S.I., Uncertainty Analysis of Transport-Transformation Models, PhD thesis, Rutgers, The State University of New Jersey, New Brunswick, New Jersey, 1999. Electronically available at http://www.ccl.rutgers.edu/~ssi/thesis/.
10. Saltelli, A., Chan, K., and Scott, E.M., *Sensitivity Analysis*, John Wiley & Sons, New York, 2000.
11. Holliday, C.H., On the Maximum Sustained Winds Occurring in Atlantic Hurricanes, ESSA Technical Memoranda, U.S. Department of Commerce, 1969.
12. Morchland, J.D. and Weber, G.G., A moments method for the calculation of confidence interval for the failure probability of a system, Proceedings of the 1972 Annual Reliability and Maintainability Symposium, IEEE, New York, 1972, 505.
13. Apostolakis, G. and Lee, V.T., Methods for the estimation of confidence bounds for the top event unavailability of fault trees, *Nuclear Engineering and Design*, 41(3), 411, 1977.
14. Martz, H.F., A comparison of methods for uncertainty analysis of nuclear plant safety system fault tree models, U.S. Nuclear regulatory Commission and Los Alamos National Laboratory, NUREG/CR-3263, Los Alamos, NM, 1983.
15. Atwood, C.L., *Handbook of Parameter Estimation for Probabilistic Risk Assessment*, NUREG/CR-6823, Washington, DC, 2003.
16. Dezfuli, H. and Modarres, M., Uncertainty analysis of reactor safety systems with statistically correlated failure data, *Reliability Engineering International Journal*, 11(1), 47, 1985.
17. Modarres, M., Kaminskiy, M., and Krivtsov, V., *Reliability Engineering and Risk Analysis: A Practical Guide*, Marcel Dekker, New York, 1999.
18. Martz, H.F. and Duran, B.S., A comparison of three methods for calculating lower confidence limits on system reliability using binomial component data, *IEEE Transactions on Reliability*, R-34(2), 113, 1985.
19. Helton, J.C. and Davis, F.J., Latin hypercube sampling and the propagation of uncertainty in analyses of complex systems, *Reliability Engineering and System Safety*, 81(1), 23, 2003.
20. Helton, J.C. and Davis, F.J., Sampling-based methods for uncertainty and sensitivity analysis, SANDIA Report, SAND99-2240, July 2000.
21. Paté-Cornell, M.E., Conditional uncertainty analysis and implications for decision making: the case of the waste isolation pilot plant, *Risk Analysis*, 19(5), 995, 1999.

22. Kaplan, S., On the method of discrete probability distributions in risk and reliability calculation — application to seismic risk assessment, *Risk Analysis Journal*, 1(3), 189, 1981.
23. Efron, B.A. and Tibshirani, R.J., *An Introduction to the Bootstrap*, Chapman and Hall, London, New York, NY, 1979.
24. Wilks, S.S., Determination of sample sizes for setting tolerance limits, *The Annals of Mathematical Statistics*, 12(1), 91, 1941.
25. Wilks, S.S., Statistical prediction with special reference to the problem of tolerance limits, *The Annals of Mathematical Statistics*, 13(4), 400, 1942.
26. Wald, A., An extension of wilks' method for setting tolerance limits, *The Annals of Mathematical Statistics*, 14(1), 45, 1943.
27. Wald, A., Tolerance limits for a normal distribution, *The Annals of Mathematical Statistics*, 17(2), 208, 1946.
28. Guba, A., Makai, M., and Pal, L., Statistical aspects of best estimate method I, *Reliability Engineering & System Safety*, 80(3), 217, 2003.
29. Nutt, W.T. and Wallis, G.B., Evaluation of nuclear safety from the outputs of computer codes in the presence of uncertainties, *Reliability Engineering & System Safety*, 83(1), 57, 2004.
30. Li, F. and Modarres, M., Characterization of uncertainty in the measurement of nuclear reactor vessel fracture toughness and probability of vessel failure, Transactions of the American Nuclear Society Annual Meeting, Milwaukee, 2001.
31. Montgomery, D. and Runger, G., *Applied Statistics and Probability for Engineers*, John Wiley & Sons, New York, 1994.
32. Verseput, R., Digging into DOE, 2001, www.qualitydigest.com
33. Khuri, A. and Cornell, J., *Response Surfaces Designs and Analyses*, Marcel Dekker, New York, 1987.
34. National Research Council, Protection against depletion of stratospheric ozone by chlorofluoro-carbons, Report by the Committee on Impacts of Stratospheric Change and the Committee on Alternative for the Reduction of Chlorofluorocarbon Emission, National Research Council, Washington, DC, 1979.

6 Identifying, Ranking, and Predicting Contributors to Risk

6.1 INTRODUCTION

Risks are unavoidable and as such the key challenge in engineering risk analysis is to identify the elements of the system or facility that contribute most to risk and associated uncertainties. To identify such contributors, the common method used is the importance ranking. One of the most useful outputs of a risk assessment, especially PRA, is the set of importance measures associated with the main elements of the risk models such as phenomena, failure events, and processes. These importance measures are used to rank the risk-significance of these elements in terms of their contributions to the total risk (e.g., expected loss or hazard) assessed in the PRA. Importance measures are either *absolute* or *relative*. The absolute measures define the contribution of each risk element in terms of an absolute risk metric (reference level), such as the conditional frequency of a hazard exposure given a particular state of the element. Relative measures compare the risk contribution of each element with respect to others. In most risk analyses, it is common to conclude that importance measures of a small fraction of risk elements contribute appreciably to the total risk. That is, often the Pareto rule applies in which less than 20% of the elements in the risk model contribute to more than 80% of the total risk. Moreover, the importance indices of risk elements usually cluster in groups that may differ by orders of magnitude from one another. As such, the importance indices are radically different such that they are generally insensitive to the precision of the data used in the risk model.

As the most important product of the risk assessment, identification of major risk contributors through importance measures provides a strong guidance for subsequent risk management efforts, for example, in allocating resources. These might include resources for reliability improvement, surveillance and maintenance, design modification, security, operating procedure, training, quality control requirements, and a wide variety of other resource expenditures. By using the importance of each risk-significant element, resource expenditure can be properly optimized to reduce the total life-cycle resource expenditures while keeping the risk as low as practicable. Alternatively, for a given resource expenditure such as for upgrades or for maintenance, the importance measure of each risk-significant element can be used to allocate resources to minimize the total system risk. This approach allows the risk manager to offer the "biggest bang for the buck." An advantage of these optimal allocation approaches is that relative risk importance measures can be used, which have generally smaller uncertainties than absolute values. Moreover, uncertainties in the importance measures can also be estimated. This subject will be discussed in more detail in this chapter.

In addition to allocating resources, the importance measures of risk elements can be used to assign optimum or allowed maintenance and repair downtimes (actually a type of resource too). Borgonovo and Apostolakis [1] categorize applications of importance measures into the following three areas:

1. *Design*: To support decisions of the system design or redesign by adding or removing components, subsystems, operating procedures, etc.
2. *Test and maintenance*: To address questions related to the system performance by changing the test and maintenance strategy for a given design.
3. *Configuration and control*: To measure the significance of the effect of failure of a component or taking a component out of service.

There are various techniques for performing sensitivity analyses. These techniques are designed to determine the importance of key assumptions and parameter values to the risk results. The most commonly used methods are the so-called "one-at-a-time" methods, in which assumptions and parameters are changed individually to measure the change in the output. These methods are very valuable and powerful, because the analyst can vary virtually any input or model assumption and observe their impact in final risk calculations. However, because they do not capture the full effect of interactions among variables, one-at-a-time methods should be used with caution. Sometimes groups of variables may be changed, but selection of the group is subjective and is largely based on the judgment of an analyst. The possibilities for sensitivity studies are almost limitless. The analyst simply has to change a parameter value (as for example a probability distribution or density model) or the form of the model and rerun the calculations. One-at-a-time sensitivity studies are relatively straightforward to perform and interpret. However, in complex PRAs and other risk models, there are often complex interactions and dependencies among the variables. For example, the occurrence of a catastrophic steam explosion in a nuclear power plant might depend on the values of two input variables, the fraction of the core slumping into the lower plenum and the amount of water present. Changing the two variables individually to their extreme conservative values might produce benign results, while changing them simultaneously to their extreme values might result in a catastrophic explosion. Currently, there are no unique rules for finding such interactions among variables and they are dependent upon the insights and expertise of the risk analysts.

Finally, derived from actual observed incidents and events, the concept of precursor analysis is also an attractive approach for identifying and predicting contributors to risk. *Precursor events* (PEs), in the risk assessment context, can be defined as events that constitute important elements of scenarios leading to accidents (or hazard exposure) in complex systems experience such as a severe core damage in a nuclear power plant, severe aviation or marine accidents, and chemical plant accidents. The significance of a PE is measured through the conditional probability of occurrence of all risk scenarios involving the actual PE, if the remaining (nonfailed) hazard barriers also fail to complete the scenarios. In other words, PEs are those events that substantially reduce the margin of safety available for prevention or mitigation of hazard exposure (risk).

In this chapter, the techniques for importance ranking, sensitivity analysis, and precursor analysis will be discussed.

6.2   IMPORTANCE RANKING IN PROBABILISTIC RISK ASSESSMENT

The total risk of a system is usually computed from multiple risk scenarios consisting of failure events, phenomena, processes, human and software errors, etc. In PRAs the risk is estimated through the frequency or probability of occurrence of scenarios represented in the form of minimal or mutually exclusive cut sets. The frequency or probability of scenario cut sets are always in the form of sum of product of terms (event probabilities). Wall et al. [2] represent the total risk, R, by representing the frequency or probability of all risk scenarios as a linear function of probability events in the form of:

$$R = aP + b \qquad (6.1)$$

where aP is simply all the cut sets (of risk scenarios) containing a specific event "P" (component failure, human error, occurrence of a phenomenon, etc.), and parameter b represents all other cut sets. That is, the first term represents all cut sets involving P, and b represents contributions from those not containing P. Clearly P can be factored out and parameter "a" is the probability (or frequency) of the remaining elements of the cut sets containing P. This formulation provides a good representation of the relationship between the total risk and probability of basic event probability, P, for which we are measuring its importance. But it should be noticed that this formulation is only useful for single component importance measures. That is, if simultaneous changes in risk elements were being considered (i.e., importance of a class of elements), this representation would not be appropriate. In the following, a brief description of some customary importance measures in the context of (6.1) and their interpretation has been presented.

6.2.1 BIRNBAUM

The Birnbaum [3] importance, I_B, is defined as the rate of change in total risk of the system with respect to changes in a risk element's basic probability (or frequency),

$$I_B = \frac{dR}{dP} = a \qquad (6.2)$$

One may also normalize it to the baseline risk of the system

$$I_B = \frac{dR/dP}{R} = \frac{a}{aP + b} \qquad (6.3)$$

When applied in PRAs the Birnbaum measure reflects the rate of change in risk (e.g., frequency of hazard exposure) as a function of changes in a risk element probability. It is a measure of functional margin in the system design (for example the defense-in-depth in the nuclear power plant designs) for a component. As such, small Birnbaum measure implies high degree of functional redundancy or reliability margin.

As is evident from (6.2) and (6.3), the Birnbaum measure does not depend on the performance of the element itself. It is rather obvious that it would be more difficult and costly to further improve the more reliable components than to improve the less reliable ones. But, since the Birnbaum measure does not consider the present or baseline performance (probability of success or failure of an element) it would be hard to use it for risk-informed decision making, since low-failure probability items are not necessarily the prime candidates for any change. To remedy this shortcoming, an extended version of this measure may be used. Called *criticality importance*, this measure considers the probability of the risk element in the Birnbaum definition as follows:

$$I_C = \frac{dR}{dP} \times \frac{P}{R} = \frac{aP}{R} \qquad (6.4)$$

It is clear that the Birnbaum importance measure is adjusted relative to the failure probability of the individual elements (components, phenomena, etc.) and as a fraction of the total risk of the whole system. Therefore, if the Birnbaum importance measure of an element is high, but the failure probability of the component is low with respect to the risk of the system, then critically importance will accordingly adjust its estimate to a lower value.

Another subset of criticality importance measure is *inspection importance* measure. This measure is defined as the product of Birnbaum importance measure times the failure probability or occurrence rate of the element. This measure is used to prioritize test and surveillance activities to ensure sufficient readiness and performance.

The Birnbaum measure, which is the result of the change in the probability of a given event, is equivalent to a sensitivity analysis and can be estimated by first calculating the total risk (frequency of a hazard exposure) with the probability of the given event set to unity and then subtracting the top event probability with the probability of the given event set to zero.

6.2.2   FUSSELL–VESELY

This measure is introduced by Vesely et al. [4,5] and later applied by Fussell [6] and is defined as the fractional contribution of a risk element, the total risk of the system of all scenarios containing that specified element. Accordingly,

$$I_{FV} = \frac{aP}{R} = \frac{aP}{aP + b} \tag{6.5}$$

Wall et al. [2] note that this measure is proportional to the basic event probability P of the specified risk element, unless it is larger than about 20%. From (6.5), we can get

$$I_{FV} = \frac{(a/b)P}{(a/b)P + 1} = \frac{kP}{kP + 1} \tag{6.6}$$

where $k = a/b = $ constant.

This relationship is linear with respect to kP for $I_{FV} < 0.2$; therefore, it expresses long-term average contribution of the risk element's performance to the overall risk. In risk elements with large Fussell–Vesely importance measures, it is imperative not to allow their long-term average probabilities to further increase. Accordingly, in an aging regime, I_{FV} can be interpreted as the amount of allowed degradation of performance as a function of risk increase. This measure also shows the importance relative to the long-term averaged performance of a component (so, it is not appropriate for measuring importance of a set of similar components instantaneously taken out of service).

6.2.3   RISK REDUCTION WORTH

The risk reduction worth (RRW) importance is a measure of the change in risk of the system when an input variable (an element's failure probability or frequency) is set to zero. That is, assuming that the component is perfect or its failure probability is zero and thus eliminating any postulated failures. This importance measure highlights the theoretical limit of the performance improvement of the system. This importance measure is defined as a ratio in failure space (although there is another version of it defined as the difference between the baseline risk and the reduced risk that will be discussed later, but is not as popular). It is the ratio of the baseline risk of the system to the new reduced risk value when probability or frequency of the specified risk element is set to zero (that is assuming that the risk element is perfect). Accordingly,

$$I_{RRW} = \frac{R}{R(P=0)} = \frac{aP+b}{b} \tag{6.7}$$

In practice, this measure may be used to identify system elements that are the best candidates for efforts leading to reducing the system risk (or improving safety).

Using (6.5) one may derive the following relationship between Fussell–Vesely measure and RRW measure:

$$\frac{1}{I_{RRW}} = 1 - I_{FV} \tag{6.8}$$

According to (6.8) I_{FV} and I_{RRW} should yield consistent ranking for risk elements in the failure space. Also it is evident that for small I_{RRW} values, I_{FV} is more sensitive and possibly more accurate for applications. However, as I_{RRW} increases toward large values, I_{FV} becomes less sensitive to changes in importance ranking.

As noted earlier there is also a differential definition of RRW given as $I_{RRW} = (aP+b)-(b) = aP$ which is not as popular as the fractional form.

6.2.4 RISK ACHIEVEMENT WORTH

The risk achievement worth (RAW) importance measure is the reverse view of RRW measure. Chadwell and Leverenz [7] discuss RAW as a measure in which input variable probability or frequency is set to unity, and the effect of this change on the system risk is measured. Therefore, RAW is the ratio of the new (increased) risk to the baseline risk of the system when the probability of the specified risk element is set to unity. Hence,

$$I_{RAW} = \frac{R(P=1)}{R} = \frac{a+b}{aP+b} \tag{6.9}$$

This measure expresses increase in risk when an element of the risk model is unavailable, out of service, failed, etc. It is the importance of an element permanently failed (or removed) or importance of one under "extreme" degradation when it results in an abrupt increase in risk. Also, I_{RAW} is the indicator of how quickly we should return a failed element to service. From (6.9) this measure may be interpreted either as an average frequency of hazard exposure versus average risk achievement over a long time, or as instantaneous frequency of hazard exposure versus instantaneous risk achievement. The latter is more meaningful because it is a measure of "existing" events rather than "occurring events."

Note that in complex systems usually total risk is not very sensitive to changes of failure probability of system elements. Therefore, one may conclude that, the RAW measure is independent of the baseline system element probability of occurrence, loss, or failure. As we discussed before, and based on the mathematical expression of this measure, it is evident that the RAW importance of a risk element is influenced by the system configuration with respect to that risk element, rather than the probability, P. As such, by not considering the logic of system or facility configuration as risk element occurs (e.g., an event happen), importance of certain risk elements may be masked.

Further, the actual procedure of the calculation would be problematic, because computer codes usually calculate the risk by considering the cut sets and setting the risk element probability to unity. Borgonovo and Apostolakis [1] caution that not considering this in a logical form (i.e., by not setting the risk element to a null condition in the logic model), this may lead to errors due to missing extra cut sets that originally (i.e., in the baseline cut sets calculation) are eliminated in the truncation process due to their small frequencies. Clearly, this will not be the case if the system logic model is used to determine the conditional frequencies given the risk element (e.g., a failure event) fails or works.

Note that similar to RRW, a less popular differential definition of RAW is also available in the form of $I_{RAW} = (a+b) - (aP+b) = a(1-P)$. Birnbaum measure is related to RAW and RRW. When these are expressed on an interval scale (absolute value), $I_B = I_{RAW} + I_{RRW}$.

6.2.5 DIFFERENTIAL IMPORTANCE MEASURE

The total variation of a function due to a small variation of its variables is expressed by the differential of the function, so the change in total risk of the system depends on how the risk element (e.g., event failure probability) is varied

$$dR = \frac{\partial R}{\partial P_1}dP_1 + \frac{\partial R}{\partial P_2}dP_2 + \frac{\partial R}{\partial P_3}dP_3 + \cdots + \frac{\partial R}{\partial P_n}dP_n \qquad (6.10)$$

The differential importance measure (DIM) introduced by Borgonovo and Apostolakis [1] is the fraction of total changes in risk (R) of the system due to a change in failure probability (P) of a certain element in the risk model. It has two mathematical expressions:

$$DIM_i = \frac{\partial R/\partial P_i}{\sum_j \partial R/\partial P_j} = \frac{a_i}{\sum_j a_j} = \frac{(\partial R/\partial P_i)x_i}{\sum_j (\partial R/\partial P_j)x_j} = \frac{a_i P_i}{\sum_j a_j P_j} \qquad (6.11)$$

DIM is inherently additive because of its definition. Thus, if we are interested in the DIM of a subset of risk elements we simply have:

$$DIM(C_1, C_2, C_3, \ldots, C_n)_i = DIM_{C_1} + DIM_{C_2} + DIM_{C_3} + \cdots + DIM_{C_n} \qquad (6.12)$$

From the definitions of DIM as in (6.11), Birnbaum as in (6.3) and criticality importance measure as in (6.4), it is clear that these measures are closely related. The first DIM definition in (6.11) produces the same ranking for individual risk elements as the Birnbaum measure (6.3), since they both report the rate of changes in system total risk with respect to individual risk element (e.g., event failure probability). Also, DIM under the second definition in (6.11) produces the same ranking as the criticality importance measure (6.4). Note, however, that Birnbaum and criticality importance measures are not additive generally, while DIM is. Thus, the importance of changes affecting multiple risk elements can be estimated in a simple way using DIM (6.12), rather than I_B or I_C.

Table 6.1 summarizes the importance measures discussed above and their relationships to each other.

6.3 IMPORTANCE MEASURES IN SUCCESS SPACE

Sometimes the objective is to obtain importance of risk elements in improving operation or safety (i.e., the success) of the system. Application of these measures in defining safety or preventive importance (as opposed to risk importance) requires slight modification in the definitions of some of these measures. To define importance measures in success space, one should calculate measures based on the total success of the system (e.g., nonexposure of hazards, reliability, and availability) instead of total risk. It is possible to start by the same structure of mathematical definitions of measures in failure domain, and by considering the

TABLE 6.1
Most Common Risk Importance Measures in Risk Assessment

Measure	Relation to Other Measures	Principle
Birnbaum	$I_B = I_{RAW_1} + I_{RRW_1}$	$R(x_i = 1) - R(x_i = 0) = a$
Risk reduction worth (differential method)	$I_{RRW_1} = I_B - I_{RAW_1}$	$R(\text{base}) - R(x_i = 0) = aP$
Fussell–Vesely	$I_{FV} = \dfrac{I_{RRW_1}}{R(\text{base})}$	$\dfrac{R(\text{base}) - R(x_i)}{R(\text{base})} = \dfrac{aP}{aP + b}$
Risk reduction worth (fractional method)	$I_{RRW_2} = \dfrac{1}{1 - I_{FV}}$	$\dfrac{R(\text{base})}{R(x_i = 0)} = \dfrac{aP + b}{b}$
Criticality importance	$I_C = I_{FV}$	$\dfrac{R(x_i = 1) - R(x_i = 0)}{R(\text{base})} x_i(\text{base}) = \dfrac{aP}{aP + b}$
Risk achievement worth (differential method)	$I_{RAW_1} = I_B - I_{RRW_1}$	$R(x_i = 1) - R(\text{base}) = a(1 - P)$
Risk achievement worth (fractional method)	$I_{RAW_2} = \dfrac{I_{RAW_1}}{R(\text{base})} + 1$	$\dfrac{R(x_i = 1)}{R(\text{base})} = \dfrac{a + b}{aP + b}$
Differential importance	DIM	$\dfrac{R}{P_i} \Big/ \sum_j \dfrac{R}{P_j}$

complement interpretations both in the success and failure spaces. The result of such an approach is summarized in Table 6.2, which leads to mathematically symmetric definition of measures in both success and failure domains.

In Table 6.2 we have exactly the same mathematical expressions for importance measures converted to a success domain from the failure domain. For instance in RRW and RAW measures, definitions are still fractional type and the concept remains the same. It is obvious that reduction of risk means achievement of success, and increasing of the total risk of the system means reduction of the system's total success probability. Fussell–Vesely importance measure in success space is simply the contribution of the specified system element's success, normalized by baseline success of the system. Ranking by this measure as defined in the success space essentially yields the same results as ranking by prevention worth (PW) importance measure introduced by Youngblood [8, 9].

Applying the mathematical expressions introduced in Table 6.2, the importance of the components can be calculated and studied, in both success and failure domains. Through a simple conceptual example in the following section, we will show that some of these definitions do not preserve their meaning in the success space, and fail to rank the components appropriately. We will then redefine a more appropriate definition of importance measures in success space that closely captures the intent and interpretations of these measures in the failed space. As such the measures yield the same importance ranking both in success and failure domains.

6.4 A COMPREHENSIVE EXAMPLE OF IMPORTANCE MEASURE APPLICATION

Figure 6.1 shows the MLD, risk scenarios, and failure probabilities of a simple conceptual example. Probability of failure or success of this simple example can be calculated by a truth table using a combinatorial method. Eight different components in the system yield 2^8 different mutually exclusive combinations of failure events (as discussed in Chapter 3). One

TABLE 6.2
Definition of Importance Measures in Success and Failure Domain

Importance Measures	Failure Space		Success Space	
	Mathematical Definition	Interpretation	Mathematical Definition	Interpretation
Birnbaum	$I_B = F_{i=0} - F_{i=1}$	The rate of system failure changes with respect to the failure of component i	$I_B = S_{i=1} - S_{i=0}$	The rate of system success changes with respect to the success of component i
Risk reduction worth	$I_{RRW} = \dfrac{F}{F_{i=1}}$	The relative improvements in system failure, realizable by improving component i	$I_{RRW} = \dfrac{S_{i=1}}{S}$	The relative improvement in system, realizable by improving component i
Risk achievement worth	$I_{RAW} = \dfrac{F_{i=0}}{F}$	Factor by which probability of system failure would increase with no credit for component i	$I_{RAW} = \dfrac{S}{S_{i=0}}$	Factor by which probability of system success would decrease with no credit for component i
Fussell–Vesely	$I_{FV} = \dfrac{F - F_{i=1}}{F}$	Fraction of system unavailability (or risk) involving failure of component i	$I_{FV} = \dfrac{S - S_{i=0}}{S}$	Fraction of system success, involving success of component i

I_i, importance measure for component I; $i=1$, the condition that element i operates successfully; $i=0$, the condition that element i has failed; S, total success of the system; F, total failure of the system.

may add probability (frequency) of those combinations leading to system failures together (or success states) to find the total system failure (or success) probability (frequency). To determine the final state for each combination one can evaluate the Boolean logic statement of the system. Probability of events (A_1, A_2, B_1, B_2, X, Y, and Z) and the frequency of initiating event, I, have been shown in Figure 6.1.

As illustrated in Figure 6.1, this simple example contains a redundant component whose failure may be regarded as an initiating event (i.e., component X).

Before calculation of the importance measures, one may determine the impact of failure probability of each component on the total risk of the system in a simple manner. Suppose failure probability of all components is constant except for the ith component. We are interested to determine the change in the total system risk as a function of this individual component probability. The result will be a line, as we expect from (6.1). By repeating the same calculation for all other components we will get an interesting preliminary evaluation of the risk elements importance ranking, and a useful visualization of the ranking.

As illustrated in Figure 6.2 risk of the system is a linear function of each component probability or frequency of failure (or occurrence of an undesirable event). The larger the slope of this line, the higher contribution (and thus more importance), the component would have toward the total failure frequency (risk) of the system. From Figure 6.2 it is evident that component X has the largest contribution to system risk, so we expect that our analysis ranks X as the most important component in this system.

As in (6.1), each line will have a slope of "a" with a y-axis intercept of "b." Clearly, the slope of each line is actually the component's Birnbaum importance measure of the component, and the y-axis intercept is the conditional risk of the system given the ith component is perfect. Also, it shows risk of the system excluding the contribution of the ith component.

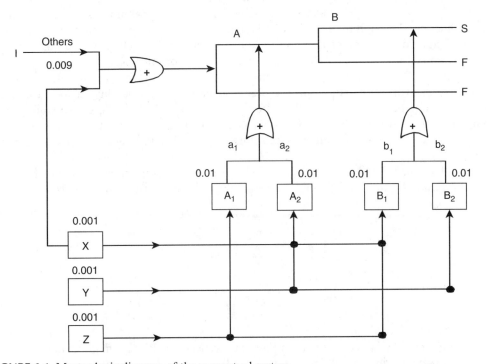

FIGURE 6.1 Master logic diagram of the conceptual system.

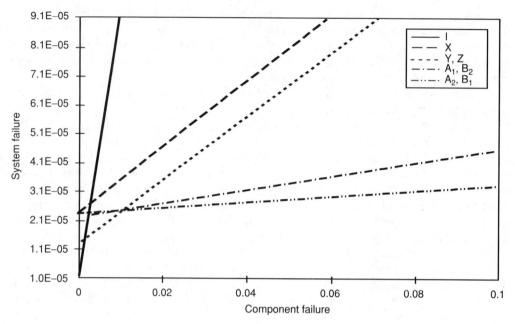

FIGURE 6.2 Linear dependency of system failure with respect to component failure.

A large y-axis intercept may be interpreted as lower contribution from that component to the total risk of the system, leading to lower importance of the component.

Since every component has a unique failure probability (or frequency of occurrence), the baseline risk point of the system will be masked by the way the lines are plotted in Figure 6.2. To remedy this problem the plots may be shown relative to the baseline risk of the system, so that all lines pass through the same baseline reference point. Further, in practical applications usually failure probability of components do not change in such a wide range, so only practical ranges around the baseline risk may actually be of interest.

Figure 6.3 is a better representation of the system. In this figure the x-axis for each component has been normalized by its baseline failure probability, so that the x-axis would actually become a probability multiplier instead of the failure probability of a component itself. In this figure the y-axis has been normalized by the total risk of the system (i.e., the baseline risk value). Accordingly, the y-axis also represents the system risk multiplier.

If the baseline risk value of the system (i.e., when all the components assume their normal probabilities) is R_b, and P_i is the failure probability of component i, (6.1) yields,

$$R' = \frac{R}{R_b}, \quad P' = \frac{P}{P_i} \qquad (6.13)$$

Therefore,

$$R' = \left(\frac{aP_i}{R_b}\right) P' + \frac{b}{R_b} \qquad (6.14)$$

As noted earlier, the slope of the lines in Figure 6.2 reflects the Birnbaum importance measure of the component, while the slopes of the lines in Figure 6.3 show the criticality importance of

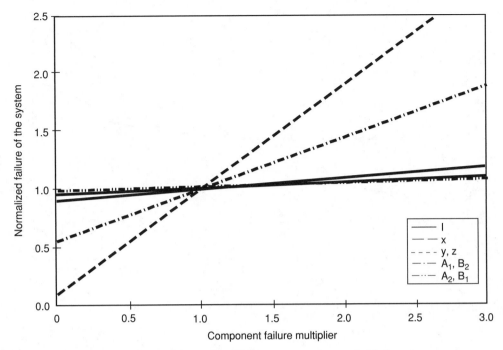

FIGURE 6.3 Dependency of system normalized failure on component multipliers.

the components (as introduced in (6.4)). The y-axis intercept of (6.14) shows the risk factor relative to the baseline risk of the system.

Figure 6.3 also shows that when failures of the components increase by several folds from their baseline probability or frequency values, failure of the system (or its risk) may change by other folds or factors from its baseline. Some components like X may drastically affect the risk, while others like A_2 and B_1 have negligible effects. Also, it should be noticed that the ranking of the components may change depending on whether Figure 6.2 or Figure 6.3 is used for ranking, and even in each figure it depends on whether the slope or the y-axis values are used for ranking.

To have a better grasp of changes caused by each component, one may plot Figure 6.3 in a logarithmic scale. Logarithmic scales highlight those components whose failure probabilities may change by several orders of magnitude, and eliminate those components with negligible effects. Figure 6.4 shows the result of converting to a logarithmic scale. This figure simply shows the fact that components X, A_1, and B_2 are the only ones having reasonable contribution to risk.

6.4.1 IMPORTANCE MEASURES IN FAILURE AND SUCCESS DOMAINS

If we estimate the importance measures based on their mathematical definition (as summarized in Table 6.1 and Table 6.2) and normalize them with respect to the largest measure for scaling purpose, Figure 6.5 shows the results of ranking the components of the sample system in Figure 6.1 based on these importance measures.

The Birnbaum method, as expected from its mathematical definition, yields the same result in both success and failure spaces, since it assumes that the success of the system is the symmetric compliment of the probability or frequency of failure. That is,

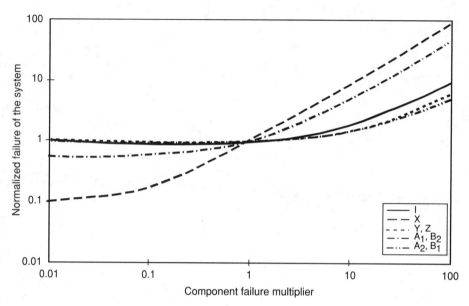

FIGURE 6.4 Dependency of system normalized failure on component multipliers (logarithmic scales).

$$I_B^S = S_{i=1} - S_{i=0} = (1 - F_{i=1}) - (1 - F_{i=0}) = F_{i=0} - F_{i=1} = I_B^F \qquad (6.15)$$

In RRW and RAW methods, their fractional definitions enable us to show the number of factors (folds) or orders of magnitude by which the system failure (success) changes due to failure (or success) of a specific component with respect to the baseline failure (or success) of the system. The definitions of RRW and RAW are ineffective and inappropriate in the success space because several factors or even orders of magnitude changes are possible in the system failure (risk) probability or frequency value, but not in system success probability. See Figure 6.5(b) and 6.5(c) for illustration of this point. As evident all components will have RRW and RAW importance measures close to unity in the success space. The results are also inconsistent with those shown in Figure 6.2 through Figure 6.4. As such one needs to provide new definitions for RRW and RAW for success space that capture the essence of what these measures intend to achieve.

Fussell–Vesely measure in the failure space has a clear concept; it ranks components based on their contribution in total failure of the system, normalized by the system failure presented in Table 6.1, it can be used exactly in the same manner in the success space. Figure 6.5(d) compares the importance of components in failure and success domains for this method. Fussell–Vesely measures in success space are consistent with its measures in the failure space, except for components A_1 and B_2. The reason for the difference is in the mathematical definition of the Fussell–Vesely method. This measure gives the same rank to redundant components in parallel configuration, since by setting one of them to a complete success one masks (or nullifies) the effect of others on the system total failure. In the success space, on the other hand, this masking problem does not work, because we set the component to a completely failed condition to estimate its contribution in total success. In this way not only we do not nullify the effects of others, but also we magnify them.

As a result of the above discussion, the masking problem in the Fussell–Vesely measure is the source of different raking results in failure and success domains, which may or may not be a problem, depending on the decision maker's point of view. For instance, consider three components A, B, and C in the system shown in Figure 6.6.

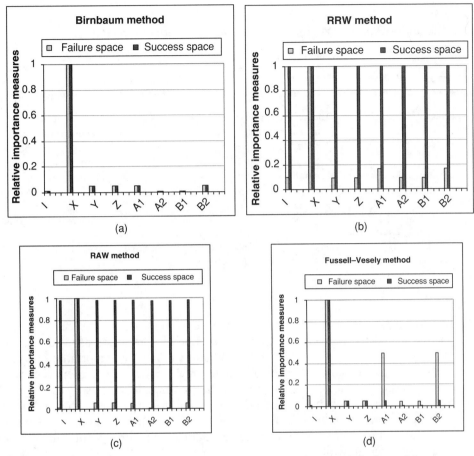

FIGURE 6.5 Importance measures in success and failure space for sample logic model.

In this simple system, regardless of the failure probability of components B and C, the Fussell–Vesely method suggests the same importance measure for these components in the failure space. While in success space, it gives a higher value to component C.

Suppose the decision maker desires to improve the system reliability by adding an extra redundant (parallel) unit to components B or C. The Fussell–Vesely measure in the failure space suggests no difference between importance of components B and C, despite the fact that overall reliability of the system improves by adding a redundant unit with a probability of failure of frequency of occurrence even lower than that of component C. Nevertheless, the use of Fussell–Vesely measure in success space appropriately highlights the difference in the importance of each of these redundant units.

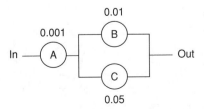

FIGURE 6.6 Example problem for illustration of Fussell–Vesely measure in success and failure spaces.

TABLE 6.3
Redefinition of RRW and RAW Importance Measures in Success Space

Importance Measures	Failure Space		Success Space	
	Mathematical Definition	Interpretation	Mathematical Definition	Interpretation
Risk reduction worth	$I_{RRW} = \dfrac{F}{F_{i=1}}$	The relative improvements in system failure, realizable by improving component i	$I_{RRW} = \dfrac{S_{i=1} - S}{S}$	The percentage of improvements in system success, realizable by improving component i
Risk achievement worth	$I_{RAW} = \dfrac{F_{i=0}}{F}$	Factor by which probability of system failure would increase with no credit for component i	$I_{RAW} = \dfrac{S - S_{i=0}}{S}$	The percentage of degradations of system success, realizable by failure of component i

$i = 1$, the condition that component i, operates successfully; $i = 0$, the condition that component i has completely failed; S, total success of the system; F, total failure of the system.

As discussed before, to have a better consistency between the results in success and failure spaces, the RAW and RRW methods should be redefined for applications in success space. While the fractional definition of RAW and RRW fails to show the importance of components in the success space, a simple "percentage change in probability or frequency of success" may remedy this shortcoming. Table 6.3 presents the modified definitions.

It should be noticed that with these modified definitions for the RRW in the success space, the results become similar to that of Fussell–Vesely method in the failure space, because the numerators of these two expressions are the same, and denominators are for normalizing the expressions. Thus one can write:

$$I_{RRW}^S \times S_b = S_{i=1} - S_b = (1 - F_{i=1}) - (1 - F_b) = I_{FV}^F \times F_b \tag{6.16}$$

where S_b and F_b are the baseline values of the system success and failure probabilities or frequencies. Figure 6.7 and Figure 6.8 show the result of importance assessment of the sample

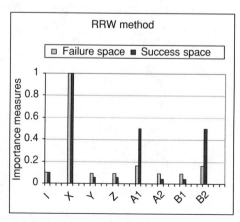

FIGURE 6.7 RRW method.

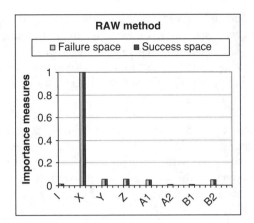

FIGURE 6.8 RAW method.

system by using the new RRW and RAW definitions in Table 6.3. Results demonstrated in Figure 6.7 through Figure 6.8 are reasonably consistent in both failure and success spaces.

RRW measure in success space gives exactly the same ranking as Fussell–Vesely in failure space, as one expects from (6.16). Accordingly, importance ranking from a measure applied in the failed space mostly yields the same results as another measure having the same (symmetric) meaning in the success space. The new definition of the RAW measure as illustrated in Figure 6.8 is also consistent with the results in the failure space. It shows that a simple percentage type definition for success space as introduced in Table 6.3 works as a powerful measure to capture the importance of components.

It is the meaning and interpretation of the importance measures that really matters, and there is no difference between success and failure space applications, once one agrees on the measures' meanings and interpretations in both spaces.

Design and operating configuration changes in complex systems and facilities such as nuclear power plants, which cause some components becoming more important (e.g., risk-significant) than others. This rise in importance of the components with respect to system configuration can be measured using suitable importance measures such as Fussell–Vesely, RRW, and criticality importance. Estimate of component's performance characteristics such as unavailability, reliability, and failure rate, along with its role in system topology will correspond to an estimate of importance (risk-significance) of the component. Under this condition, a unique ranking of components can be determined.

6.5 CONSIDERATION OF UNCERTAINTIES IN IMPORTANCE MEASURES USED FOR RISK-BASED RANKING

Information used to estimate component performance and other elements of risk are associated with some levels of uncertainty. Uncertainties associated with, for example, component performance characteristics should be propagated through the system model (system topology, risk or configuration changes) leading to uncertainties associated with system performance characteristics or the final risk value. As such when performing importance calculations, for example, for ranking risk contributors, the result will also involve uncertainties. In risk analysis, an important issue is how to rank components and events in the risk model in the presence of these uncertainties. For example, determining uncertainty associated with importance measure affects the ranking of the components (e.g., as compared to ranking based on only mean values of importance measures as is done in practice today). If the importance measure is used as a basis for ranking, then the "importance measure distribution" could potentially affect ranking. This section focuses on the issues related to importance and risk-based ranking under uncertainty. It also presents some examples to clarify the approach used and highlights significance of uncertainty consideration in any ranking process.

There are two ways to consider the effect of basic data and other uncertainties in PRAs on component ranking. First, by finding the uncertainty in *ranking order* associated with an element of the PRA (component, initiating event, subsystem, human action, etc.). Second, by calculating uncertainty associated with the *importance measure* of that element with respect to the whole facility or system. In the remainder of this section we refer to "component" as a basic element of a PRA model (i.e., component, initiating event, subsystem, human action, etc.).

6.5.1 RANKING ORDER UNCERTAINTY

In the presence of uncertainty, there is a probability associated with goodness of a given component rank order. Also, there may be some component rank orders that are equally probable. So the logical approach for ranking under uncertainty is to generate a *ranking distribution*, i.e., a histogram of ranks for each component in many simulations of assessing ranking of the components.

Assume that there are n components (failure events, etc.) in the PRA model. So each component can take a rank order from 1 to n, with 1 being the highest rank (the most risk-significant contributor). Monte Carlo simulation or other uncertainty propagation techniques can be used to generate a large number of random samples from the probability distributions representing risk elements such as component unavailabilities, rate of occurrence failures, or failure rates. Each of these samples yield P_j as the probability or frequency of failure of component i (using any of the known measures). Similarly, each sample gives a unique system performance, risk value or frequency of a loss, S_j. Importance measure is then a function of P_j and S_j, i.e., for any risk element (component) i. For example, the Birnbaum importance of component i is $(I_B)_{i,j}$ in the Monte Carlo iteration j, regardless of the specific measure used. Each simulation iteration j results in a single value of importance for the risk element i. These importance measures (e.g., Birnbaum) are used as an index to rank risk elements (components) and provide a ranking order $(R_B)_{i,j}$. Each of these R's give a particular rank to each of the components or risk elements. Here, $(R_B)_{i,j}$ is the rank order generated by using the Birnbaum importance measure for the ith component or risk element in the jth Monte Carlo trial. The R values range from 1 to n for any simulation iteration j. The ranks assigned to each risk element may vary from one ranking order to another. If a large number of simulation runs are made, R_i may be considered as a random variable representing ranking order of the component or risk element i. For simplicity the distribution of R_i may be represented as a histogram plot. The ranking order can be done either in relative terms or according to some absolute preset criteria. This topic will be discussed later in this section.

6.5.2 UNCERTAINTY ASSOCIATED WITH THE IMPORTANCE MEASURES

As an alternative to the R_i, one can also develop the distribution of importance measures of a risk element first and then calculate the rank order from such distributions. For example, considering a RAW, the importance of risk element i, (I_{RAW}^i) is calculated from (6.9). Since variables a, b, and P each would be represented by a pdf (obtained from the propagation of individual risk element uncertainties), then I_{RAW}^i would also be represented by a pdf. Again,

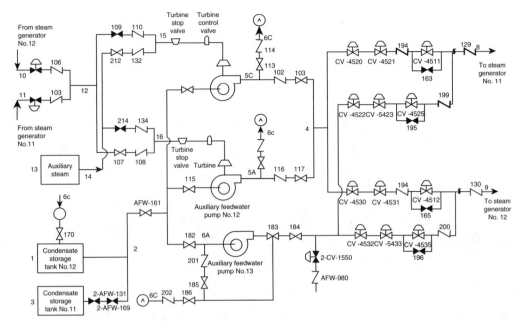

FIGURE 6.9 Auxiliary feedwater system simplified P&ID.

Monte Carlo simulation can be used to find the distribution of I^i_{RAW}. If other importance measures such as RRW or Fussell–Vesely are used, a distribution representing uncertainty of the importance I^i can be equally obtained.

6.5.3 ILLUSTRATIVE EXAMPLE

To demonstrate the methodology described in this section, a simplified auxiliary feedwater system in a pressurized water reactor (PWR) will be analyzed. The system is used for emergency cooling of the steam generators when secondary cooling is lost. Consider the simplified piping and instrumentation diagram (P&ID) of a typical system like the one shown in Figure 6.9. To further simplify this system, the reliability block diagram representation of this P&ID may be used as shown in Figure 6.10.

If we assume that all components are in standby mode; all components are periodically tested, and the failure rates are uncertain and represented by lognormal distributions, the unavailability of each block can then be predicted.

Table 6.4 shows the failure and other operating characteristics and data for this example. In this table, λ is failure rate (per hour), T is test interval (h), T_R is the average repair time (h), T_t is the average test duration (hour), f_r is the frequency of repair/test per interval, and T_0 is operating time ($T_0 = T - T_R - T_t$).

Monte Carlo simulation technique may be used to sample from the component (block) distributions and calculate the importance measures for each trial and determine the rank order of each block (component). As such many ranking orders may be generated.

Fussell–Vesely, RAW, and RRW importance measures are used to simulate the rank order of each block. A total of 5000 samples (i.e., 5000 rank orders) are used. The results are summarized in Table 6.5, Table 6.6, and Table 6.7. Each column shows the probability that a given block (shown in the first column) takes the rank order associated with that column. For example, according to Table 6.5, Block A took the ranked order 9 in 48.3% of times (9th most important unit with 1 being the most important), it took rank order 10 in 51.2% of trials, and

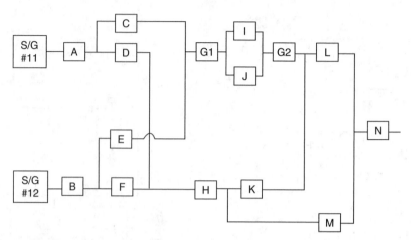

FIGURE 6.10 Simplified auxiliary feedwater system of a PWR.

rank 11 in 0.2% times, and finally, rank order 12 in 0.3% of simulations trails. Therefore, each row of Table 6.5 can be viewed as a histogram showing the distribution of the rank order of each block (risk element). Note that in this case the rank order is a random variable.

It is important to observe that all three importance measures predict close, but somewhat different ranking orders as they can be interpreted differently. Clearly, since they have different meanings as discussed in Sections 6.2 and 6.3, care must be taken to use appropriate measures. Therefore, ranking depends upon the type of the importance measure chosen.

A completely different approach is to first estimate the uncertainties with the importance measures themselves. Then these measures and their uncertainties, are used to rank order

TABLE 6.4
Failure Data for Blocks of Figure 6.2

Block Name	Lognormal Distribution for Failure Rate λ		Frequency of Repair f_r	Average Test Duration T_t (h)	Average Repair Time T_R (h)	Test Interval T (h)
	Mean (μ)	SD (λ)				
A	1×10^{-7}	5.0×10^{-8}	9.2×10^{-3}	0	5	720
B	1×10^{-7}	5.0×10^{-8}	9.2×10^{-3}	0	5	720
C	1×10^{-6}	5.0×10^{-7}	2.5×10^{-2}	0	10	720
D	1×10^{-6}	5.0×10^{-7}	2.5×10^{-2}	0	10	720
E	1×10^{-6}	5.0×10^{-7}	2.5×10^{-2}	0	10	720
F	1×10^{-6}	5.0×10^{-7}	2.5×10^{-2}	0	10	720
G (G1&G)	1×10^{-7}	5.0×10^{-8}	7.7×10^{-4}	0	15	720
H	1×10^{-7}	5.0×10^{-8}	1.8×10^{-4}	0	24	720
I	1×10^{-4}	5.0×10^{-5}	6.8×10^{-1}	2	36	720
J	1×10^{-4}	5.0×10^{-5}	6.8×10^{-1}	2	36	720
K	1×10^{-5}	5.0×10^{-6}	5.5×10^{-1}	2	24	720
L	5×10^{-7}	2.5×10^{-7}	4.3×10^{-3}	0	10	720
M	3×10^{-4}	1.5×10^{-4}	1.5×10^{-1}	0	10	720
N	1×10^{-7}	5.0×10^{-8}	5.8×10^{-4}	0	5	720

TABLE 6.5
Fussell–Vesely Rank Order of Components Using Monte Carlo Simulation

Block Name								Rank Order						
	1	2	3	4	5	6	7	8	9	10	11	12	13	14
A	0	0	0	0	0	0	0	0	0.483	**0.512**	0.002	0.003	0	0
B	0	0	0	0	0	0	0	0	**0.513**	**0.484**	0.001	0.002	0	0
C	0	0	0	0	0	0	0	0	0	0	0.073	0.069	**0.434**	**0.423**
D	0	0	0	0	0	0	0	0	0.002	0.003	**0.429**	**0.418**	0.076	0.073
E	0	0	0	0	0	0	0	0	0	0	0.075	0.074	**0.422**	**0.428**
F	0	0	0	0	0	0	0	0	0.002	0.001	**0.419**	**0.434**	0.068	0.076
G	0	0	0	0	0	0	0.261	**0.739**	0	0	0	0	0	0
H	0	0	0	0	0	0	**0.739**	0.261	0	0	0	0	0	0
I	0	0.011	0.095	**0.645**	0.249	0	0	0	0	0	0	0	0	0
J	0	0	0.011	0.095	**0.645**	0.249	0	0	0	0	0	0	0	0
K	0	0.004	0.029	0.259	0.089	**0.619**	0	0	0	0	0	0	0	0
L	0	0.224	**0.68**	0	0.012	0.084	0	0	0	0	0	0	0	0
M	**0.46**	**0.54**	0	0	0	0	0	0	0	0	0	0	0	0
N	**0.54**	0.221	0.185	0.001	0.005	0.047	0	0	0	0	0	0	0	0

components. The distribution associated with FV, RAW, and RRW importance measures is calculated using the same Monte Carlo simulation methodology discussed earlier. For example, the importance measure uncertainty distribution for Fussell–Vesely and RAW measures for block I of the example under study is shown in Figure 6.11.

If we use the mean, median, or 95th percent quantiles of the importance measure distribution of each block and the corresponding rank order of the blocks based on these quantities, then one may arrive at the rank orders of the components as shown in Table 6.8.

TABLE 6.6
RAW-Based Rank-Order Monte Carlo Simulation Results

Block Name							Rank Order							
	1	2	3	4	5	6	7	8	9	10	11	12	13	14
A	0	0	0	0	0	0.004	0.091	0.129	**0.387**	**0.389**	0	0	0	0
B	0	0	0	0	0	0.003	0.093	0.133	**0.387**	**0.384**	0	0	0	0
C	0	0	0	0	0	0	0	0	0	0	0.059	0.07	0.417	**0.454**
D	0	0	0	0	0	0	0	0	0	0	**0.44**	**0.434**	0.076	0.05
E	0	0	0	0	0	0	0	0	0	0	0.054	0.07	**0.433**	**0.443**
F	0	0	0	0	0	0	0	0	0	0	**0.447**	**0.425**	0.075	0.053
G	0	0	0.087	**0.913**	0	0	0	0	0	0	0	0	0	0
H	0	0	**0.913**	0.087	0	0	0	0	0	0	0	0	0	0
I	0	0	0	0	0	0.086	**0.359**	0.323	0.114	0.118	0	0	0	0
J	0	0	0	0	0	0.077	**0.367**	0.335	0.112	0.109	0	0	0	0
K	0	0	0	0	**0.741**	0.253	0.004	0.002	0	0	0	0	0	0
L	0	**1**	0	0	0	0	0	0	0	0	0	0	0	0
M	0	0	0	0	0.259	**0.576**	0.085	0.079	0	0	0	0	0	0
N	**1**	0	0	0	0	0	0	0	0	0	0	0	0	0

TABLE 6.7
RRW-Based Rank-Order Monte Carlo Simulation Results

Block Name	

Block Name	1	2	3	4	5	6	7	8	9	10	11	12	13	14
							Rank Order							
A	0	0	0	0	0	0	0	0	**0.571**	0.424	0.002	0.002	0	0
B	0	0	0	0	0	0	0	0	0.425	**0.572**	0.001	0.002	0	0
C	0	0	0	0	0	0	0	0	0	0	0.094	0.06	**0.575**	0.27
D	0	0	0	0	0	0	0	0	0.002	0.002	**0.507**	0.344	0.093	0.051
E	0	0	0	0	0	0	0	0	0	0	0.058	0.086	0.278	**0.578**
F	0	0	0	0	0	0	0	0	0.002	0.001	0.337	**0.506**	0.054	0.101
G	0	0	0	0	0	0	0.261	**0.739**	0	0	0	0	0	0
H	0	0	0	0	0	0	**0.739**	0.261	0	0	0	0	0	0
I	0	0.011	0.095	**0.645**	0.249	0	0	0	0	0	0	0	0	0
J	0	0	0.011	0.095	**0.645**	0.249	0	0	0	0	0	0	0	0
K	0	0.004	0.029	0.259	0.089	**0.619**	0	0	0	0	0	0	0	0
L	0	0.224	**0.68**	0	0.012	0.084	0	0	0	0	0	0	0	0
M	**0.46**	0.54	0	0	0	0	0	0	0	0	0	0	0	0
N	**0.54**	0.221	0.185	0.001	0.005	0.047	0	0	0	0	0	0	0	0

The results are different from those tabulated in Tables 6.5 through 6.7. It can be argued that once we determine the importance measure distribution, ranking can only be based on a given quantile or the mean value. The drawback of this approach is that in doing so we compare the values corresponding to the specific quantile (e.g., 50% or 95%) for all of the importance distributions. The probability that all the components assume a specific quantile level is extremely small. So this ranking approach would give erroneous results. However, from this example it can be concluded that the mean values of component performance yield a reasonably accurate ranking, despite some moderate uncertainties in the component performance values. This conclusion, however, may not hold if uncertainties associated with component performance values are very large.

It is important to know how the results of the ranking obtained by using the various importance measures under uncertainty. The ranking process is often performed for one of the two reasons:

- To order components, systems, human actions, etc. relative to each other.
- To determine whether contribution of components, systems, human, actions, etc. to risk meets an absolute preset criterion.

This issue is further discussed in the context of this example in the following sections.

6.5.4 RELATIVE RANKING BASED ON UNCERTAIN IMPORTANCE MEASURES

When a relative ranking is performed based on a given importance measure, the ranking order of each risk element (phenomena, component, system, human action, etc.) relative to each other is desirable. Uncertainty associated with the unavailability or other performance measures associated with the components of the PRA may affect the ranking order. Consider the AFW system. The relative ranking order for components N, M, L, K, H, G, B, and A considering the uncertainty associated with the unavailability of these component blocks is the same as the mean ranking. On the other hand, ranking of the components I, J, K, D,

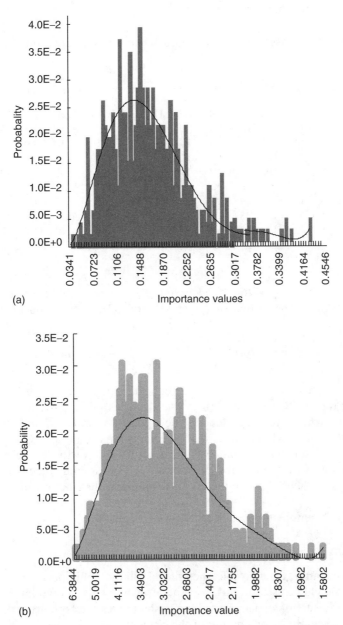

FIGURE 6.11 Importance measure distribution for block 1 of AFW systems. (a) FV and (b) RAW.

F, C, and E is demonstratively different. A relative ranking can be performed using Figure 6.12.

In Figure 6.12 the relative ranking order can be visually inspected as: I, H, G, and A with I being at the highest rank. The selection may not be always as straightforward as the one shown in Figure 6.12. For example, consider Figure 6.13. To clearly determine whether A ranks higher than B or the reverse applies, a probabilistic measure of exceedance between the two distributions may be needed.

Suppose R_A and R_B represent two random variables describing the ranking order of component blocks A and B (note that there are n such component blocks). We are interested

TABLE 6.8
Ranking of Block A–N Based on Various Importance Measure Distributions

	Quantile	Rank Order												
		1	2	3	4	5	6	7	8	9	10	11	12	13	14
FV	Mean	N	M	L	K	I, J	I, J	H	G	A, B	A, B	D	F	C, E	C, E
	50%	N	M	L	K	I, J	I, J	H	G	A, B	A, B	D, F	D, F	C, E	C, E
	95%	N	M	L	I, J	I, J	K	H	G	A, B	A, B	D	F	E	C
RAW	Mean	N	L	H	G	K	M	J	I	B	A	F	D	C	E
	50%	N	L	H	G	K	M	J	I	B	A	D	F	E	C
	95%	N	L	H	G	K	M	I	J	B	A	F	D	E	C
RRW	Mean	N	M	L	I, J	I, J	K	H	G	A, B	A, B	D, F	D, E	C, E	C, E
	50%	N	M	L	I, J	I, J	K	H	G	A, B	A, B	D, F	D, F	C, E	C, E
	95%	N	M	L	I, J	I, J	K	H	G	A, B	A, B	D, F	D, F	C, E	C, E

in the probability that $R_A > R_B$. If $f(R_A)$ and $f(R_B)$ represent the truncated distribution functions for ranking order of the components A and B between ϕ and 17 then

$$\Pr(R_A > R_B) = \int_0^n f(R_A) \int_0^{R_A} f(R_B) d(R_B) d(R_A) \qquad (6.17)$$

If $\Pr(R_A > R_B)$ is greater than 0.5 it may be concluded that ranking order of A exceeds B. Conversely, if $\Pr(R_B > R_A)$ is greater than 0.5, then the ranking order of B exceeds A. As a practical matter, higher values than 0.5 may be used, for example 0.7 and above, with values in the range of 0.3–0.7 may be considered as an equally important rank. Generally truncated a normal distribution for $f(R_i)$ would be a reasonable choice. In this case, the mean and standard deviation of the sample for the ranking level of a component can be used as the mean and standard deviation parameters of a binomial distribution or its equivalent the normal distribution.

For the AFW example, the results are shown in Table 6.9.

6.5.5 ABSOLUTE RANKING BASED ON UNCERTAIN IMPORTANCE MEASURES

When an absolute ranking is desirable, then the distribution of importance measures for each system element should be used. In this case, the probability that a random variable representing importance measure of an item exceeds an absolute preset acceptance criterion is used as a measure of the acceptability or ranking of the element.

For example, consider the case where absolute ranking of certain components are based on the preset absolute acceptance criteria as shown in Table 6.10.

Again, in this case ranking decision should consider the quantitative uncertainty in the calculated importance measures as depicted in Figure 6.14.

By qualitative inspection, if a large part of the density associated or probability function with importance of a given component or risk element falls within any of the qualitative ranking orders (high, medium, and low discussed in Table 6.10) then such a ranking can be selected. If the distribution spreads over two or more ranking criteria, then it would be possible to use probabilistic exceedance analysis. For example, by assuming a lognormal distribution for the random variable I_i (where I is the importance of item i), then $\Pr(b > I_i > a)$ is the probability that importance measure of component i falls between two acceptance criteria a and b represents the probability that the selected qualitative ranking is adequate.

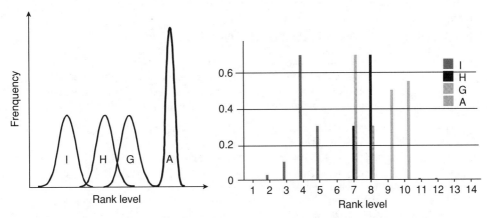

FIGURE 6.12 Relative ranking results based on order obtained from Fussell–Vesely.

Otherwise, other qualitative ranking measures such as medium–high may be used, if a distribution reasonably spreads over both the "medium" and "high" regions.

Consider the example in this section using the criteria discussed earlier. The ranking results shown in Table 6.11 can be obtained.

A final observation based on the AFW example is that the mean ranking yields reasonable results and captures most of the uncertainties in the PRA model.

6.6 UNCERTAINTY IMPORTANCE MEASURES

Beside the importance of each element of the risk model to the total system risk (or performance), one may also seek contribution of uncertainty of each risk element to the uncertainty of total system risk (or performance). This notion is called uncertainty import-

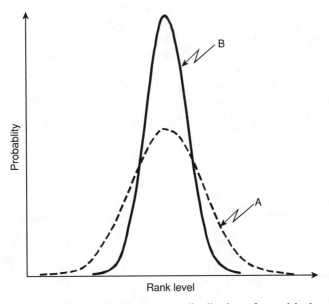

FIGURE 6.13 A typical case of comparing importance distribution of two risk elements.

TABLE 6.9
Relative Ranking of the Blocks in the Example Problem

R_M	Mean Ranking	Probability that Component's Rank Exceeds the Next Mean Rank Component	Probability of Exceeding Next Two Mean Rank Component	Final Risk-Based Ranking	R_f
1	N	Pr(N > M) = 0.5404	Pr(N > L) = 0.7674	M	1
2	M	Pr(M > L) = 1.0	Pr(M > K) = 1.0	N, M	2
3	L	Pr(L > K) = 0.9078	Pr(L > I) = 0.9044	L	3
4	K	Pr(K > I) = 0.2816	Pr(K > J) = 0.2816	I, J	4
5	I, J	Pr(I > J) = 0.5	Pr(I > H) = 1.0		
6	I, J	Pr(J > H) = 1.0	Pr(J > G) = 1.0	K	5
			Pr(K > H) = 1.0		
7	H	Pr(H > G) = 0.7394	Pr(H > B) = 1.0	H	6
8	G	Pr(G > B) = 1.0	Pr(G > A) = 1.0	G	7
9	A, B	Pr(B > A) = 0.5158	Pr(B > D) = 0.996	A, B	8
10	A, B	Pr(A > D) = 0.9954	Pr(A > F) = 0.9956	A, B	9
11	D	Pr(D > F) = 0.5062	Pr(D > C) = 0.8518	D, F	10
12	F	Pr(F > C) = 0.8558	Pr(F > E) = 0.8518		
13	C, E	Pr(C > E) = 0.497	—	C, E	11
14	C, E	—	—		

ance. The importance measures that we discussed in Section 6.2 are no longer valid for this purpose. We are interested to know the elements that highly affect the uncertainty of the total risk value (e.g., the tails of the distribution of the risk value) rather than the contributors to the risk value itself. Therefore, in general, the uncertainty importance measure focuses on the contribution of a particular input variable to uncertainty of the output risk value.

The method mostly used for uncertainty ranking is to find contributors from the input elements and variables to variance of the total system risk. In particular, if the variance of a particular input variable goes to zero, we can measure the degree to which the variance in the output distribution (e.g., the total risk value of system performance) is reduced. Using a Monte Carlo simulation approach (using random sampling or LHS), the uncertainty importance can be readily estimated by fixing the input variable at its mean or point estimate value and repeating the sampling analysis to recalculate the output distribution. However, because of the instability in estimating the expected values of skewed distributions and sensitivity to the shapes of the tails of the distributions, the calculation is often performed on a logarithmic scale. That is, the calculations will be done with respect to the variance in the logarithm of the output variable.

TABLE 6.10
Acceptance Criteria for Risk-Significant Components

FV Importance Value	Qualitative Rank
$I_{FV} > 0.01$	High
$0.001 < I_{FV} < 0.01$	Medium
$I_{FV} < 0.001$	Low

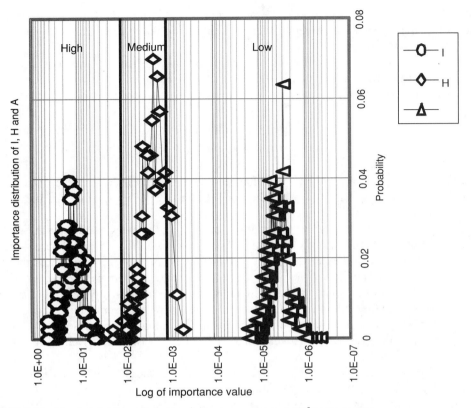

FIGURE 6.14 Absolute ranking based on importance measures of components.

To measure the uncertainty importance of the dependent variable $y = f(x_i)$, if coefficient of variation of the independent variable x_i is reduced to zero, but all other x_i's are kept unchanged, then depending on the uses of the uncertainty importance, one or more of the following expressions may be used as a measure of uncertainty importance of x_i:

$$I_{xi} = CVy_{base}/CVy_{new} \tag{6.18}$$

TABLE 6.11
Absolute Ranking of Blocks in the Example Problem Using FV Importance Measure

Rank	Mean Ranking	Probability	Blocks	Qualitative Rank Order
High	I, J, K, L, M, N	$Pr(I_i > 0.01) = 1.0$	I, J, K, L, M, N	High
Medium	H	$Pr(I_H < 0.001) = 0.0477$ $Pr(0.001 < I_H < 0.01) = 0.941$ $Pr(I_H > 0.01) = 0.0113$	H	Medium
	G	$Pr(0.001 < I_G < 0.01) = 0.8446$ $Pr(I_G < 0.001) = 0.1554$	G	
Low	A, B, C, D, E, F	$Pr(I_i < 0.001) = 1.0$	A, B, C, D, E, F	Low

where I_{xi} is the uncertainty importance of x_i, and $CV_{y\text{base}}$, and $CV_{y\text{new}}$ are coefficient of variation of the base value of y (e.g., total risk) and the coefficient of variation of y when CV_{xi} is set to zero, respectively.

Two variations of (6.18) may also be used for cases when we are interested in the sensitivity of the mean value of y and the tails of the distributions of y to the uncertainty of the risk element or parameter x_i. These are:

$$I_{xi} = \mu_{y\text{base}}/\mu_{y\text{new}}$$
$$I_{xi} = L_{y\text{base}}/L_{y\text{new}} \qquad\qquad (6.19)$$

where $\mu_{y\text{base}}$ and $\mu_{y\text{new}}$ are the base (no change) and new (with CV_{xi} set to zero) mean of y, respectively. Also, $L_{y\text{base}}$ and $L_{y\text{new}}$ are the upper bound (e.g., 95% quantile) of the distribution of base and new distributions of y.

Equation (6.19) for estimating uncertainty importance involves examining the response of fixed quantiles of the output distribution, such as mean, the 0.05 and 0.95 quantile, to changes in the uncertainties in the input parameters. This approach calculates the ratios described by (6.19) and shows the effect of input variance on the mean and quantiles of the output distribution, when the individual input uncertainties are held at their mean values. These quantiles allow the analyst to consider how the overall output distribution may change (e.g., become narrower) as a result of eliminating the selected epistemic uncertainties, especially if a risk management or policy decision is driven by the tails of the output distribution.

Several other uncertainty importance measures exist in the literature. Bier [10] has presented an uncertainty importance measure as percentage change in the total risk variance caused by a given percentage change in the variance of the input elements of the risk model. Iman and Hora [11] have developed a measure by using the Taylor series expansion, which yields the total contribution to the variance or the frequency change of the top event that can be assigned to the corresponding quantities of each basic event. Another measure by Rushdi [12] assesses the expected reduction in the system variance V due to probability of each event x_i and its variance. A theoretically complete formulation of an uncertainty importance measure is given by Iman and Hora [11]. This measure is derived as an expected reduction in variance of the top event frequency with respect to each input parameter of interest. A decision theoretic framework has been proposed by Chun et al. [13], which provides an estimate of the risk elements that would be most valuable, from the decision-making point of view, to procure more information.

Some useful importance uncertainty measures similar to (6.18) and (6.19) are shown in Table 6.12. They are expressed in terms of variance of the total risk value, R, or error factor (for lognormal distribution) of each basic element of the risk model. The method as addressed above assesses the expected percentage reduction in the variance V of the output risk, R, due

TABLE 6.12
Some Uncertainty Importance Measure

Iman	$\dfrac{V_{xi}[E(R\|x_i)]}{V} \times 100\%$
Iman and Hora	$\dfrac{V_{xi}[E(\log R\|x_i)]}{V(\log R)}$
Cho and Yum	$\dfrac{\text{Uncertainty of } \ln R \text{ due to the uncertainty of } \ln x_i}{\text{Total uncertainty of } \ln R}$

to fixing an input variable (element i) value to its mean value μ_{xi}. To evaluate this measure, the base variance V needs to be computed. For large-sized risk models, it is very complicated to analytically compute V, and, therefore, V is often estimated by using a Monte Carlo simulation. This, however, is difficult because of instability in the value of V.

To obtain a stable uncertainty importance measure of each basic event, Iman and Hora [11] suggest the second measure listed in Table 6.12. This measure represents the percentage variance of logarithm of R explained by setting variance of input element i to zero and use its mean μ_{xi}. The third measure by Cho and Yum [14] assesses the mean effect of the uncertainty involved in a log-transformed basic event probability, on the uncertainty of $\ln(R)$, under the assumption that all input elements are independent of each other and they distribute lognormally.

6.7 PRECURSOR ANALYSIS

Risk analysis may be carried out by completely hypothesizing scenarios of events, which can lead to exposure of hazard, or may be based on actuarial scenarios of events. Sometimes, however, certain actuarial scenarios of events may have occurred without leading to an exposure of hazard, but involve a substantial erosion of margin of hazard prevention or mitigation. These scenarios are considered as precursors to accidents (exposure of hazard). We are often interested to know significance and importance of these events.

Precursor events or simply PEs, in the context of risk assessment, can be defined as those operational events that constitute important elements of risk scenarios leading to consequences (or hazard exposure) in complex systems or facilities such as a severe core damage in a nuclear power plant, severe aviation or marine accidents, chemical plant accidents, etc. The significance of a PE is measured through the conditional probability that the actual event or scenarios of events would result in exposure of hazard. In other words, PEs are those events that substantially reduce the margin of safety (or hazard exposure barriers) available for prevention or mitigation of risk value and losses.

The precursor analysis methodology considered in this section is mainly based on the methodology developed for nuclear power plants [15], nevertheless, its application to other complex systems seems to be straightforward.

6.7.1 BASIC METHODOLOGY

Considering a risk scenario in a system given as one following the HPP, the MLE for the rate of occurrence of risk events, λ, can be written as

$$\hat{\lambda} = \frac{n}{t} \tag{6.20}$$

where n is the total number of risk exposure events observed in nonrandom exposure (or cumulative exposure) time t. The total exposure time can be measured in such units as reactor years (for nuclear power plants), aircraft hours flown, vehicle miles driven, etc.

Because a severe risk event such as an accident is a rare event (i.e., n is quite small), estimator (6.20) cannot be applied, so one must resort to postulated events, whose occurrence would lead to severe accident. The marginal contribution from each PE in the numerator of (6.20) can be counted as a positive number less than 1. For nuclear power plants Apostolakis and Mosleh [16] have suggested using conditional core damage probability given a PE in the numerator of (6.20). Obviously this approach can be similarly used for other complex systems.

Considering all such PEs that have occurred in exposure time t, the estimator (6.20) is replaced by

$$\hat{\lambda} = \frac{\sum_i p_i}{t} \tag{6.21}$$

where p_i is the conditional probability of the risk event given PE i.

The methodology of precursor analysis has two major components — screening, i.e., identification of PEs with anticipated high p_i values, and quantification, i.e., estimation of p_i and λ, and developing corresponding trend analysis, as an indicator of the overall system's risk trend, which are discussed below.

6.7.2 Categorization and Selection of Precursor Events

The conditional probabilities of risk due to hazard exposure given PEs i $(i = 1, 2, \dots)$, p_i, are estimated based on the data collected on the observed operational events to identify those events that are above a threshold level. These events are known as significant PEs. The process of estimating the p_i's is rather straightforward. Events are mapped onto an event tree (since presumably a partial part of a scenario modeled in the event tree has actually happened as part of the PE), and other failures, which eliminate the remaining hazard barriers, are postulated to fail so as to complete a risk scenario. The precursor analysis event trees are developed the same way as in regular PRA methods. In fact regular PRAs are used for precursor studies. The probabilities that such postulated events occur are multiplied to estimate the conditional probability of a risk (loss) event of interest.

The process of mapping an event i onto event trees and subsequently calculating the conditional probability p_i turn out to be time-consuming. However, because majority of the events are rather minor, only a small proportion of events — those that are expected to yield high p_i values (meet the preset qualitative screening criteria) — need to be analyzed. On the other hand to estimate the rate of occurrence of hazard exposure events, $\hat{\lambda}$, using (6.21), it would be advisable to include the risk-significance of all PEs because the more frequent but less significant events are not considered. For example, in a system having no events that meet some precursor selection criteria, (6.21) yields a zero estimate for $\hat{\lambda}$. However, provided the system may had some other incidents with potentially small p_i values which do not meet the selection criteria chosen, the zero value underestimates the system is true rate of occurrence of losses due to hazard exposure events, $\hat{\lambda}$. Therefore, a background risk correction factor that collectively accounts for these less serious incidents is sometimes introduced [15].

6.7.3 Properties of Precursor Estimator for the Occurrence Rate of Hazard Exposure Events and Their Interpretation

Because the set of hazard exposure events (e.g., accidents scenarios) corresponding to the observed PEs usually overlap, it has been shown (see Rubenstein [17], Cooke et al. [18], Bier [19], Abramson [20], Modarres et al. [15]) that there is overcounting of conditional probability of exposure given the observed event.

To assess the appropriateness of using (6.21) as an estimator of the rate of occurrence of hazard exposure events λ, it is essential to evaluate the statistical properties of this estimator. To do this, one needs a probabilistic model for the number of PEs and a model for the magnitude of the p_i values. Usually it is assumed that the number of precursors observed in exposure time t follows the HPP with a rate (intensity) μ, and p_i is assumed to be an independently distributed continuous random variable having a truncated (due to the threshold mentioned above) pdf $h(p)$. For the U.S. nuclear power plants examples considered below, the lower truncation value p_0, as a rule, is 10^{-6}.

Under these assumptions the estimator (6.21) can be written as

$$\hat{\lambda} = \frac{\sum_{i=1}^{N(t)} p_i}{t} \tag{6.22}$$

where the number of items in the numerator $N(t)$ has the Poisson distribution with mean μt, and the conditional probabilities p_i are all independently and identically distributed according to pdf $h(p)$. Suppose now that $N(t) = n$ precursors have occurred in exposure time t, thus, $\hat{\mu} = n/t$. As it was mentioned, the exposure time t may be cumulative exposure time. For example, for the U.S. nuclear power plants, for the period 1984 through 1993, $n = 275$ precursors were observed in $t = 732$ reactor year of operation (see Modarres et al. [15]); thus, $\hat{\mu} = 0.38$ precursors per reactor year.

There exist numerous parametric and nonparametric methods that can be used to fit $h(p)$, based on the available values of p_i. Some parametric and nonparametric approaches are considered in Modarres et al. [15].

For an appropriately chosen (or fitted) distribution $h(p)$, one is interested in determining the corresponding distribution of the estimate (6.22), from which one can then get any moments or quantiles of interest, such as the mean or 95th quantile. In general, it is difficult analytically to determine the distribution of $\hat{\lambda}$, therefore; Monte Carlo simulation is recommended as a universal practical approach.

The HPP model considered can be generalized by using the NHPP model (introduced in Chapter 4) with intensity $\mu(t)$ for $N(t)$, which allows one to get an analytical trend for λ.

Another approach is based on the use of a truncated nonparametric pdf estimator of $h(p)$ [21] and Monte Carlo simulation to estimate the distribution of $\hat{\lambda}$. This approach is known as the smooth bootstrap method.

An alternative but similar model can be obtained through the use of the extreme value theory. An analogous example for earthquakes is considered in Castillo [22], in which the occurrence of earthquakes is treated as the HPP, and severity (or intensity in geophysical terms) of each earthquake is assumed to be a positively defined random variable. It is clear that the conditional probability of hazard exposure p_i given a precursor considered is analogous to earthquake severity given the occurrence of an earthquake.

To further illustrate the application of extreme value theory, suppose that we are interested in the distribution of the maximum value of conditional probability of risk events, which we denote by P_{max}, for exposure time t, based on random sample of size n precursors that occur in t. Let $H(p)$ denote the cdf corresponding to $h(p)$. The distribution function of P_{max} for a nonrandom sample of size n is given by $H_n(p)$. Consider n has the Poisson distribution with parameter μt, the cdf of P_{max} becomes

$$H_{max}(p,t) = \sum_{n=0}^{\infty} \frac{\exp(-\mu t)(\mu t)^n H_n(p)}{n!}$$

Using the MacLaurin expansion for an exponent, this relationship can be written as

$$H_{max}(p,t) = \sum_{n=0}^{\infty} \exp\{-\mu t[1 - H(p)]\} \tag{6.23}$$

Correspondingly, the probability that the maximum value is greater than p (probability of exceedance) is simply $1 - H_{max}(p, t)$. Equation (6.23) can be generalized for the case of the NHPP with the rate $\mu(t)$ as

$$H_{\max}(p,t_1,t_2) = \exp\left\{\int_{t_1}^{t_2} \lambda(s)\,ds[1-H(p)]\right\}$$

Using the corresponding sample (empirical) cdf for the PEs to estimate $H(p)$, it is possible to estimate the probability of exceeding any value p in any desired exposure time t. The corresponding example associated with nuclear power plant safety problems is given in the following section.

6.7.4 APPLICATIONS OF PRECURSOR ANALYSIS

From the discussion above it is obvious that the precursor analysis results can be used as follows:

1. To identify and compare safety significance of operational events, which are then considered as major precursors.
2. To show trends in the number and significance of the PEs selected.

Some examples of precursor analysis for the nuclear power plant data for the 1984 through 1993 period [15] are considered below. In the framework of nuclear power plant terminology "sever accident" is referred to accidents leading to a *core damage*; correspondingly the term *conditional probability of core damage* is used as a substitute of conditional probability of severe accidents (conditional risk).

The results of analysis of precursor data for the 1984 through 1993 period are given in Table 6.13. The table gives a breakdown of important precursors but it does not show trends in the occurrence of precursors as an indicator of the overall plant safety. Figure 6.15 represents one such indicator. In this figure, the conditional core damage probabilities p_i of the precursors for each year are summed to calculate a value which is then used as an indicator of the overall safety of plants.

Provided the bias in (6.22) is constant or approximately constant, one can use the estimator to analyze an overall trend in the safety performance. The accumulated precursor data for 1984 through 1993 are used at the end of each subsequent year to sequentially estimate the intensity of the HPP μ for the occurrence of the precursors.

TABLE 6.13
Analysis of Nuclear Power Plant Precursor Data for the 1984 Through 1993 Period

Year	Cumulative Reactor Years	Cumulative Number of Precursors, n_i	Cumulative p_i	Rate of Occurrence of Core Damage
1984	52.5	32	0.00579	1.1×10^{-4}
1985	114.2	71	0.02275	2.0×10^{-4}
1986	178.1	89	0.02857	1.6×10^{-4}
1987	248.6	122	0.03268	1.3×10^{-4}
1988	324.7	154	0.03509	1.1×10^{-4}
1989	400.7	184	0.03741	9.3×10^{-5}
1990	481.4	212	0.04124	8.6×10^{-5}
1991	565.4	238	0.05124	9.0×10^{-5}
1992	649.1	262	0.05358	8.1×10^{-5}
1993	732.0	275	0.05440	7.2×10^{-5}

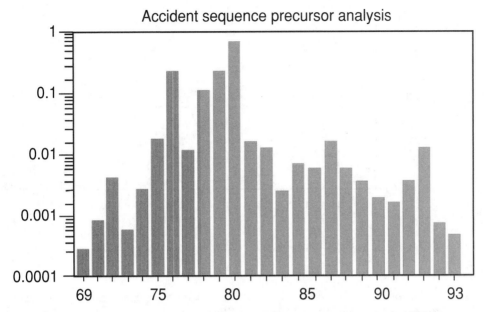

FIGURE 6.15 Annual sum of precursor in terms of conditional core damage probabilities.

Figure 6.16 illustrates the trend obtained from an approach based on the truncated lognormal distribution of the conditional core damage probabilities p_i which was fitted using the method of moments and the sample of 275 values of p_i.

Having this distribution estimated, the distribution of $\hat{\lambda}$ in (6.20) may be estimated using Monte Carlo simulation from which the mean and upper 95% quantile were calculated. The maximum for 1985 is associated with the outlying PEs observed in this year for which $p_i = 0.011$.

Finally, Figure 6.17 shows the trend based on the extreme value approach (6.23). The probabilities that p_{max} exceeds the two indicated values (0.01 and 0.001) are plotted based on the same precursor data. Note that the results in Figure 6.17 indicate the same general trend as in Figure 6.16.

6.7.5 DIFFERENCES BETWEEN PRECURSOR ANALYSIS AND PROBABILISTIC RISK ASSESSMENTS

The precursor analysis originated with the objective of validating (independently) the PRA results. However, the two approaches are fundamentally the same but with different emphasis. For example, both approaches rely on event trees to postulate risk scenarios and both use system or facility-specific data to obtain risk values (e.g., failure probability or severe accidents such as core damage in the case of nuclear power plants). The only thing that differentiates the two approaches is the process of identifying significant events. Readers are referred to Cook and Goossens [23], which conclude that PRA and precursor analyses are only different in the way these analyses are performed; however, both approaches use the same models and data for the analysis. Therefore, the results from the two methods cannot be viewed as totally independent, and one cannot validate the other.

Another small difference between the two approaches is the way dependent failures are treated. Dependent failures such as common-cause failures are considered in precursor analysis because a PE may include dependent failures. This is a favorable feature of precursor analysis. One can also estimate the contribution that common-cause or other events to make

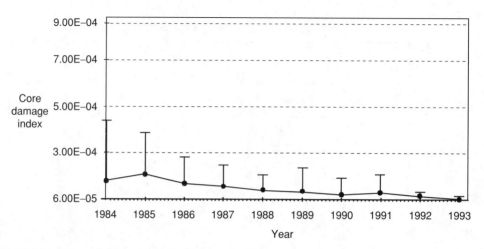

FIGURE 6.16 Truncated lognormal distribution for $h(p)$.

the overall rate of occurrence of severe accidents. Common cause failures are explicitly modeled in PRA as the same way discussed in Chapter 4.

The last difference is that PRAs limit themselves to a finite number of postulated events. However, some events that are not customarily included in PRA (due to unknown events not considered and known at the time of risk assessment by the analysts) may occur as PEs, and these may be important contributions to risk. This is certainly an important strength of precursor methodology to reduce the so-called unknown–unknown events discussed in Table 5.1.

Exercises

6.1 For the containment spray system below calculate the Birnbaum and Vesely–Fussell importance measures for all events using a fault tree analysis assuming "No H_2O spray" from the system as the top event. Assume all events have a probability of failure 0.01 per demand and failure rate of 1×10^{-4} per h of operation.
   • There are no secondary failures.
   • There is no test and maintenance.
   • There are no passive failures.

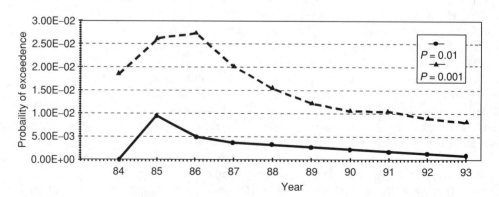

FIGURE 6.17 Safety trends based on Equation (6.23). Probability that P_{max} exceeds P.

- There are independent failures.
- One of the two pumps and one of the two spray heads is sufficient to provide spray (only one train is enough).
- One of the valves sv_1 or sv_2 is opened after demand. However, sv_3 and sv_4 are always normally open.
- Valve sv_5 is always in the closed position.
- There is no human error.
- SP_1, SP_2, sv_1, sv_2, sv_3, and sv_4 use the same power source P to operate.

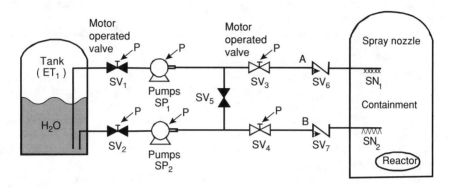

6.2 Calculate RAW and RRW (fraction form) of each component C_i ($i = 1, 2, 3$) with respect to the risk (R) if

$$1 \cdot R = I(1 - C_1)(1 - C_3)C_2$$

and $I = 10^{-3}$ per year, $\Pr(C_1) = 0.001$, $\Pr(C_2) = 0.008$, and $\Pr(C_3) = 0.005$, interpret the results.

6.3 A scenario is characterized by the following cutest.

$$S = a \cdot b + a \cdot c + d$$

Probability of a and b can be obtained from an exponential model such as

$$\Pr_a(t) = 1 - \exp(-\lambda_a t),$$

but probability of c and d can be obtained from a Weibull model such as

$$\Pr_c(t) = 1 - \exp\left[-\left(\frac{t}{\alpha}\right)^\beta\right]$$

If $\lambda_a = \lambda_b = 10^{-4}$ per h and $\alpha_c = \alpha_d = 2250$ h and $\beta_c = \beta_d = 1.4$. Determine the importance of parameters, α, β, and λ at a given time T in modeling this scenario. Discuss pros and cons of using various important measures for this purpose.

REFERENCES

1. Borgonovo, E. and Apostolakis, G.E., A new importance measure for risk-informed decision making, *Reliability Engineering and System Safety*, 72(2), 193, 2001.
2. Wall, I.B., Haugh, J.J., and Wortledge, D.H., Recent applications of PSA for managing nuclear power plant safety, *Progress in Nuclear Energy*, 39(3–4), 367, 2001.

3. Birnbaum, Z.W., *On the Importance of Different Components in a Multicomponent System, Multivariate Analysis II*, Krishnaiah, P.R., Ed., Academic Press, New York, 1969.
4. Vesely, W.E. et al., Measures of Risk Importance and their Applications, NUREG/CR-3385, Washington, D.C., 1983.
5. Vesely, W.E. and Davis T.C., Evaluations and Utilization of Risk Importances, NUREG/CR-4377, Washington, D.C., 1985.
6. Fussell, J., How to hand calculate system reliability and safety characteristics, *IEEE Transaction on Reliability*, R-24(3), 169, 1975.
7. Chadwell, G.B. and Leverenz, F.L., *Importance Measures for Prioritization of Mechanical Integrity and Risk Reduction Activities*, AICHE Loss Prevention Symposium, Houston, TX, March 15, 1999.
8. Youngblood, R.W., Risk significance and safety significance, *Reliability Engineering and System Safety*, 73, 121, 2001.
9. Youngblood, R.W. and Worrell R.B., Top event prevention in complex systems, in *Proceedings of the 1995 joint ASME/JSME Pressure Vessels and Qualitative Risk Assessment System (QRAS)*, Original version 1997–1998, Extended version 1999–2001, Reliability Engineering Program, University of Maryland at College Park.
10. Bier, V.M., A measure of uncertainty importance for components in fault trees, *Transaction of American Nuclear Society*, 45, 384, 1983.
11. Iman, R.L. and Hora, S.C., A robust measure of uncertainty importance for use in fault tree system analysis, *Risk Analysis*, 10(3), 401, 1990.
12. Rushdi, A.M., Uncertainty analysis of fault tree outputs, *IEEE Transactions on Reliability*, R-34, 458–462, 1985.
13. Chun, M.H., Han, S.J., and Tak, N.I., An uncertainty importance measure using a distance metric for the change in accumulative distribution function, *Reliability Engineering and System Safety*, 70, 313, 2000.
14. Cho, J.G., and Yum, B.J., Development and evaluation of an uncertainty importance measure in fault tree analysis, *Reliability Engineering and System Safety*, 57(2), 143, 1997.
15. Modarres, M., Martz, H., and Kaminskiy, H., The accident sequence precursor analysis: review of the methods and new insights, *Nuclear Science and Engineering*, 123(2), 238, 1996.
16. Apostolakis, G. and Mosleh, A., Expert opinion and statistical evidence. An application to reactor core melt frequency, *Nuclear Science and Engineering*, 70(2), 135, 1979.
17. Rubenstein, D., Core Damage Overestimation, NUREG/CR-3591, U.S. Nuclear Regulatory Commission, 1985.
18. Cooke, R.M., Goossens, L., Hale, A.R., and Von Der Horst, J., Accident Sequence Precursor Methodology: A Feasibility Study for the Chemical Process Industries, Technical University of Delft, 1987
19. Bier, V.M., Statistical methods for the use of accident precursor data in estimating the frequency of rare events, *Reliability Engineering and System Safety*, 41(3), 267, 1993.
20. Abramson, L.R., Precursor Estimated of Core Damage Frequency, PSAM-II Conference, San Diego, CA, March 20–25, 1994.
21. Scott, P.K. and Fehling, K.A., Probabilistic Methods To Determine Risk-Based Preliminary Remediation Goals, Presented at 13th Annual Meeting of the Society of Environmental Toxicology and Chemistry, Cincinnati, OH, November 8–12, 1992.
22. Castillo, E., *Extreme Value Theory in Engineering*, Academy Press, San Diego, CA, 1988.
23. Cook, R. and Goossens, L., The accident sequence precursor methodology for the European post-severe era, *Reliability Engineering and System Safety*, 27(1), 117, 1990.

7 Representation of Risk Values and Risk Acceptance Criteria

7.1 INTRODUCTION

Risk acceptance criteria are used to compare risk assessment results against figures of merit for making decisions. As such, it is important to first develop a meaningful and consistent expression of risk values. Ambiguities in the expression of risk and improper definitions hamper acceptability of risk. In this chapter, a number of ways in which risk can be expressed and methods to arrive and convey acceptance criteria have been discussed.

Two most important points of view of risk measures are related to individual (e.g., a person, a system, or a facility) and societal risks (e.g., group of people, a fleet of aircrafts, all nuclear plants). The first point of view is that an individual (or system/facility owner) is able to undertake an activity, weighing the risks against the direct and indirect benefits. This point of view leads to acceptable individual risk, usually defined as the "frequency" or "probability" by which an individual may be exposed to a given level of hazard- or experience-specific consequences from such an exposure. The specified level of consequence may be expressed in terms of frequency of loss of life, loss of system, or suffering from certain illnesses. The societal point of view attempts to measure risk (exposure or consequence) of a large population such as loss of life expectancy or changes in the number of cancers due to exposure to a carcinogen for a given total population. Commonly, the notion of risk in a societal context is expressed in terms of the total number of casualties such as the relation between frequency and the number of people affected from a specified level of consequence in a given population from exposure to specified hazards. The specified level of consequence may be modeled by the so-called Farmer's curves (frequency of exceedance curve) of the consequences as discussed in Chapter 2. While in this chapter, we refer to an individual as a person, the concepts discussed equally apply to a facility (e.g., a plant or a process) or a system (an aircraft). Similarly, societal risk applies to a group of facilities or a fleet.

Individual risk is one of the most widely used measures of risk and is simply defined as the fraction of the exposed population to a specific hazard and subsequent consequence per unit time. A further refinement is possible if we make individual risk specific to certain subgroups. For example, the individual risk of death among coal miners or elderly and chronically ill population. It is important to realize that in this subgroup each individual puts in about the same number of hours, so that all of the members are roughly equally exposed. This could not be the case for activities where the individual's annual participation varies enormously, as in transportation.

Once the risk measure (individual or societal) is defined, an upper or lower limit may be placed on certain aspects of risk value (cumulative or continuous limits) A limit may be applied to:

1. The maximum frequency or probability of occurrence of an event (such as failure of a hazard barrier) considered as a significant precursor to individual or societal consequences of fatality, morbidity, or economic loss.

275

2. The maximum rate, duration, or frequency of exposure to a hazardous condition or material (such as exposure to radiation or carcinogens) leading to individual or societal consequence of fatality, morbidity, or economic loss.
3. The maximum individual risk of early death (as in accidents), or delayed death (as in cancer or other major causes of death).
4. The maximum societal risk of early or delayed death.
5. The minimum performance level of hazard barriers designed to prevent, protect, or mitigate exposure to hazards.
6. The maximum risk-aversion level associated with infrequent risks involving large consequences (such risk scenarios involving numerous casualties, environmental, or economical disasters).
7. A risk characterization factor (maximum or minimum) such as importance measure of a risk contributor or uncertainty measure.

Setting specific acceptance criteria is generally a very complex and subjective process. In this chapter, we will discuss in more detail the methods, pitfalls, and strengths of using risk acceptance criteria. As an example, consider Table 7.1 which can be used as a simple guideline for establishing *absolute* individual risk acceptance limits.

There are several alternative strategies for developing risk acceptance criteria which are *relative* in nature. The following three are the important ones:

1. Comparing risk, frequency, or exposure with historical average levels of the same or similar hazards.
2. Comparing actual observed events with estimated risk levels for exposure to the same hazard.
3. Comparing with other general risks in society.

The most practical relative risk level is to compare a specific risk with general risks in the society, mainly due to lack of relevant statistics. When comparing with the general risk in everyday life, it is normal to use the natural fatality risk for the age group with the lowest individual fatality risk (typically young and healthy group). For example, accidental risk of 1×10^{-4} fatality per year for the age group 10–14 years leads to a maximum acceptable risk

TABLE 7.1
Considerations in Establishing Involuntary Individual Risk Acceptance Limits

Fatality Risk Level per Year	Considerations
10^{-3}	This level is unacceptable to everyone. Accidents exposing hazards at this level are difficult to find. When risk approaches this level, immediate action is taken to reduce the hazard.
10^{-4}	This level is acceptable; however, individuals would like to see that public money is spent to control a hazard exposure at this level (e.g., placing traffic signs).
10^{-5}	While acceptable, the individuals in public recognize this level of risk, but generally feel comfortable about it. At this level, individuals may accept inconveniences to avoid this level of risk, such as avoiding air travel.
10^{-6}	Generally acceptable with no great concern. Individuals may be aware of the risks at this level, but feel that they cannot happen to them. They view this level of risk as "an act of God."
Less than 10^{-6}	Acceptable. Individuals are not generally aware of this level of risk.

level for death rate of 1×10^{-4} per year. This approach was advocated in the landmark Reactor Safety Study [1], which estimated the risks from operation of nuclear power plants in the U.S. and established the formal PRA methodology.

Finally, an important consideration in risk acceptance criteria is the way the risk results have been presented. This topic will be further discussed in Section 7.2. However, besides numerical risk value such as the expected loss measure, variations of the Farmer's curves can be used to express the risk, consequences, and frequency results. For example, Figure 7.1 shows, in general, the ways the results are displayed.

In general, the coordinates of the curve in Figure 7.1 can be in form of the examples below:

- Risk vs. consequence
- Probability (frequency) vs. consequence (e.g., economic damages)
- Risk vs. frequency
- Risk vs. time
- Risk vs. space (e.g., distance)
- Magnitude of exposure vs. time
- Performance vs. time

Further, the form of the curve can be continuous, discrete, or a combination of both. Limits may be set on "risk versus a variable" curve. For example, on the value of risk itself, exposure to hazards or performance (plotted on the y-axis), and the risk model input variable (plotted on the x-axis) as shown in Figure 7.1. As an example, consider consequences, c (e.g., in terms of cost), of a risk scenario with frequency, f. Suppose that an expression of the acceptability bounds, as shown in Figure 7.1 have the power law form

$$f \cdot c^\beta = k \tag{7.1}$$

where k is a constant. For a situation where the risk recipient is risk-averse, parameter β is less than 1 (e.g., 0.6). This parameter along with the constant k is selected by the policy-maker or decision-maker, depending on the risk-averseness of the recipient. Because one may argue that for setting risk acceptance criteria, one should judge the degree to which the recipient would

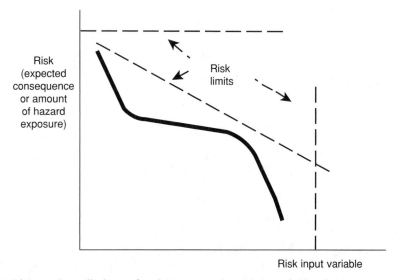

Risk (expected consequence or amount of hazard exposure)

Risk limits

Risk input variable

FIGURE 7.1 Risk results or limits set for risk versus a dependent variable of risk.

react to the likelihood of being exposed to a hazard or facing losses, it is natural for the decision-maker to understand recipient's point of view before selecting β and k. For a risk neutral–risk acceptance criterion, $\beta = 1$ is adequate. Limits on f and c can be placed by the decision-maker.

7.2 REPRESENTATION OF RISK RESULTS

As discussed earlier, the results of risk assessment can be presented for individual or societal risks, and exceedance limits may be placed on such estimates. The most popular methods include the following:

Loss of Life Expectancy. A simple way of showing the results of risk assessment and corresponding risk acceptance criteria is expressing and comparing the results of a specific risk exposure or loss situation in terms of the loss of life expectancy measures (loss of human life, but also reduction in life of systems such as an aircraft). Since risks involving fatality causes "premature" fatalities, it ultimately affects the overall life expectancy of a peculiar societal group exposed to the risk. Since life expectancy is a strong aggregate index of the overall societal health and safety, it is an excellent measure to use for representing the societal risk of an activity in terms of its impact on the overall life expectancy.

Cohen and Lee [2] have developed the measure of loss of life expectancy due to various causes. The idea is roughly like this: Define the population *mortality rate*, h (probability of dying due to all causes per person for year i) as a function of age as follows:

$$h(i) = \frac{\text{Number who die (or fail) in their } i+1 \text{ th year of life}}{\text{Number who live (function) in age } i} \qquad (7.2)$$

Equation (7.2) is the discrete version of the so-called hazard rate in reliability analysis. The probability $\Pr(n)$ of death at age n is then $\Pr(n) = \Pr(\text{death at age } n | \text{alive at age } n-1) \times \Pr(\text{alive at age } n-1)$. Therefore

$$\Pr(n) = h(n) \prod_{i=1}^{n-1} [1 - h(i)] \qquad (7.3)$$

and the life expectancy is

$$LE = \sum_{i=1}^{\infty} i \Pr(i) \qquad (7.4)$$

The mortality rate due to cause j, h_j is defined as

$$h_j(i) = \frac{\text{Number at age } i+1 \text{ who die (or fail) of } j}{\text{Number who live (function) in age } i} \qquad (7.5)$$

Define $h(\Delta_j) = h - h_j$, and use this in (7.3) and (7.4) to compute LE_j. Cohen and Lee [2] define the loss of life expectancy due to cause j, LLE_j as

$$LLE_j = LE_j - LE \qquad (7.6)$$

Example 7.1
Table 7.2 and Table 7.3 show the death rate from all causes and from diseases of the circulatory system (mainly heart diseases and stroke) registered in New Zealand in the year 2000. Calculate the loss of life expectancy due to diseases of the blood circulatory system.

TABLE 7.2
Death Rates by Total Population from All Causes (Rates per 100,000 Population per Year)

			Age-Specific Rate			
0–1	2–14	15–24	25–44	45–64	65–74	75+
632.6	24.6	68.7	112.1	495.4	2117.3	7488.9

Solution

Using (7.2) and (7.5) with the data in Table 7.2 and Table 7.3 yields the following mortality rates for each of the age-specified groups.

Ranges of i	h	h_j
0–1	6.326E–03	5.286E–05
2–14	2.460E–04	6.101E–06
15–24	6.870E–04	2.076E–05
25–44	1.121E–03	1.961E–04
45–64	4.954E–03	1.497E–03
65–74	2.117E–02	7.557E–03
75+	7.489E–02	3.738E–02

Using the mortality rates above along with (7.3) and (7.4) for $i = 1$ to a large number (say 120) and assuming that h and h_j linearly increase starting from the age of 45 years (with the values of h in the table falling in the center of the range), the LE for all causes is estimated as 79.2 years. Similarly LE_j values using (7.3) and (7.4) can be calculated, assuming that the mortality rate is defined as $h(\Delta_j) = h - h_j$. Accordingly, the $LE_j = 86.3$ years. Therefore, the loss of life expectancy due to the diseases of the circulatory system according to (7.6) is measured as 7.1 years, or about 2600 days.

Another way to determine the loss of life expectancy is by measuring the actual age at which a loss of life due to a cause occurs, and subtracting it from the prevailing life expectancy at that time. For example, Hibbert et al. [3] obtained data on the age distribution of ex-coal workers, who died from pneumoconiosis during 1974 and calculated the total loss of life expectancy suffered by those victims. He found that 449 individuals had collectively suffered 4871 years loss of life expectancy. Reasoning that these deaths were incurred among a group of workers exposed over a period of years to conditions prevailing in the industry. Hibbert estimated the total groups at risk to be 450,000. He then calculated the average loss of life expectancy as 4871 years divided by 450,000 workers, which yields 3.95 days lost per worker exposed. Examples of similar societal risks are shown in Table 7.4.

Equivalent Annual Risk. This method introduced by Wilson [5] computes the intensity at which a given activity or exposure to hazard should occur in order to increase the annual

TABLE 7.3
Death Rates by Total Population from Diseases of the Circulatory System (Rates per 100,000 Populations per Year)

			Age-Specific Rate			
0–1	2–14	15–24	25–44	45–64	65–74	75+
5.29	0.61	2.08	19.61	149.69	755.73	3738.50

TABLE 7.4
Loss of Life Expectancy for Some Activities Involving Fatality Risk [4]

Activity or Risk	Loss of Life Expectancy (Days)
Living in poverty	3500
Being male (versus female)	2800
Cigarette smoking (male)	2300
Being unmarried	2000
Being black (versus white)	2000
Working as a coal miner	1100
30 pounds overweight	900
Grade school dropout	800
Suboptimal medical care	550
15 pounds overweight	450
All accidents	400
Motor vehicle accidents	180
Occupational accidents	74
Married to smoker	50
Drowning	40
Speed limit: 65 versus 55 mph	40
Falls	39
Radon in homes	35
Firearms	11
All electricity, nuclear (based on Union of Concerned Scientists Data)	1.5
Peanut butter (1 Tbsp/day)	1.1
Hurricanes, Tornadoes	1
Airline crashes	1
Dam failures	1
All electricity, nuclear (based on Nuclear Regulatory Commission Data)	0.04

probability of death of 1 in 1-million. For example, the following activities or exposures have a risk of 10^{-6}:

- Living 2 days in New York City or Boston (exposure to pollutants)
- Smoking 1.4 cigarettes (exposure to smoke)
- Flying 1000 miles by jet (accidental crash)

This representation is used for easy comparison with other risks and for risk communication. In general, comparison of risks which are widely different is highly discouraged because it can be misleading, as in Chapter 9 we will further elaborate on this subject and its adverse influences on risk communication. Table 7.5 shows some examples of this representation.

The problem with this measurement is that it is often difficult or impossible to compute or measure; and one should also be suspicious of some of these computations at low exposure duration or amount, since the relationships between exposure and response at these situations are not linear. This means that if, for example, high-dose exposures show that 10,000 person-rem of radiation exposure cause one statistical death, it does not mean that 0.01 person-rem (i.e., 10 millirem of exposure) which is the amount of exposure one receives in a single chest x-ray or in a round trip flight from New York to Los Angeles, has a one-in-one-million chance of death.

For purposes of cross-activity comparison, the rates in Table 7.5 may be treated as constant and estimated as an average mortality rate, say from activity j per unit of time or exposure amount

TABLE 7.5
Risk Exposures that Increase Chance of Death by 1 in 1,000,000 per Year [5]

Nature of Risk Exposure	Cause of Death
Spending 1 hour in a coal mine	Black lung disease
Spending 3 hour in a coal mine	Accident
Traveling 10 miles by bicycle	Accident
Traveling 300 miles by car	Accident
Traveling 10,000 miles by jet	Accident
Having chest x-ray taken in a good hospital	Cancer caused by radiation
Living 50 years within 5 miles of a nuclear plant	Cancer caused by plant emissions

$$\lambda_j = \frac{\text{Number of deaths (or failure) during activity } j}{\text{Number of hours spent or exposure amount in activity } j} \qquad (7.7)$$

This measure is only meaningful if the deaths *caused* by activity j actually occur while doing j. For example, risks from automobile driving or commercial aviation accidents can be represented in this way, but not the risks from smoking.

We illustrate how the data in Table 7.5 are estimated by means of the following example. In the U.S., 627 billion cigarettes are made annually. This translates to 3000 cigarettes smoked per smoker per year, or a little less than half a pack a day per smoker. It is estimated that 30% of all smokers die from lung and other cancers, or from heart diseases due to smoking. This figure can be expressed as an average lifetime mortality risk of 0.15 per smoker. Dividing this by a 70-year life expectancy, it gives a yearly risk of 0.002; dividing it again by 3000 (the average number of cigarettes smoked per year), the risk per cigarette smoked of 0.7×10^{-6} is estimated. Therefore, smoking 1.4 cigarettes is equivalent to a fatality risk of 1 in 1-million. Of course, as stated earlier, these estimations are only to use as figure of merit for comparing risks rather than as a realistic estimate of risk itself.

Risk per unit of time (or space) per unit of population exposed. This method is a simple and effective way of showing the risk results. While it is mainly used for high frequency events, it can equally be used for rare events too. Table 7.6 and Table 7.7 show examples of this type of representation for automobile accidents and some other accidental causes of death in the U.S.

A variant of the *risk per year per unit time per unit population* measurements is to use small relative units of time (per single unit of total population exposed). For example, the number

TABLE 7.6
Risk of Various Activities per Year per Million Population Exposed [6]

Cause	Annual Rate (Deaths per Million)
Asbestos exposure in schools	0.005–0.093
Whooping cough vaccination (1970–1980)	1–6
Aircraft accidents (1979)	6
High school football (1970–1980)	10
Drowning (ages 5–14)	27
Motor vehicle accident, pedestrian (ages 5–14)	32
Home accidents (ages 1–14)	60
Long-term smoking	1200

TABLE 7.7
Risk of Fatalities and Injuries from Automobile Accidents per Year per Unit Population Exposed [7]

National Rates: *Fatalities*

Fatalities per 100 million vehicle miles traveled	1.5
Fatalities per 100,000 population	15.26
Fatalities per 100,000 registered vehicles	19.56
Fatalities per 100,000 licensed drivers	22.23

National Rates: *Injured Persons*

Injured persons per 100 million vehicle miles traveled	120
Injured persons per 100,000 population	1187
Injured persons per 100,000 registered vehicles	1522
Injured persons per 100,000 licensed drivers	1729

of deaths of bicyclists may be expressed relative to the number of hours spent cycling or alternatively to the distance traveled. As an example, Table 7.8 illustrates this format for presenting risk data reported by the U.K. Royal Society [7].

Similarly, the risk results may be expressed in the form of risk per unit of space (e.g., distance traveled) per exposed population unit. Table 7.9 compares risk per unit of travel distance (100 km) for various activities for different periods.

Exposed population size that leads to occurrence of a single loss (event) per unit of time. This method shows the risk in terms of "odds." It is possibly the oldest and simplest method of showing risk results. An example for this method is shown in Table 7.10.

Relative contribution to total risk. This method is not a risk representation *per se*, but a method of describing important risk contributors. For example, Doll and Peto [8] describe the data listed in Table 7.11 to describe the contribution from various sources of cancer-deaths.

7.3   HEALTH AND SAFETY RISK ACCEPTANCE CRITERIA

Health and safety risk criteria can be divided into two categories of acceptance criteria that are designed for individual risk recipient, and those designed for group (population) appli-

TABLE 7.8
Hourly Mortality Rates of Various Activities per Person in the Exposed Population [7]

Activity	Hourly Mortality Rate
Plague in London in 1965	1.5×10^{-4}
Rock climbing (on rock face)	4.0×10^{-5}
Civilian in London Air-Raids, 1940	2.0×10^{-6}
Jet travel	7.0×10^{-7}
Automobile travel	1.0×10^{-7} (at 70 mph)
Bicycling	1.0×10^{-7} (at 30 mph)
Coal mining accidents (accidents)	4.0×10^{-7}
Background	1.0×10^{-6}

TABLE 7.9A
Deaths per 100 km Traveled in U.K. [7]

Sector	Class	1967–1971	1972–1976	1986–1990
Rail travel	Passengers from train accidents	0.65	0.45	1.10
Air travel	Airline passengers on scheduled service	2.30	1.40	0.23
Road Travel	Public service vehicles	1.20	1.20	0.45
	Private cars or taxis	9.00	7.50	4.40
	Cyclists	88.00	85.00	50.00
	Motorcycle driver	163.00	165.00	104.00
	Motorcycle passenger	375.00	359.00	104.00
	Pedestrian, based on 8.7 km walk per week	110.00	105.00	70.00

TABLE 7.9B
Travel Risks on per Mile and per Hour Basis [7]

	Death Rate	
	Passenger Hours (per 10^9)	Passenger Miles (per 10^9)
Air		
Bus		
U.S.A.	80	1.7
U.K.	30	1.3
Rail		
U.S.A.	80	1.3
U.K.	50	1.3
Car		
U.S.A.	950	24
U.K.	570	19

cations. Further, these criteria can be absolute or relative. Finally, the hazard exposure can be continuous or accidental. Implications, methods, and examples of these criteria will be discussed in the following subsections.

When we are considering the chance of a large-scale industrial accident from a facility or system, the important question for each individual would be: what is the risk to me and to my family? The way this is answered in risk estimation is to calculate the risk to any individual who lives within a particular distance from the system or facility; or who follows a particular pattern of life that might subject him to the consequences of an accident. The consequences of a major accident, however, are wider than injury to particular people. Perhaps, because each one of us has to die somehow, sometime, we commonly take large numbers of individual accidental deaths far less seriously than we do a single event killing a similarly large number of people.

Farmer's curves discussed earlier can provide useful insight into the degree of risks from a facility or hazardous process to the employees on the plant site and to the community located beyond the plant boundaries. Assessing the risk beyond the facility or system boundaries requires a definition of the population at risk (industrial, residential, school, hospital, etc.), likelihood of people being present, and reliability of mitigation factors (people evacuation procedures, etc.).

TABLE 7.10
Odds of Certain Activities per Year

Activity/Accident	Approximate Chance per Year
A fire killing 10 or more people	1
A railway accident killing or seriously injuring 100 or more people	1 in 20
An aircraft accident killing 500 people	1 in 1,000
Airplane crashing into full football stadium	1 in a 100 million
A house catches fire	1 in 200
A person dies from heart disease	1 in 280
A person dies of cancer	1 in 500
Automobile fatality	1 in 6,000
Killed by lightning	1 in 1.4 million
Killed by flood or tornado	1 in 2 million
Killed in hurricane	1 in 6 million
Drowning	1 in 20,000
Car crash	1 in 5,300
Choking	1 in 68,000

7.3.1 INDIVIDUAL RISK ACCEPTANCE CRITERIA

7.3.1.1 Absolute and Relative Risk Acceptance Criteria for Individuals

Several authors have applied the notion of loss of life expectancy as a means of expressing risk limits. For example, consider the acceptable level of risk expressed as the individual loss of life expectancy of 10^{-5} per year. This level of risk implies a rate of individual loss of life expectancy that is roughly balanced by the observed rate of increase in life expectancy with date of birth. For more discussions on individual risk acceptance, see Vrouwenvelder et al. [9].

Individual risks sometimes are expressed as fatal accident rates (FAR) such as those shown in accidents, fatalities, and rates [10], NHTSA [11, 12]. They can be expressed as an annual fatality probability or as the probability per unit time of a person (or system) being

TABLE 7.11
Cancer Risk Contributors

Factor or Class of Factors	Percent of All Cancer-Deaths	
	Best Estimate	Range of Estimate
Tobacco	30	25–40
Alcohol	3	<1–24
Diet	35	10–70
Food additives	<1	Unknown
Reproduction and sexual behaviors	7	1–13
Occupation	4	2–8
Pollution	2	<1–5
Industrial products	<1	<1–2
Medicines and medical procedures	1	0.5–3
Geophysical factors	3	2–4
Infection	10	1 — unknown

killed (or failed) when actually performing a specific activity. An almost unavoidable risk is the probability of dying from natural causes. In developed countries, this probability for a person under 60 years of age is about 10^{-3} per year. The probability of losing life in normal daily activities, such as driving vehicles or working in a factory, is in general one or two orders of magnitude lower than the normal probability of dying. Activities such as mountain climbing involve a much higher risk. These probabilities may be reflected as an implicit risk acceptance model. Of course, people do not have those numbers actually in mind, but by participating in such activities they "reveal" their preferences and risk taking patterns. For example, actual observation of such revealed activities shows that for activities considered as attractive and done voluntarily much higher risks are accepted as for involuntary activities. This means that we may use these numbers to set up limits. The FAR numbers indicate that in highly developed countries, 10^{-4} per year appears to be an acceptable voluntary risk limit. The nuclear industry, especially in the U.S. and U.K., has worked toward establishing goals of this kind. For some other more intensive hazard exposure situations, one might then have the absolute risk acceptance criteria such as the ones shown in Figure 7.2 for living or working near or at the nuclear power plants in the U.K.

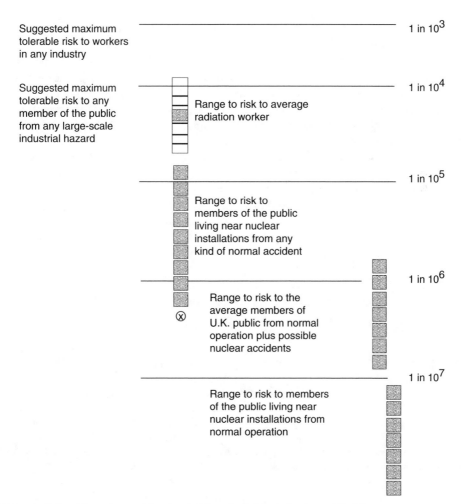

FIGURE 7.2 Tolerable and actual levels of risk to individuals in the public [7].

According to these guidelines, frequency of early fatality for an imaginary individual at a reference point from a nuclear plant due to all possible causes added together may be, for example, set to be less than 1 in 10-million/year.

In establishing relative acceptable risk limits, it is prudent to take into account societal bias about certain classes of risk, without letting it drive the final decision. The so-called U.S. Nuclear Regulatory Commission (U.S. NRC) quantitative safety goals [13] may be viewed for establishing these levels. The individual risk criterion (there is also a societal risk that will be described in the next section):

The risk to an individual in the vicinity of a nuclear power plant, of fatalities resulting from reactor accidents should not exceed one-tenth of one percent (0.1%) of the sum of prompt fatality risks resulting from other accidents to which members of the U.S. population are generally exposed.

Clearly, since the risk acceptability is based on other risks that the individuals are exposed to, this risk acceptance criterion is relative in nature. There are 97,900 accidental deaths in the U.S. in 2003 (NHTSA [11,12]). This makes the individual risk in the U.S. with an estimated population of 285 million in 2003 as 3.4×10^{-4} per person per year. Assuming that this risk estimate describes the accidental risk of individuals around nuclear power plants, the acceptable annual risk level for an individual would be 0.1% of this risk or 3.4×10^{-7}.

Since this reference level is a relative measure for the "de minims" individual risk limit, if the society manages to reduce accident rate, then some plants that are close to the limit may now violate the safety goal.

Bedford and Cooke [14] discuss similar risk acceptance safety goals as shown in Table 7.12.

It is also possible to express individual risk limit in the form of a line depicting the relationship between the frequency of occurrence and the magnitude of the consequences for a spectrum of events as proposed by Farmer [15] (see Figure 7.3). Interpretation of this representation in individual acceptance criteria can be extended by expressing the consequences of the number of casualties.

The individual risk limits expressed in the form shown in Figure 7.3 may be expressed in terms of limits over the "space" as opposed to "time." For example, in a case of limiting activities around a toxic pipeline, a limit such as the one shown by Whittaker et al. [16] in Figure 7.4 may be used. Clearly for an individual risk of 10^{-6}, an exclusion zone of 1.0 km or more may be considered acceptable.

TABLE 7.12
Examples of Risk Acceptance Levels for Various Cases [14]

Technology	Risk Goal
Marine structures	Failure probability for different accident classes 10^{-3}–10^{-6}
Aviation, air planes	Catastrophic failure per flight hour: less than 10^{-9}
Space vehicles	Catastrophic consequence for a crew transfer vehicle (CTV) smaller than 1 in 500 CTV missions
Process industry (Netherlands)	Consequence of more than 10^n fatalities must be smaller than probability 10^{-3-2n}. Individual risk of less than 10^{-6} per plant year
IEC 61508 (for all technologies) Electrical/electronic safety systems with embedded software	Average probability of failure per demand: 10^{-1} to 10^{-5} for different safety levels 1–4 and low demand mode of operation

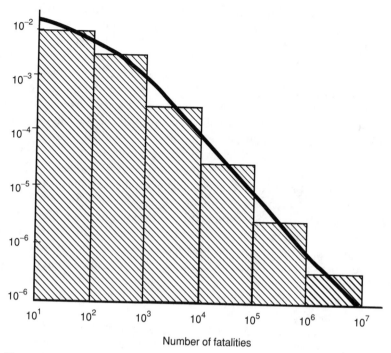

FIGURE 7.3 Farmer's curve used as acceptance criterion.

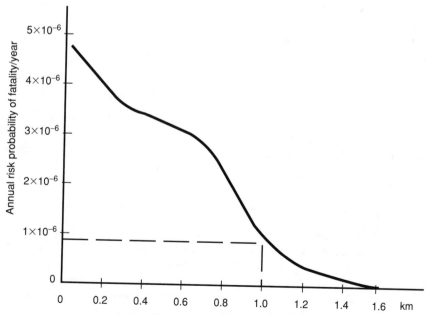

FIGURE 7.4 Risk versus distance from the release site [16].

Since the nature of risk acceptance and perception of risks due to continuous exposure to hazards may be different, this subject will be further elaborated in Chapter 9.

Many industrial facilities emit toxic and other pollutants to the environment. A good example of this involves emission of radioactive materials into air by research and nuclear facilities. For this purpose, the maximum cumulative doses allowed for workers exposed to radiation and for members of the public are described in various national and international rules. In order to minimize the consequential risks of cancer and organ damage, a strict measure of control on the total annual dose allowed is specified. In the case of workers in the U.S., this limit is currently at 0.05 Sv/year (Sv stands for "Sievert" which is a unit of harmful equivalent radiation dose received by an individual and can be used to calculate the health consequences such as cancer). Also, the dose limit to any individual organ or tissue other than the lens of the eye is 10 times higher, which is 0.5 Sv (for eye lens, it is 0.15 Sv). Similarly, for the members of the public, it is 50 times less than the workers' limit or 1 mSv (milliSievert). These provisions apply in practice to members of the public near enough to a nuclear plant possibly to be affected by discharges to air or water. Assuming the estimated increase in one cancer incident is 5×10^{-5}/mSv (obtained by extrapolating from high-dose exposures), then acceptable risk of continuous exposure to radiation is 5×10^{-5} per year. The readers should also be cautioned as the topic of health effects of exposure to low levels of radiation is still evolving, and most scientists believe that consequences obtained from radiation exposure extrapolated from high dose to low dose, yield overly conservative results. Note that 1 mrem/year (0.01 mSv/year) $= 0.5 \times 10^{-6}$ fatality/year-individual and 10 mrem/year (0.1 mSv/year) $= 5 \times 10^{-6}$/year-individual risk.

Similarly, radioactive waste disposal sites must be so designed such that the maximum risk to any individual not exceeds a released dose of 0.1 mSv/year (i.e., a conservative acceptable cancer risk of 5×10^{-6} per year). In practice, the systems of control that are applied make it fairly rare for any worker to receive regular annual doses approaching 0.050 Sv, and the average for workers at nuclear installations varies between about 1 mSv/year up to 5 mSv at a few installations.

A limited but important aspect of the long-term exposure risk concerns is the development of a criterion for cases that lead to the delayed death of an individual. This includes exposure to low-level ionizing radiation and to certain chemicals. The effects of exposure to chemicals in this context are much less well understood than those of radiation.

7.3.2 Societal Risk Acceptance Criteria (Exposure, Consequence, Precursor)

Societal risks are those risks that affect a substantial number of people (or systems such as a fleet). In general, the principles applied to individual risk acceptance criteria apply to societal risk criteria too. Recently, the societal risk acceptance criterion for any accidental risk to human life is presented in a general form by ISO 2394 [17] as

$$P(N_d > n) < An^{-k} \qquad (7.8)$$

This *absolute* criterion should hold for all n. Here N_d is the number of people killed in 1 year in one accident. The value of parameter A may range from 0.001 to 1/year and the value of k ranges from 1 to 2. Figure 7.5 shows examples of plots of (7.8).

High values of k express the social aversion to large disasters. The probability or frequency $P(N_d > n)$ depends on the nature and frequency of events, and on the factors that determine the number of fatalities in case of a hazard exposure. In some cases, exceeding the upper line in Figure 7.5 may be considered as unacceptable and below the lower limit may exhibit a negligible risk level.

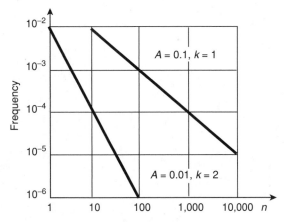

FIGURE 7.5 Examples of (7.8) criterion presented as "*F–N* curves."

In the case of possible nuclear accidents, consider any accident involving radiation releases bigger than the so-called design basis accident radiation release. Societal criteria may be used to place some limits (absolute or relative) for radiation exposures risk to a large population. For example, the U.S. NRC [13] proposes nuclear plant societal safety goal and is expressed as:

> *The risk, to the population in the area near a nuclear plant, of cancer fatalities resulting from nuclear power plant operation should not exceed one-tenth of one percent (0.1%) of the sum of cancer fatality risks resulting from all other causes.*

Clearly, this is a *relative* societal risk acceptance criterion. According to the American Cancer Society in 2003, there were 556,500 cancer-deaths in the U.S. For a population of 285 million in 2003, the risk of cancer-death is 0.0195 per person per year. One-tenth of a percent of this value amounts to 1.95×10^{-5} long-term acceptable probability of cancer-death.

A limited but important aspect of the long-term exposure risk is the development of a criterion for consequences in form of delayed mortality. This includes exposure to low-level ionizing radiation and exposure to certain chemicals. The effects of exposure to chemicals in this context are less understood than those of radiation.

In attempting to propose a criterion for the risk of delayed death (such as societal exposure to low-level radiation), one must be sensitive to the possibility that the public may be indifferent to the latency of a death risk, if its cause can be traced to a particular industrial activity. A criterion based on such a view would then be the same as that for immediate deaths. However, few people would argue that death at some time in the future is preferable to immediate death. Examples include criteria proposed by Kinchin [18] in the form of separate Farmer-type frequency lines for early and delayed deaths, with the frequency of occurrence proposed for the delayed death criterion being a factor of 30 greater than that for the early deaths. Figure 7.6 describes an example of these acceptance criteria. Note that anywhere on the *F–N* lines of acceptance criteria in Figure 7.6 represent an equivalent annual risk of 1×10^{-4} per person for delayed risk, and 3.33×10^{-5} per person for early risk.

Similar examples of this representation are societal nonresidential radiation exposures at high-dose exposure levels caused by major accidents set as having a risk acceptance criterion of less than 10^{-6} per year. Haugom and Rikheim [19] show the corresponding *F–N* curves as in Figure 7.7 for three areas of "acceptance" (below 10^{-6} risk value for $N = 10$), "ALARP"

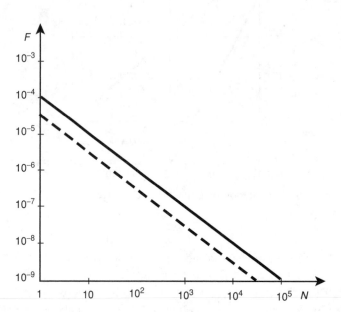

FIGURE 7.6 Proposed risk acceptance criterion for delayed death risk (solid line) compared with early deaths (broken line) both in cumulative form.

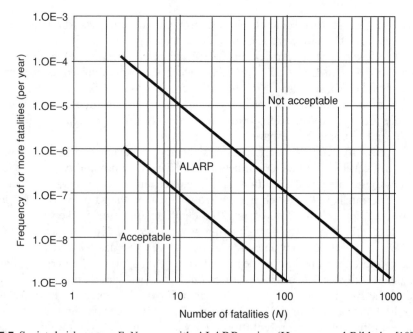

FIGURE 7.7 Societal risk curve, F–N curve with ALARP region (Haugom and Rikheim [19]).

or as low as reasonably practicable (between 10^{-4} and 10^{-6} for $n = 10$) and "unacceptance" (above 10^{-4} for $N = 10$). If the calculated risk is above the curve, the risk must be reduced.

7.4 ECONOMIC RISK AND PERFORMANCE ACCEPTANCE CRITERIA

Risk acceptance criteria can be defined for performance of hazard barriers, precursors to risks, and economic losses. For example, the U.S. Nuclear Regulatory Commission has set other risk levels based on reactor vessel and containment performance measures as follows:

(1) Consistent with the traditional defense-in depth approach and the accident mitigation philosophy requiring reliable performance of containment systems, the overall mean frequency of a large release of radioactive materials to the environment from a reactor accident should be less than 1 in 1,000,000 per year of reactor operation.

(2) The frequency of core damage should be less than 1 in 10,000 per year of reactor operation.

Among the countries using light water reactors in the western world, beside the U.S., only France has announced a quantitative safety goal for its nuclear power plants. It has required that the probability of cancer due to all causes for radiological exposure not to exceed 10^{-6} per reactor per year. In connection with the proposed Sizewell B nuclear reactor in the U.K., the Central Electricity Generating Board proposed similar design safety guidelines that have been accepted by the Nuclear Installations Inspectorate as reliability targets for licensing design. Italy has announced the safety goal for future nuclear reactors to be 10^{-5} per reactor per year for serious core damage accidents, with each risk scenario leading to such an event being less than 10% of the total risk.

Similar to fatality risk limits in form of *F–N* curves, economic risk limits may also be imposed. Figure 7.8 shows an example of this approach. In this case, the cost of certain consequences (such as cancer risk) is measured against the cost of control, mitigation, or protection from such exposures. These kinds of risk criteria are used as risk management decision criteria.

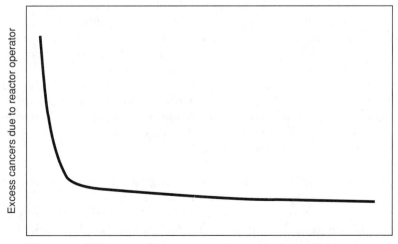

Control cost (current value of capital and operating costs)

FIGURE 7.8 Exceedance cost versus control cost.

TABLE 7.13
Median Costs of Life-Saving Measures per Sector

Sector	Cost per Statistical Life ($)
Medical	19,000
Injury reduction	48,000
Toxin control	2,800,000
Overall	42,000

Government agencies are frequently faced with the problem of prioritizing some life-saving measures. Cost–benefit analysis uses the notion of the "value of human life" to judge which life-saving measures are economical (this topic will be discussed in Section 7.4.6). The U.K. government, for example, uses figures from the Ministry of Transport in several policy areas. Table 7.13 shows an example of such life-saving cost criteria.

If government authorities want to choose an amount to spend on saving lives, the appropriate value of a human life used is usually determined on the basis of surveying members of the public. The survey gathers information on amount of willingness-to-pay (WTP) to avert a death and willingness-to-accept (WTA) a potentially fatal exposure. Typically, the amount people are willing to pay to avoid exposure to a risk is considerably less than what they would accept to be exposed.

7.4.1 OTHER RISK ACCEPTANCE CRITERIA IN FORM OF FIGURES OF MERIT

One of the most important aspects of a risk assessment application is the identification of the appropriate figure of merit to be estimated and evaluated including the corresponding decision criteria. The term "figure of merit" refers to the quantitative analytical risk-related measures to be utilized. In some cases, more than a single figure of merit may be appropriate. The term "decision criteria" refers to the specific quantitative screening or acceptance criteria applied to a figure of merit to evaluate the acceptability of the results.

Depending upon their scope, risk assessments can provide a wide variety of figures of merit in applications. Figures of merit can be categorized by the two types of application: evaluation of risk significance and risk-based prioritization/ranking.

For regulatory applications involving assessment of risk significance, the primary figures of merit consider both short-term and long-term health effects. No single figure of merit can effectively capture both aspects of public health and safety. This could include use of the risk of early fatalities and injuries and latent fatalities as the figures of merit.

In nuclear plant applications, core damage frequency (CDF) is a prevalent figure of merit. Large, early release frequency (LERF) is another preferred figure of merit. In combination, these figures address both prevention (CDF) and mitigation (LERF) events leading to the risk of radiation exposure. For the purposes of facilitating decisions using these figures of merit, the following general definitions are provided:

- *Core Damage.* Uncovering and heating of the reactor core to the point where prolonged clad oxidation and severe fuel damage are anticipated.
- *Large, Early Release.* A radioactive release from the containment which is both large and early. Large is defined as involving the rapid, unscrubbed release of airborne aerosol fission products to the environment. Early is defined as occurring before the effective implementation of the off-site emergency response and protective actions.

In estimating the core damage figures of merit, the PRAs of nuclear plant should develop scenarios typically involving the following conditions:

- Collapsed liquid level below top of active fuel (pressurized water reactors)
- Collapsed liquid level less than one-third of the core height (boiling water reactors)
- Core peak temperature > 1800F
- Core exit thermocouple reading > 1200F
- Core maximum fuel temperature > 2200F

Similarly for large, early release, the PRA scenarios should consider conditions such as:

- Unscrubbed containment bypass pathway occurring with core damage or
- Unscrubbed containment failure pathway of sufficient size to release the contents of the containment within 1 h, which occurs before or within 4 h of reactor vessel breach

7.4.2 ACCEPTABLE RISK LEVELS BASED ON CHANGES IN FIGURES OF MERIT

In some situations, applications involving risk acceptance directly utilize changes in figures of merit such as change in nuclear plant CDF or LERF due to changes in the input to the risk model. Sometimes instead of acceptable levels of RAW and RRW, importance measures are used. Figure 7.9 shows an example of risk acceptance regions based on changes

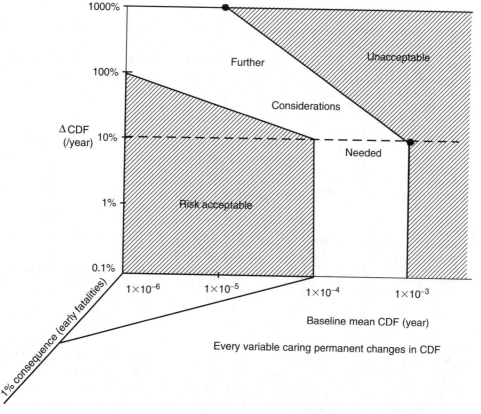

FIGURE 7.9 An example of risk acceptance based on changes in the figure of merit (see EPRI [20]).

in the figure of merit (CDF of a nuclear plant) resulting from changes due to input to the risk model. These types of acceptance criteria are used to determine acceptability of changes to the system or facility design and operation that affect the risk model (for example, eliminating or modifying operation of a hazard barrier). While changes in the figures of merit are important to set the acceptability limit, their base value should also be considered in the decision process. For example, in cases where the figure of merit itself is small (i.e., corresponds to small risk and below the acceptable limit), there is more room to increase the risk due to proposed changes and, therefore, the risk acceptance criteria should be less restrictive.

7.4.3 USING FIGURES OF MERIT IN SINGLE EVENT RISK

Consider a maintenance activity of a unit that results in additional risk in an operating system or facility. That is the case where, while performing a maintenance action, a safety barrier will become inactive or degraded. If the unit is taken out of service for duration of d, the risk increases by ΔR, and the risk, r, of this maintenance activity can be measured by

$$r = \Delta R \cdot d \qquad (7.9)$$

where d is the duration of risk exposure.

If this activity occurs frequently or periodically, as is the case in preventive maintenance, the total risk in terms of the frequency of the maintenance event, f, is $R = f \cdot r$ or

$$R = f \cdot \Delta R \cdot d \qquad (7.10)$$

where f is the frequency of the maintenance activity resulting in a temporal risk change ΔR.

According to (7.10), if a risk acceptance criterion for any single risk contributing event (e.g., maintenance event) is r_c and over a long period (such as a year) is R_c, then the criterion for the acceptable length (in time) of the maintenance would be

$$
\begin{cases}
r \leq r_c \quad d \leq \dfrac{r_c}{\Delta R} \\
\qquad\qquad\quad \text{Criterion } d \leq \min\left[\dfrac{r_c}{\Delta R}; \dfrac{R_c}{\Delta R \cdot f}\right] \\
R \leq R_c \quad d \leq \dfrac{R_c}{\Delta R \cdot f}
\end{cases}
\qquad (7.11)
$$

This concept has been illustrated in Figure 7.10 for a nuclear plant in which a maintenance activity on a safety system, while the plant is in operation increases the CDF which is considered as a figure of merit, during the maintenance time d. Note that the total risk would be expected to reduce, as the maintenance improves the performance of the safety system.

Example 7.2
A nuclear power plant has a nominal CDF of 2.5×10^{-5}/year. Calculate the maximum allowable out of service time for a specific safety component in the plant, if the CDF increases by a factor of 40 when this component is taken out of service. Assume risk acceptance criterion for maintenance is $r_c = 1 \times 10^{-6}$ per event. Consider two preventive maintenance per year, and the annual risk acceptance criterion of $R_c = 1 \times 10^{-5}$.

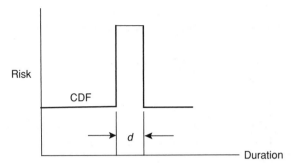

FIGURE 7.10 Illustration of a change in CDF due to a maintenance activity of duration d.

Solution
For the risk increases, according to (7.11),

$$d < \min\left[\frac{1 \times 10^{-6}}{40 \times 2.5 \times 10^{-5}}; \frac{1 \times 10^{-5}}{2 \times 40 \times 2.5 \times 10^{-5}}\right]$$

$$d < \min[0.001; 0.005]$$

$$d < 0.001 \text{ years or } 8.76 \text{ h}$$

Then the maintenance has to be planned in order to finish it in less than 8.76 h to meet the acceptance criteria.

Since the CDF level would be decreased as the result of the maintenance (as the performance of the units or hazard barriers improve), the basis for making a decision to proceed with such an action may be based on the overall amount of risk reduction anticipated. For example, Table 7.14 shows risk acceptance criteria based on the amount of reduction in CDF.

7.4.4 INSTANTANEOUS ABSOLUTE VALUE LIMITS

Absolute figure of merit limits may be imposed on frequencies such as CDF and LERF in a nuclear power plant. Upper limits are intended to screen out those time-dependent risk situations that lead to risk "spikes" (such as the one shown in Figure 7.10), i.e., excessively high instantaneous risk levels. Lower limits are intended to screen out events that are not risk significant and should not require detailed quantitative analyses.

7.4.5 RISK ACCEPTANCE BASED ON IMPORTANCE MEASURES

Another way to set risk acceptance criteria is through setting limits on risk importance measures. For example, for hazard barriers and other units in a complex system, one may set

TABLE 7.14
Amount of Reduction in CDF Due to Maintenance and
Subsequent Actions

Estimated Reduction in CDF (per Year)	Action
$>10^{-4}$	Proceed to cost–benefit analysis
10^{-4}–10^{-5}	Calculate large, early release
$\leq 10^{-5}$	No further analysis needed

the following limits proposed by the nuclear industry BWR Owners' Group Report [21] on the RRW or RAW of a safety system or component with respect to the total risk (or figure of merit):

Hazard Barrier or Safety System Importance Level Risk Criteria

(i) $I_{RRW} > 1.05$, and

(ii) $I_{RAW} > 2.00$.

At the level of components of a risk barrier (such as a safety unit), lower values may be necessary. For example:

Component Level Importance Risk Criteria

(i) $I_{RRW} > 1.005$, and

(ii) $I_{RAW} > 2.00$.

For specific hazard barriers such as passive barriers (piping, pressure vessels, etc.), more stringent criteria may be set. For example:

Passive Structure Importance Risk Criteria

(i) $I_{RRW} > 1.001$, and

(ii) $I_{RAW} > 1.005$.

7.4.6 RISK ACCEPTANCE BASED ON VALUE OF HUMAN LIFE

Another way to establish risk acceptance criteria for risks involving human death is by placing an acceptable monetary value over human life. While a difficult, controversial and even by some accounts an unethical approach, nevertheless for societal policy decision making, a value of life can be broadly suggested and used as a risk acceptance criterion.

Key aspects of the value of life and safety have been discussed by several researchers. The first method is concerned with lost gross output based on goods and services which a person could have produced. Sometimes gross loss of productivity is reduced by an amount representing consumption (i.e., net output). The output approach usually gives a small value for life, especially if discounted consumption is used.

The livelihood approach to value of life, which is not fundamentally different from the output approach, assigns valuations in direct proportion to income. The present value of future earnings of an individual is estimated and reduced by an amount equal to discounted consumption. This would give the net value of an individual to a family. This method also gives a small value for life. As in the case of output approach, deduction of consumption is to some extent unethical, but is used in some estimates. The livelihood method normally favors males over females, working persons over retired, and higher paid over lower paid persons, and does not reflect individual or societal preferences.

The third approach assumes that if an individual has a life insurance policy for known amount, then that amount implies values of he/she places on his/her life. Data from insurance companies on these amounts are readily available. In adopting the insurance method, there are two drawbacks. First, a decision as to whether or not to purchase insurance, and the amount of insurance purchased is not necessarily made in a manner consistent with one's best judgment of the value of one's life. This decision largely depends upon the premium the insured person can afford. Second, purchasing an insurance policy does not affect the mortality risk of an individual; this action is not intended to compensate fully for death or to reduce the risk of accidental death. Hence insuring life is not exactly a trade-off between mortality risks and costs.

The fourth method for assessing value of life involves court awards of compensation to beneficiaries of a deceased person. Here again, collection of necessary data is not a problem. Assessment of values of life could also be expected to be reasonably accurate since legal professionals rely on professional expertise. The objective of such an analysis is to discover whether the risk could have been reasonably foreseen, and whether the risk was justified or unreasonable.

There are, however, some problems in using court awards for valuing human life. The courts should ideally be concerned with the assessment of adequate sums as compensation for an objective loss, e.g., loss of earnings of the deceased and damages to spouse and children. In some countries, damages can include a subjective component for pain and suffering of the survivors, but certain courts are generally against such compensation to persons who are not themselves physically affected. It is also difficult to value the quality of a lost life. People who themselves suffer severe personal injury, of course, qualify for substantial damages. Resource costs such as medical and hospital expenses are significantly higher for obvious reasons in serious injury cases than in fatal cases; hence, awards for subjective losses tend to be much larger and more important in serious nonfatal cases than in fatal cases.

The fifth approach, discussed earlier in Chapter 6, is the one widely adopted for valuing life. This approach is the WTP, which is based on the money people are prepared to spend to increase their safety or to reduce a particular source of mortality risk. It is difficult to differentiate between the benefit from increasing peoples' feeling of safety and that from reducing the number of deaths. According to VanDoren [22], because government policies reduce risks of death rather than eliminate specific individual deaths, the correct benefit value is society's WTP for the reduction in risk. VanDoren argues that, if a regulation would reduce risk by 1 in 1-million to everyone in a population of 1 million, then the risk reduction policy or regulation would save one statistical life. If the average WTP for that risk reduction is US\$ 6 per person, then the value of a statistical life is US\$ 6 million. Using data on wages, economists have estimated people's trade-offs between money and fatality risk, thus establishing a revealed value of lives. Recent estimates for averaged risks of death at work imply that, in 2003 dollars, workers appear to receive premiums in the range of US\$ 600 to face an additional annual work-related fatality risk of 10^{-4}. This yields an approximate statistical life value of US\$ 6 million.

The implied value of life revealed by a WTP criterion would depend on a number of factors. The acceptable expenditure per life saved for involuntary risks is likely to be higher than the acceptable expenditure for voluntary risks, as people are generally less willing to accept involuntarily the same level of risk they will accept voluntarily. The sum people are prepared to pay to reduce a given risk will also depend on the total level of risk, the amount already being spent on safety, and the earnings of the individuals.

The WTP method connects with the principle of "consumer sovereignty," that goods should be valued according to the value individuals put on them. This consumer preference approach treats safety as a commodity like any other, so that when a government carries out projects or devise regulations to reduce risk, it should estimate costs and benefits as people do. This method would provide a level of safety expenditure that people could be expected to accept or bear, thereby avoiding the disadvantages of compulsory regulations. This approach is the most appropriate when we are considering the expenditure of government money of its citizens.

Following the WTP criterion, people may be asked to specify the amounts they are willing to spend to avoid different risks. However, surveys carried out in this connection have shown variability and inconsistencies in the responses to questionnaires; quite simply, individuals have difficulty in answering questions involving very small changes in their mortality risks. Due to insufficient knowledge about the risk, most people find it difficult to accurately quantify the magnitude of a risk. Also, the benefits are often intangible, e.g., enjoyment, peace of mind. It is difficult to put a monetary value on these factors. As literature on compensating wage differential indicates, individual WTP can be estimated by methods other than direct questioning of individuals.

Graham and Vaupel [23] have compared the costs and benefits of 57 life-saving programs. Quoting surveys of expressed WTP for small reductions in the probability of death, these authors have shown that values of a life ranged from US$ 50,000 to US$ 8 million (in 1978 dollars). Nine labor market studies of wage premiums have produced a narrower, but still disparate range of values spread from US$ 300,000 to US$ 3.5 million. Graham and Vaupel conclude that within a broad range, the monetary value assigned to the benefits of averting a death usually does not alter the policy implications and decisions.

Exercises

7.1 Consider the risk of fatality a person traveling from Washington DC to Chicago faces due to air and automobile transportation as well as natural causes.
   a. Calculate risk of driving using published data. Search for data, for example, from NHTSA (www.nhtsa.gov) or other sources of your choice.
   b. Calculate risk of flying (use FAA or other sources for historical experience). In calculating this risk, consider the risk of driving to the airport (one-way for 20 miles), risk of airplane crash, risk of driving from the airport to the hotel (15 miles).
   c. Compare the risks in parts 1 and 2 and use a formal method to compare the two options. (Assume the cost of the two options is the same, considering the value of the time saved by flying offsetting the lower expense of driving.)
   d. Use a set of risk conversion factors in determining risk perceptions of options 1 and 2.
   e. Use "loss of life expectancy" to compare fatality risk due to general total air travels with auto travels in the U.S. Assume the whole U.S. population is exposed to both risks.

7.2 Lave and Seskin find from a statistical study that an additional microgram per cubic meter of mean annual SO_2 concentration is associated with increased mortality of 0.039 per 10,000 population per year. In London, the annual mean concentration of SO_2 in the atmosphere is about 115 $\mu g/m^3$ (average). Calculate the extra death risk per million population as compared with EPA standards which require U.S. SO_2 average concentration of 80 $\mu g/m^3$.
   (a) Discuss the results.
   (b) Compare your results with those you find in literature.
   (c) Make a rough estimate of the effect of SO_2 in London on the life expectancy of males (assume a 72-years life expectancy for males). How does this compare with other causes of life expectancies due to cigarette smoking?
   (d) Compare the results with the risk of traveling 10 miles by bicycle which has a risk of 10^{-6}/person/year.

7.3 If the risk profile in terms of loss of life in person-years in a new technology is transferred (using regression analysis) to the following equation:

$$L(t) = 10 - 4(1 - t), \ 0 < t < 10{,}000 \ (t = \text{person-years})$$

in a population of 10,000, what is the expected life reduction because of the above risk?

7.4 Consider the following plot representing probability of car accidents per year as a function of the driver's age.
(a) Calculate mean age of drivers involving accident.
(b) Calculate the annual conditional probability that a person who is 55 year and older will be involved in an accident.

(c)

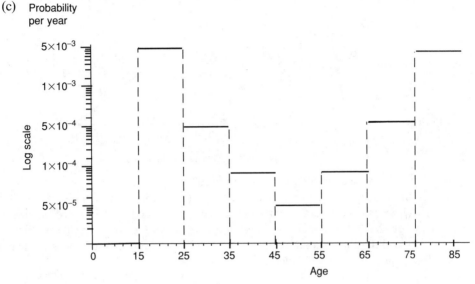

7.5 Describe the disadvantages of using life expectancy as a method of measuring and evaluating risk significance of various societal activities.

7.6 According to the following table, what is the *average life expectancy gained* due to adaptation of a safety measure?

Age Group	Fraction of Individuals Whose Lives Are Saved
0–10	0.001
10–20	0.008
20–30	0.007
30–40	0.004
40–50	0.003
50–60	0.001
60–80	0.001
80–90	0.001

Assume life expectancy of 73 years for both male and female.

7.7 If acceptable risk is considered as the area below the following curve:

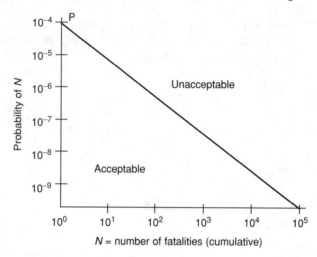

What is the acceptable risk criterion used?

7.8 It has been proposed that NASA adopts an acceptable risk limit of 5×10^{-5} cancer fatalities per year due to each accident scenario caused by the Ulysses nuclear powered space vehicle. Since there is a spectrum of accident scenarios, each scenario associated with a different radioactive release and different frequency of release, one could plot the results in the form below.

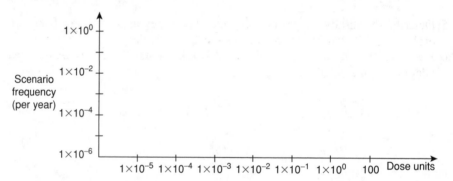

If the dose–cancer relation is 5×10^{-2} fatality/dose unit, draw a curve representing "frequency" as a function of "dose units."

7.9 If the acceptable risk is 0.1% of all cancers unrelated with Ulysses accidents, how does this acceptable level compare with the frequency versus dose units shown above? Assume that there are 500,000 cancer deaths per year in the U.S.

7.10 Consider the risk acceptance criterion below.

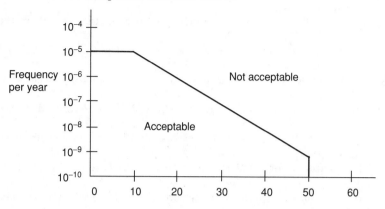

Design system configurations for systems A and B (composed of a number of units configured in parallel–series form). Systems A and B in the event tree below act as risk barriers so that above criterion is not violated.

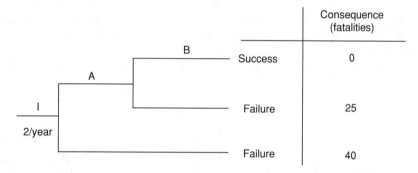

Each unit in system A has a failure probability of 5×10^{-2}, and for system B each unit has a failure probability of 1×10^{-2}. (Neglect dependent failures between the units.)

REFERENCES

1. U.S. Nuclear Regulatory Commission, Reactor Safety Study: An Assessment of Accidents in U.S. Commercial Nuclear Power Plants, U.S. Regulatory Commission, WASH-1400, Washington DC, 1975.
2. Cohen, B.L. and Lee, I.S., A catalog of risks, *Health Physics*, 36, 707, 1979.
3. Hibbert, M.E., Lung function values from a longitudinal study of healthy children and adolescents, *Pediatric Pulmonary Journal*, 7, 101, 1989.
4. Cohn, B., *Nuclear Energy Option*, Plenum press, New York, NY, 1990.
5. Wilson, R., Analyzing the daily risks of life, *Technology Review*, 41, 1979.
6. Mossman, B.T., Bignon, J., Corn, M., Seaton, A., Gee, J.B.L., Asbestos: scientific developments and implications for public policy, *Science*, 247, 1990.
7. U.K. Royal Society, *Royal Society Study Group on Risk Assessment*, U.K. Royal Society, 1983.
8. Doll, R. and Peto, R., The causes of cancer: quantitative estimates of the avoidable risk of cancer in the U.S. today, *Journal of the National Cancer Institute*, 66, 1191, 1981.
9. Vrouwenvelder, T., Lovegrove, R., Holicky, M., Tanner, P., Canisius, G., Risk assessment and risk communication in civil engineering, *Safety, Risk, Reliability — Trends in Engineering*, 2001.
10. U.S. National Highway Traffic Safety Administration, Motor Vehicle Traffic Crash Injury and Fatality Estimates 2002 Early Assessment, DOT HS 809 586, Technical Report, 2003.

11. National Highway Traffic Safety Administration, Accidents and Accident Rates, NHTSA Classification, 1984–2003, 14 CFR 121, 2004.
12. National Highway Traffic Safety Administration, Passenger Injuries and Injury Rates, NHTSA, 1984–2003, 14 CFR 121, 2004.
13. U.S. Nuclear Regulatory Commission, Safety Goals for the Operations of Nuclear Power Plants, NRC Federal Register Notice 51 FR 28044, 1986.
14. Bedford, T. and Cooke, R., *Probabilistic Risk Analysis*: *Foundations and Methods*, Cambridge Press, Cambridge, 2001.
15. Farmer, F.R., *Siting Criteria — A New Approach*, IAEA SM-89/34; Reprinted in Nuclear Safety, 8, 539, 1967.
16. Whittaker, J., Angle, R., Wilson, D., and Choukalos, M., Risk-based zoning for toxic gas piplines, *Risk Analysis*, 2 (3), 1982.
17. International Organization for Standardization, General Principles on Reliability for Structures, ISO 2394, 1998.
18. Kinchin, G.H., Assessment of hazards in engineering work, *Proceeding of the Institute of Civil Engineers*, 64, 431, 1978.
19. Haugom, G.P. and Rikheim, H., Hydrogen Applications Risk Acceptance Criteria and Risk Assessment Methodology, DNV, The Research Council of Norway, 1990.
20. Electric Power Research Institute, PSA Applications Guide, EPRI TR-105396, Final Report Dated August 1995.
21. BWR Owners' Group Report, Risk-Based Ranking, ESBU/WDG-95-034, Wash. Elec. Corp., 1995.
22. VanDoren, P., Cato Handbook for Congress: Policy Recommendations for 108th Congress, Cato Institute Report, 2003.
23. Graham, J.D. and Vaupel, J., The value of life: what difference does it make, *Risk Analysis*, 89, 1981.

8 Decision Making Techniques Using Risk Information

8.1 INTRODUCTION

There are two classes of decision making methodologies: economic and noneconomic. Another useful division is descriptive and normative theories of decision making. Descriptive theory of decision making represents preferences of individuals as they are. However, normative theories model individual's beliefs structure following certain rules of consistency. As these are controversial views, we only discuss economic and noneconomic division in this book. The economic methodologies make the decisions based on a value (actual measurable monetary value or subjective value) assigned to risk (or loss). This value can be absolute or relative. Examples of economic-based methods that will be addressed in this chapter are cost–benefit, cost-effectiveness, and risk-effectiveness methods. The noneconomic methodologies include those that assign a value by experts or lay people to risk and other risk management attributes such as cost and time. These values can be real measurable ones (such as losses incurred by natural events), or values perceived by people as to the frequency or consequence of certain events. Examples of these methods are loss of life expectancy, comparison to natural risks, and expressed preference methods. Similar to the economic methodologies, the values assigned to risk can be either relative or absolute. The noneconomic methods may involve value judgments that assign (mostly subjective) values to the risk (frequency and consequence) and benefits. Noneconomic methods that will be discussed in this chapter are exceedance analysis, value analysis, decision tree, and analytical hierarchy process (AHP).

The types of risk-based and risk-informed decisions can also be grouped as: risk acceptability, risk reduction, and risk alternative. In the risk acceptability decisions, the value of loss (or any figure of merit of interest, including performance measures) is usually compared to either an absolute or relative risk acceptance level, and determined whether the risk exceeds this limit, or estimating the probability of exceedance. This level can also be in the form of a single value, a range of values, or a distribution of values. Risk reduction decisions are those involving reduction of the frequency, consequence, or both associated with a risk by comparing a number of options and selecting the best alternative. Finally, risk alternative decisions involve comparing several events and their risks (frequency and consequence) and selecting the best.

In this chapter, several decision making methods along with examples of their applications have been discussed.

8.2 ECONOMIC METHODS IN RISK ANALYSIS

Risks are controlled (through risk aversion) by reducing the frequency that specific risk scenarios will occur or by limiting the consequence by averting or limiting exposure pathways. The process requires having a risk assessment model, with which the method of reducing the frequency or consequence can be applied. In the economic-based methods, the decision relies on the economic viability of the proposed risk-controlling approach. There are various methods to measure the risk viability that will be discussed in this section.

Examples of risk control strategies are reducing the frequency of an initiating event, such as removing concentration of a corrosive agent in a pipe, thus reducing the possibility of a pipe break due to stress corrosion cracking failure mechanism in safety critical pipes in a facility. Another control strategy might include addition of a new barrier to contain the release of contaminants into the environment due to the occurrence of a scenario. In health risk situations, quitting smoking to avoid cancer or using filtered cigarettes to reduce amount of cancer-causing agent are also examples of risk-controlling measures.

A less practical approach that most economists find having a great theoretical appeal is the externality tax [1]. This requires that a person who operates a hazardous system or facility that imposes some societal loss must pay a tax roughly equal to the cost of that loss. One of the few contexts in which this approach has seen a major application is in the control of pollution in the Rhine River valley in Germany, where industries are taxed by the amount of pollution they discharge into the river. Similarly, some suggest levying additional taxes on the cigarettes to the extent to cover the total societal losses attributed to smoking in terms of medical and loss of productivity costs.

The economic-based methods for selecting a risk-controlling approach are all based on assigning monetary values to the net gains in terms of amount of risk reduction (actual money or other economic-based indices) and determining whether such gains are adequate, desirable, and practical.

8.2.1 Benefit–Cost Analysis

In this approach, the risk in terms of reducing the frequency of occurrence of an undesirable event and associated consequences are appraised using some form of monetary value. Similarly the expected benefits by adopting each risk control approach are also assessed using the same or similar forms of monetary value. Then exceedance of benefit value from that of cost would constitute a condition for acceptance of the risk control measure. The benefit and cost of a risk control alternative may be measured in terms of real dollars or by using the so-called utility functions. The utility function allows for expressing the value of the money spent on controlling risk using other than monetary values, especially value judgments. For example, in a risk-averse situation, small amount of money may be judged to correspond to low utility (as compared to the face value of the money). But higher values, for example above a given limit, may correspond to exponentially higher values than their actual face values.

The benefit may be direct and indirect. For example, a direct benefit may be reduction in release of a pollutant or reduction in frequency of occurrence of an unsafe event. By direct, it means that the risk recipient directly gains the benefits of such a risk control action. The benefits may also be indirect. In this case, the benefit to the recipient is realized through third parties. For example, a risk control measure to reduce the radiation exposure to workers in a nuclear power plant would benefit the electric power consumers in terms of lower electric costs, as the plant owners directly benefit from potentially reduced health costs and other liabilities that are ultimately passed to the consumers.

The cost of implementing the risk control measure may also be direct or indirect. The direct costs are usually associated with the expenditure needed to implement a risk control measure. This cost is often paid by the actual recipient of the risk. For example, if a plant owner is investing in its plant by installing a protective measure to reduce the corrosion rate of its piping to reduce risks of pipe breaks, the direct cost in this case is absorbed by the direct beneficiary. On the other hand, if the taxes are raised by a state government to pay for the cost of cleaning a polluted area, tax payers who are not in the vicinity of the affected area may have to bear the indirect costs of this clean up process. In benefit–cost analyses aimed at controlling risks, it would be highly desirable, if at all possible, to identify the direct and indirect recipients of risk, benefits, and costs.

The decision criteria based on the direct and indirect costs and benefits have been summarized in Table 8.1. In this table C_D, B_D, C_I, and B_I are direct and indirect costs and benefits, respectively.

In most applications, it is difficult to separate the indirect and the direct costs and benefits, and thus only direct costs are considered, or an attempt to consider the total cost and benefits is made.

For comparing the effectiveness of multiple risk control measures, sometimes the benefit–cost ratio is used. The ratio is defined as

$$R_{b-c} = B/C, \tag{8.1}$$

where B is the benefit (direct, indirect, or total) and C is the cost (direct, indirect, or total).

Example 8.1

Raheja [2] describes the problem of risk reduction strategies in a transportation fatality risk situation attributed to pickup trucks. The case in question involves a scenario involving fuel tank side impacts in traffic accidents involving a particular design of pickup truck that may lead to explosions and fire-related injuries. The manufacturer is considering three risk reduction options. Determine the benefit-to-cost ratios for each design option. The data apply to reduction or prevention. The following risk reduction options are considered:

Option 1: Install a protective steel plate. Cost US$ 14. This will effectively prevent all explosions.
Option 2: Install a Lexan plastic plate. Cost US$ 4. This will prevent 95% of explosions.
Option 3: Install a plastic lining inside the fuel tank. Cost US$ 2. This will prevent 85% of explosions.

The following risk and cost data apply to this vehicle when no risk reduction option is implemented:
Possible fatalities from vehicles already shipped: 180
Expected cost per fatality: US$ 500,000
Number of injuries expected (no fatality): 200
Cost per injury: US$ 70,000
Expected number of vehicles damaged (no injury): 3000
Cost to repair the vehicle: US$ 1200
Number of vehicles to be manufactured: 6,000,000

TABLE 8.1
Decision Criteria in Benefit–Cost Analysis

Case	Direct Balance	Indirect Balance	Decision
1	$C_D < B_D$	$C_I < B_I$	Acceptable
2	$C_D > B_D$	$C_I > B_I$	Unacceptable
3	$C_D < B_D$	$C_I > B_I$	Unacceptable (unless allowed by regulation)
4	$C_D > B_D$	$C_I < B_I$	Unacceptable (unless subsidized)

Solution:
The cost for each option is the cost of implementing the change. The benefits are in terms of lives saved and avoidance of injury and damage. The analysis is performed as the direct cost and benefit to the manufacturer. Clearly the consumer risk and indirect manufacturer's risk may also play a major factor in the final decision.
Option 1:

$$\text{Cost} = \text{US\$ } 14 \times 6{,}000{,}000 \text{ vehicles} = \text{US\$ } 84{,}000{,}000$$

$$\text{Benefits} = (180 \text{ lives saved} \times \text{US\$ } 500{,}000) + (200 \text{ injuries prevented} \times \text{US\$ } 70{,}000)$$
$$+ (3000 \text{ damaged vehicles} \times \$1200) = \text{US\$ } 107{,}600{,}000$$

Using (8.1)

$$R = \text{US\$ } 107{,}600{,}000 / \text{US\$ } 84{,}000{,}000 = 1.28$$

Option 2:

$$\text{Cost} = \text{US\$ } 4 \times 6{,}000{,}000 = \text{US\$ } 24{,}000{,}000$$

$$\text{Benefits} = (95\% \text{ accidents prevented}) \times [(180 \text{ fatalities} \times \text{US\$ } 500{,}000)$$
$$+ (200 \text{ injuries} \times \text{US\$ } 70{,}000) + (3000 \text{ damaged vehicles} \times \text{US\$ } 1200)]$$

$$\text{Benefits} = 0.95 \times \text{US\$ } 107{,}600{,}000 = \text{US\$ } 102{,}220{,}000$$
$$R = \text{US\$ } 102{,}220{,}000 / \text{US\$ } 24{,}000{,}000 = 4.25$$

Option 3:

$$\text{Cost} = \text{US\$ } 2 \times 6{,}000{,}000 = \text{US\$ } 12{,}000{,}000$$

$$\text{Benefits} = (85\% \text{ accidents prevented}) \times [(180 \text{ fatalities} \times \text{US\$ } 500{,}00)$$
$$+ (200 \text{ injuries} \times \text{US\$ } 70{,}000) + (3000 \text{ damaged vehicles} \times \text{US\$ } 1200)]$$

$$\text{Benefits} = 0.85 \times \text{US\$ } 107{,}600{,}000 = \text{US\$ } 91{,}460{,}000$$
$$R = \text{US\$ } 91{,}460{,}000 / \text{US\$ } 12{,}000{,}000 = 7.62$$

Relatively speaking, option 3 appears to present the highest benefit–cost ratio. As noted earlier, the decision should not be solely based on this figure of merit, as other indirect factors such as the manufacturer's reputation should also be considered. The indirect costs and benefits to the manufacturer in some cases may even be more important.

Some researchers such as Kelman [3] argue that there may be instances where certain decisions might be right even though its benefits do not exceed its costs, especially when indirect nonmarket benefits such as environmental and public health benefits are involved. This view is rejected by other researchers such as Butter et al. [4].

8.2.1.1   Value of Money

To overcome the difficulties of estimating the actual monetary amounts of cost and benefit, a subjective estimate of the value of money may be used instead. Because of consideration of many indirect and direct factors, the value of money may not be linear for the risk recipient and it may be necessary to use a nonlinear relationship between the actual money and the equivalent "value" of the money to the risk recipient. For example, consider the three cases of risk-averse, risk-neutral, and risk-seeking recipient, depicted in Figure 8.1 with respect to such recipients. To better understand the concept of risk aversion, consider Bernstein's [5] explanation. Imagine that you have a choice between a gift certificate worth US$ 100 and an opportunity of winning US$ 400 with a chance of 25%. These two have the same expected amount of return. A risk-averse person takes a definite US$ 100 gift over an uncertain US$ 400 gift. A risk-neutral person would take either case. A risk seeker would select the US$ 400 gift with a chance of 25%. Similarly a project with a known cost of US$ 500,000 is preferred by a risk-averse decision maker to a similar project with uncertain cost estimated in the range of US$ 300,000 to US$ 700,000.

The value is a relative scale metric that is subjectively assigned to the amount of the actual money. A risk recipient who is risk-averse places increasingly higher relative values to the amount of money than it is actually worth. Reversely, a risk-seeking individual or organization places lower risk values (i.e., it is willing to take more risk) as the actual amount of money increases. The risk-neutral situation adopts a linear relationship between the risk value and the actual monetary amount. Hammond et al. [6] define a value system such as the ones shown in Figure 8.1 that relate the value of consequences from "worst" to "best".

VanDoren [7] argues that the opponents of economic-based methods in regulatory and policy decision making have long seized on the moral issues by claiming that their views protect individual health and that less consequential concerns such as cost should not interfere with risk management efforts. VanDoren, however, indicates the fallacy of such thinking is that high-cost, low-benefit safety decisions and regulations divert organization's and society's resources from a mix of expenditures that would be more health enhancing than the allocations dictated by the health and safety regulations. Finally, VanDoren states that "agencies that make an unbounded financial commitment to safety frequently are sacrificing individual lives in their symbolic quest for a zero-risk society."

Example 8.2

A contestant on a TV game show just won US$ 50,000. The host offers him a choice: quit now and keep the US$ 50,000, or play again. If the contestant decides to play again, there is a 50% chance that he will win again, and increases his money to US$ 100,000. On the other hand, if he decides to play again and lose, he loses all his US$ 50,000. Which option should he choose?

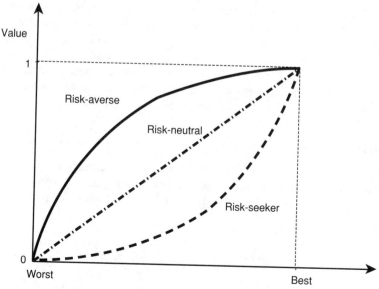

FIGURE 8.1 Value of money or consequences.

Solution:
The expected monetary value of the two choices is equal. If the contestant is risk-averse, he will quit playing. If the contestant is a risk seeker, he will play and if risk-neutral, the choice would be random.

8.2.2 COST-EFFECTIVENESS ANALYSIS

In this approach, the relationship between cost to the recipient for taking the risk (resulting in losses) and for controlling the risk (resulting in reducing the risk) is shown. In the cost-effectiveness analysis, the basic assumption is that expenditures in risk reduction operations (e.g., for life saving) indicate that residual risk is acceptable. The analysis is often performed to compare the various options for risk control, or determining the optimum risk control approach. Figure 8.2(a) illustrates this concept. As shown in this figure, the cost of loss (e.g., fatalities, material, and time losses) increase as the risk increases. This can be linear, or as shown in this figure, an exponentially increasing function. At the same time, the cost of control (leading to reduction of the risk) would decrease, as the willingness to accept the risk by the recipient increases. This is clearly obvious since for reducing the risk to very small values, one requires far more expensive risk control methods.

While a given risk scenario that relies on a specific risk control method may represent a point on the cost vs. risk graph displayed in Figure 8.2, the optimum risk control solution may also be found as the point that the cost and risk together impose the least possible net loss to the recipient. This method could lead to a model-based approach to regulating and accepting the risks. For example, all the risk control solutions to the left of the optimum point shown in Figure 8.2(a) can be considered as acceptable risk reduction techniques.

Sometimes for risk management purposes, only the cost of control is considered. For example, the area of curve in Figure 8.2(b) that is near a 45° tangent line to the cost vs. risk curve determines the optimum risk control solution (see Figure 8.2(c)). This point is referred to as the most cost-effective point. An area slightly above and below this optimum point is usually referred to as-low-as-practicable (ALAP), which determines the region within which the decision maker, for example a regulatory agency, accepts a risk reduction strategy and views the risk reduction methods as cost-effective. While more risk reduction solutions are possible, these come at an exorbitant price and are often associated with sophisticated technologies. The region that encompasses such technologies is referred to as best available technologies (BATs) as shown in Figure 8.2(b) and is not necessarily the best choice. A highly risk-averse recipient, however, would prefer this technology over the ALAP approaches. Since the risk can never be reduced to the zero level, the BAT defines a level of practically zero risk. A clear advantage of the cost-effective approach is that risk need not be converted to its equivalent monetary value as is the case in the benefit–cost analysis. This conversion usually is difficult and often highly subjective.

8.2.3 RISK-EFFECTIVENESS ANALYSIS

Risk-effectiveness is a measure of cost of a risk reduction (or control) solution per unit of risk reduction. The solution can be a risk-mitigative device, a regulatory approach, an operating or maintenance procedure, etc. Accordingly, the cost-effectiveness for a set of risk scenarios can be expressed as

$$RE = \frac{S}{\sum_{i=1}^{n} F_i C_i - \sum_{i=1}^{n} F_i' C_i'} \tag{8.2}$$

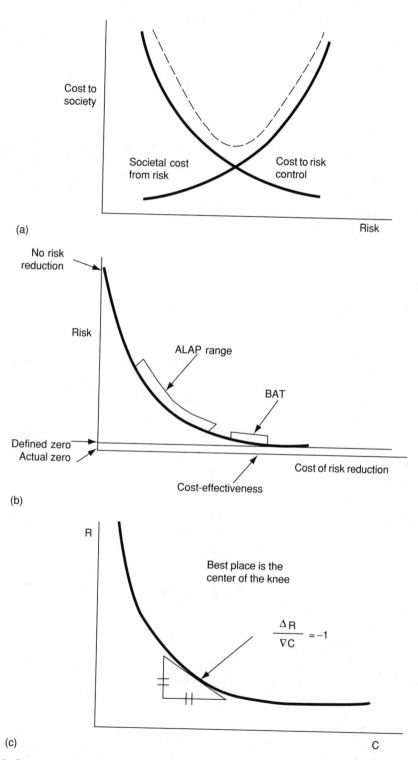

FIGURE 8.2 Cost vs. risk for both cost of risk and cost of control (a); and regions of risk vs. cost for cost of control (b).

where RE represents the risk-effectiveness measure, S the annualized cost of a risk reduction solution, F_i the frequency of scenario i without implementation of the risk reduction solution, C_i the consequence of the scenario i, and F_i' and C_i', respectively, the frequency and consequence value after implementation of the risk control solution.

One very useful application of this approach is to measure the risk reduction ability of regulations per life saved. Tengs et al. [8] present cost-effectiveness of various governmental regulations using this approach. (see Table 8.2 adopted from Tengs et al.)

Example 8.3
Addition of a diverse preventive solution for risk reduction at a complex facility costs US$ 1.5 million per year. It reduces the frequency of an undesirable event (fatalities) with actual annual frequency of 1×10^{-5} per year by an order of magnitude. Further, it reduces the consequences in terms of fatalities due to hazard exposure from 1500 to 100 cases. Determine the cost-effectiveness of this approach.

TABLE 8.2
Data on Three Modes of Transportation [8]

Regulation	Year	Health or Safety	Baseline Mortality Risk per Million Exposed	Cost per Premature Death Averted (US$ millions 1990)
Unvented space heater ban	1980	CPSC	1,890	0.1
Aircraft cabin fire protection standard	1985	FAA	5	0.1
Auto passive restrain/seat belt standards	1984	NHTSA	6,370	0.1
Steering column protection standard	1967	NHTSA	385	0.1
Underground construction standards	1989	OSHA-S	38,700	0.1
Trihalomenthane drinking water standards	1979	EPA	420	0.2
Aircraft seat cushion flammability standard	1984	FAA	11	0.4
Alcohol and drug control standard	1985	FRA	81	0.4
Auto fuel-system integrity standard	1975	NHTSA	343	0.4
Standards for servicing auto wheel rims	1984	OSHA-S	630	0.4
Aircraft floor emergency lighting standards	1984	FAA	2	0.6
Concrete and masonry construction standards	1988	OSHA-S	630	0.6
Crane-suspended personnel platform standard	1988	OSHA-S	81,000	0.7
Passive restraints for trucks and buses	1989	NHTSA	6,370	0.7
Side-impact standards for autos (dynamic)	1990	NHTSA	NA	0.8
Children's sleepwear flammability ban	1973	CPSC	29	0.8
Auto side door support standards	1970	NHTSA	2,520	0.8
Low-altitude windshear equipment and training	1988	FAA	NA	1.3
Electrical equipment standards (metal mines)	1970	MSHA	NA	1.4
Trenching and excavation standards	1989	OSHA-S	14,310	1.5
Traffic alert and collision avoidance (TCAS)	1988	FAA	NA	1.5
Hazard communication standard	1983	OSHA-S	1,800	1.6
Side-impact standards for trucks and MPVs	1989	NHSTA	NA	2.2
Gain dust explosion prevention standards	1987	OSHA-S	9,450	2.8
Rear lap/shoulder belts for autos	1989	NHSTA	NA	3.2
Standards for radio nuclides in uranium mines	1984	EPA	6,300	3.4
Benzene NESHAP (original: fugitive emissions)	1984	EPA	1,470	3.4
Ethylene dibromide drinking water standards	1991	EPA	NA	5.7

TABLE 8.2
Data on Three Modes of Transportation [8]—(*Continued*)

Regulation	Year	Health or Safety	Baseline Mortality Risk per Million Exposed	Cost per Premature Death Averted (US$ millions 1990)
Benzene NESHAP (revised: coke by-products)	1988	EPA	NA	6.1
Asbestos occupational exposure limit	1972	OSHA-S	3,015	8.3
Benzene occupational exposure limit	1987	OSHA-S	39,600	8.9
Electrical equipments standards (coal mines)	1970	MSHA	NA	9.2
Arsenic emission standards for glass plants	1986	EPA	2,660	13.5
Ethylene oxide occupational exposure limits	1984	OSHA-S	1,980	20.5
Arsenic/copper NESHAP	1986	EPA	63,000	23.0
Hazardous waste listing for petroleum refining sludge	1990	EPA	210	27.6
Cover/move uranium mil tailings (inactive sites)	1983	EPA	30,100	31.7
Benzene NESHAP (Revised: transfer operations)	1990	EPA	NA	32.9
Cover/move uranium mil tailings (active sites)	1983	EPA	30,100	45.0
Acrylonitrile occupational exposure limit	1978	OSHA-S	42,300	51.5
Coke ovens occupational exposure limit	1976	OSHA-S	7,200	63.5
Lockout/tagged	1989	OSHA-S	4	70.9
Asbestos occupational exposure limit	1986	OSHA-S	3,015	74.0
Arsenic occupational exposure limit	1978	OSHA-S	14,800	106.9
Asbestos ban	1989	EPA	NA	110.7
Diethylstilbestrol (DES) cattlefeed ban	1979	FDA	22	124.8
Benzene NESHAP (revised waste operation)	1990	EPA	NA	168.2
1,2-Dichloropropane drinking water standard	1991	EPA	NA	653.0
Hazardous waste land disposal ban (1st to 3rd)	1988	EPA	2	4,190.4
Municipal solid waste landfill standards (proposed)	1988	EPA	<1	19,107.0
Formaldehyde occupational exposure limit	1987	OSHA-S	31	86,201.8
Atrazine/alachlor drinking water standard	1991	EPA	NA	92,069.7
Hazardous waste listing for wood preserving chemicals	1990	EPA	<1	5,700,000.0

Solution:
Using (8.2) cost-effectiveness of

$$RE = \frac{US\$\ 1.5 \times 10^6}{(1 \times 10^{-5} \times 1500) - (1 \times 10^{-6} \times 100)} \cong 1 \times 10^8 \ US\$/\text{life saved}$$

Acceptability of this estimate is a policy decision and could be based on additional consideration, such as the public perception of the risk itself, or an acceptable value of the life (discussed in Section 7.4.6).

8.3   NONECONOMIC TECHNIQUES

In this section, four noneconomic techniques will be discussed. They include the probability of exceedance method, structured value analysis, analytic hierarchy process, and the decision tree method. Noneconomic techniques are particularly important for multiattribute decisions. In risk assessment, these decisions involve cases concerning attributes such as risk, performance, cost, and time.

8.3.1 PROBABILITY OF EXCEEDANCE METHOD

This is the simplest approach to risk-based and risk-informed decision making. The process involves using risk acceptance criteria such as acceptable values of the frequency of a hazard exposure event, consequences of the event, or the value risk itself and comparing them with the estimated values. The process also involves determining the degree to which the estimated risk meets the criteria over time and space of interest.

There are four possible situations. The estimated risk (frequency, consequence, or risk value) is represented by a probability distribution function and the decision criterion is also shown in form of probability distribution function. In this case, the probability that the estimated risk exceeds the criterion is calculated, similar to Equation (4.94). The decision would be based on the value of the probability. For example, if one would like to be highly conservative, he or she would choose very low probabilities for the estimated risk to exceed the risk criterion (e.g., a decision probability of 0.01 indicates a 1% chance that the estimated risk exceeds the acceptance limit). Of course, the exceedance probability should be estimated over all possible times and spaces of interest, and the maximum or combined probability of exceedance be used for the decision.

The second case is similar to the first, with the exception that the acceptance criteria are represented in terms of point estimates, not probability distribution functions. The mathematics is clearly simpler in this case, as the probability of exceedance from a deterministic point can be easily calculated. Similar to the first situation, the probability of exceedance would be the decision index of interest. If the risk is changed over time and space, the exceedance probability should be estimated for all points in time and space.

The third situation is a rare case where the actual risk is represented as a point estimate, but the acceptance criterion is a distribution function. In this case, the probability of exceedance is estimated similar to case 2 discussed above, and the decision is made based on the estimated probability of exceedance.

Finally, both risk estimation and acceptance criterion may be represented by point estimates. This is a simple case where the actual exceedance (i.e., exceedance of the minimum acceptable risk limit from the estimated risk) determines acceptability.

Sometimes, the acceptance criteria may be expressed in terms of more than one risk metric. An example of this includes the case where acceptable frequency–consequence relationship (similar to the Farmer curves) is available. In this case, the probability of exceedance for all risk metrics (for example, frequency and consequence) over the entire time and space of interest is calculated and the total probability of exceedance is measured based on the probability of exceedance of either risk metrics. For this purpose, Equation (A.17) may be used.

8.3.2 STRUCTURED VALUE ANALYSIS

Because we are often dealing with the evaluation of imprecise and intangible values in risk management, to avoid monetarizing risk, cost, and benefits, an approach discussed by Rowe [9, 10] is especially useful when the decision must consider multiple diverse risk acceptance criteria. In this approach, the risk management parameters such as the risk, consequence, cost, exposure amount, and time are each assessed based on a value function and a normalized weight of importance of the parameter. An aggregate of the overall value of all parameters are then calculated and is used as an index for decision making.

Consider the case that we are interested to evaluate the capability of some risk reduction or control options based on acceptance criteria characterized by parameters i (such as cost, risk, and consequence). Further, assume that the value functions, similar to Figure 8.1 associated with each parameter i, are F_i. These functions are subjectively assigned by the decision makers (or experts) depending on their preferences and values. Finally, assume that the normalized weight of the parameter i is W_i (often normalized to 1). The weight represents the relative importance of each parameter to the decision. This weight is a subjective value expressed by the decision maker directly or through expert elicitation. Then the linear aggregate weight of all the parameters would be:

$$V = \sum_i F_i W_i \tag{8.3}$$

Rowe [9] discusses other forms of (8.3) depending on the nature of decision. However, (8.3) is adequate for most situations. The value of V for all risk reduction solutions is estimated, and the one with the best value (either the highest or lowest, depending on the situation) is selected.

This method also allows for consideration of uncertainties in both F_i and W_i values, by propagating these uncertainties to express the values of V in terms of probability distributions. The decision for selecting the best option would be similar to the problems discussed in Section 8.3.1.

Example 8.4

Rowe [9] defines a problem of selecting the best alternative means of transportation among the three options of motorcycle, automobile, and bus. Table 8.3 shows three parameters that enter the decision for the choice of the best solution. These parameters are time of travel (from home to work and back), total cost, and accident risks. Further, data gathered for each mode of transportation are associated with some uncertainties. Suppose these uncertainties are shown by a normal distribution function with a coefficient of variation of 10% as listed in Table 8.3. Furthermore, Figure 8.3 to Figure 8.5 describe the subjective value functions assigned by the decision maker to each of the parameters. We are interested in identifying the best transportation choice under this situation.

A subjective weight of 0.4 has been selected for the parameter "time" (as it translates to the quality of life) and 0.3 for "risk" (as it related to the quantity of life) and 0.3 for cost.

Solution:
According to (8.3)

$$V = 0.4F_T \text{ (time)} + 0.3F_R \text{ (risk)} + 0.3F_C \text{ (cost)}$$
$$V_{\text{Motorcycle}} = 0.4(0.85) + 0.3(0.2) + 0.3(0.8) = 0.64$$
$$V_{\text{Auto}} = 0.4(0.65) + 0.3(0.3) + 0.3(0.6) = 0.53$$
$$V_{\text{Bus}} = 0.4(0.5) + 0.3(0.75) + 0.3(0.75) = 0.65$$

TABLE 8.3
Decision Parameters and Their Values for Transportation Options

System	Time (min) F_1	Risk per 100,000 miles F_2	Cost per Year (US$) F_3
Motorcycle	12	200	250
Automobile	45	50	750
Bus	90	3	200

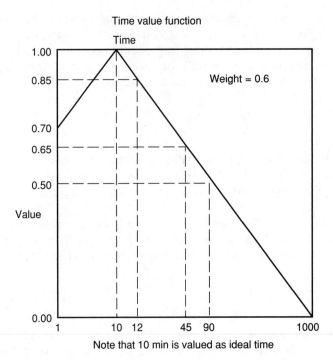

FIGURE 8.3 Value function for the parameter "time."

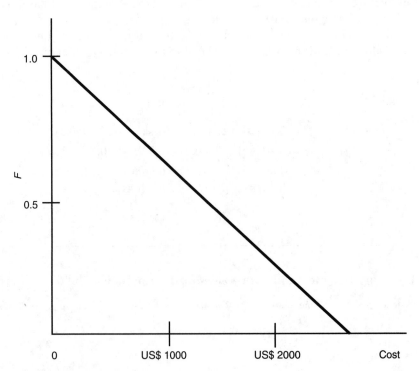

FIGURE 8.4 Value function for cost.

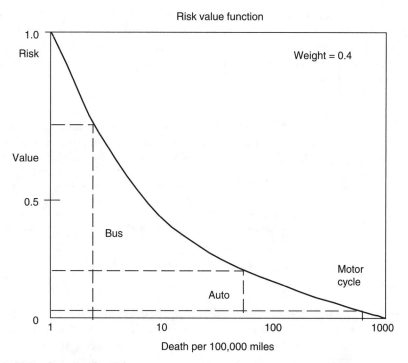

FIGURE 8.5 Value function for risk.

The uncertainties are calculated using the Taylor expansion approach (see Section 5.4):

$$\hat{V} \approx f(x_1, x_2, \ldots, x_n) = 0.4x_1 + 0.3x_2 + 0.3x_3$$

If the uncertainties are shown by a normal distribution function with a coefficient of variation of 10%,

	μ_{time}	σ_{time}	μ_{risk}	σ_{risk}	μ_{cost}	σ_{cost}
Motorcycle	0.85	0.085	0.20	0.020	0.80	0.080
Auto	0.65	0.065	0.30	0.030	0.60	0.060
Bus	0.50	0.050	0.75	0.075	0.75	0.075

The variance (as a measure of uncertainty) of the system performance characteristic V is calculated from (5.7). We have

$$\frac{\partial f(V)}{\partial x_1} = 0.4; \ \frac{\partial f(V)}{\partial x_2} = 0.3; \ \frac{\partial f(V)}{\partial x_1} = 0.3;$$

by substituting the derivatives and S^2 into (5.7),

$$\text{Var}\left(\hat{V}_{\text{Motorcycle}}\right) = 0.4 \cdot 0.085^2 + 0.3 \cdot 0.02^2 + 0.3 \cdot 0.08^2 = 0.00493$$
$$\text{Var}\left(\hat{V}_{\text{Auto}}\right) = 0.4 \cdot 0.065^2 + 0.3 \cdot 0.03^2 + 0.3 \cdot 0.06^2 = 0.00304$$
$$\text{Var}\left(\hat{V}_{\text{Bus}}\right) = 0.4 \cdot 0.05^2 + 0.3 \cdot 0.075^2 + 0.3 \cdot 0.075^2 = 0.00438$$

Figure 8.6 shows the results. Based on the point estimate results, options "Bus" and "Motor-cycle" exhibit the same values, with "Automobile" having the least value.

Considering the results including propagated uncertainties, the three options are close and overlap with the Bus and Motorcycle being marginally on the higher side, and the Bus option being marginally the best option.

Example 8.5

Consider a company's sales to its customers in the fiscal years 2003 and 2004. We are interested to rank the value of each customer based on these data. Three parameters con-sidered as contributing to the decision have been listed in the table below:

	Forecasted Sales and Customer Potential	Sales This Year	Rate of Change in Sales Prior to This Year
Parameters	F_1	F_2	F_3
Weights of each parameter	W_1	W_2	W_3

Suppose that the company places the following weights (subjectively) to each parameter. $W_1 = 0.4$, $W_2 = 0.3$, $W_3 = 0.3$. Figure 8.7(a)–(c) shows the value functions, subjectively assigned by the company executives to represent their value of each parameter.

Consider Customer X with expected sales of US$ 70 million in the immediate future, present sales of US$ 40 million, and most recent average annual sales increase of 20%. Compare this customer with Customer Y with expected sales of US$ 70 million in the immediate future, present sales of US$ 70 million, and the sale change rate of 5%.

Solution:

Customer X:

Expected 2004 sales (US$ 70 million) $V_{F_1} = 0.7$(from V_{F_1} Curve)

2003 sales (US$ 50 million) $V_{F_2} = 0.8$

2002 sales (US$ 40 million) $V_{F_3} = 0.2$

(find annual rate of change, e.g., 20%)

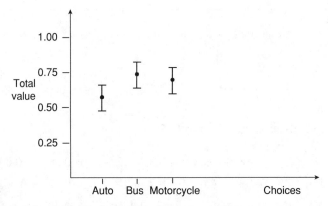

FIGURE 8.6 Results of the risk–cost–time evaluation.

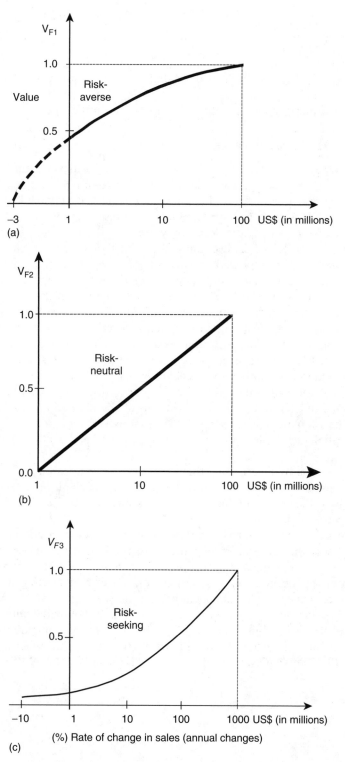

FIGURE 8.7 (a) Value of forecasted sales to a customer *i*. (b) Value of present sales to a customer. (c) Value of rate of change in sales.

$$V_X = 0.4 \times 0.7 + 0.3 \times 0.8 + 0.2 \times 0.3 = 0.58$$

Similarly, for Customer Y,

$$V_Y = 0.4 \times 0.7 + 0.3 \times 0.85 + 0.3 \times 0.1 = 0.565$$

Customer X is a better customer for the company based on the value system described.

8.3.3 ANALYTICAL HIERARCHY PROCESS

AHP was developed by Saaty [11] in the early 1970s and expanded and applied to many applications involving decision making, including risk-based or risk-informed decision making. The AHP is often used for prioritizing alternatives when multiple criteria (factors or parameters) must be considered. It is a systematic procedure for representing the elements of any problem hierarchically. It organizes the basic decision by breaking it down into its smaller constituent parts. The process then guides the decision makers through a series of pairwise comparisons and qualitative judgments to express the relative importance, strength, or intensity of impact of the element in the hierarchy. These qualitative judgments are then translated to quantitative indices for analytical ranking. The AHP includes procedures and principles used to synthesize the many judgments to derive priorities among criteria and subsequently for alternative solutions.

The AHP is one of the most popular decision theoretic approaches to multiattribute or criteria decision making and has been utilized to wide variety of applications (for example, R&D project prioritization, assessing the risks of complex systems, strategic planning in the military, and selection of manufacturing technologies).

In the simplest form, the hierarchy of a decision task is comprised of three levels: the goal, the criteria, and the alternatives. At the top of the hierarchy is the overall objective (goal) and the decision alternatives are at the bottom. Between the top and bottom levels are the relevant attributes of the decision problem that provide significant impacts on the decision process.

The process begins by determining the relative importance, strength, or intensity of the criteria in meeting the goal. Secondly, the extents to which each of the criteria is achieved are measured. Finally, the results of the two previous analyses are combined to compute the degree to which the goal is met.

In this process, the decision maker carries out pairwise comparison judgments which are then used to assess the importance of various decision elements. These comparisons are derived from a set of judgments, either verbal or numerical. They express the relative importance, strength, or intensity of one item vs. another in meeting a criteria or goal. Table 8.4 represents one such measurement scale in the AHP.

If a factor has one of the numbers in Table 8.4 (e.g., 3) compared with a second factor, then the second factor has the reciprocal value (i.e., 1/3) when compared to the first. The pairwise comparisons are reduced to a square matrix, in which an array of numbers is arranged as in the following example of a 4×4 matrix (four factors: S_1, S_2, S_3, and S_4):

	S_1	S_2	S_3	S_4
S_1	1	2	3	4
S_2	1/2	1	5	6
S_3	1/3	1/5	1	7
S_4	1/4	1/6	1/7	1

TABLE 8.4
Qualitative and Quantitative Guidelines for Pairwise Comparison

Numerical Rating	Definition	Explanation
9	Extremely preferred (or more important, stronger, more intense, etc.)	The evidence favoring one factor over another is of the highest possible order of affirmation
7	Very strongly preferred	A factor is strongly favored and its dominance is demonstrated in practice
5	Strongly preferred	Experience and judgment strongly favor one factor over another
3	Moderately preferred	Experience and judgment slightly favor one factor over another
1	Equally preferred	Two factors contribute equally to the objective
2, 4, 6, 8	For additional levels of discrimination	When compromise is needed

In making the pairwise comparisons the following kinds of questions are asked. In comparing factor S_1 with factor S_2:

1. Which has greater impact, by how much?
2. Which is more likely to happen, by how much?
3. Which is more preferred, by how much?
4. Which is better, by how much?
5. Which is stronger, by how much?

The matrix is entered from left to right, with the horizontal items being compared with vertical items. For example, in the matrix above S_1 is equally to moderately better than S_2, and a value of 2 expresses this judgment. Its reciprocal (1/2) is assigned when S_2 is compared to S_1. A value of 1/2 indicates that S_2 is equally to moderately worse than S_1. Clearly the diagonal cells are all 1, as they represent comparison of one factor to itself.

If there are n factors involved, a total of $n(n-1)/2$ judgments are needed to develop this matrix. As an example, when $n = 4$, only 6 judgments would be needed to fill the 16 cells in the corresponding 4×4 matrix. When the n is large, generally 8 or more, several available methods can be used to reduce the numbers of judgments.

AHP provides a useful mechanism for checking the degree to which an analyst's judgments are consistent. An example of perfect consistency is when A is twice better than B, B is three times better than C, and A is six times better than C. In this example, inconsistency occurs, for example, when C is expressed as better than A. Perfect consistency is not necessary and is sometimes difficult to attain. AHP reflects this concern and allows a range of inconsistency levels. The overall consistency of judgments is measured by the mean of consistency ratio (CR). In general, a CR of 1 or less is considered acceptable. If it is more than 1, the judgments may be somewhat random. Participants or decision makers should study the problem and revise their judgments. In Example 8.7, we discuss the methods to judge the adequacy of the consistency in a pairwise matrix and how to correct inconsistencies.

The decision problem is simple in the case of a single decision making when the decision problem is the selection of the most preferred alternative according to one's preferences. For the group decision making, the analysis must be able to account for the conflicts among different individuals who have different goals, objectives, criteria, etc.

Group decision making with AHP can and has been carried out in two ways:

1. Debate the judgments and vote until the consensus or compromise is reached.
2. Take the geometric mean of individual judgments to form an aggregate judgment.

To solve the AHP problems an eigenvalue problem should be solved. For example, if the pairwise comparison matrix A has been developed, then

$$\overline{A}\,\overline{W} = \lambda \overline{W} \tag{8.4}$$

where $\overline{A}$ is the binary importance matrix, $\overline{W}$ is the vector of weights of objectives, and λ is the eigenvalue.

For example, for a problem consisting of six decision factors or parameters shown in the hierarchy of Figure 8.8, matrix A is represented by:

$$\overline{A} = \begin{bmatrix} W_1/W_1 & W_1/W_2 & \cdots & W_1/W_n \\ W_2/W_1 & W_2/W_2 & \cdots & W_2/W_n \\ \vdots & \vdots & \ddots & \vdots \\ W_n/W_1 & W_n/W_2 & \cdots & W_n/W_n \end{bmatrix}$$

where $a_{ij} = W_i/W_j$; accordingly $a_{ji} = W_j/W_i$. The eigen value problem can be written in the form:

$$\begin{bmatrix} 1 & a_{12} & \cdots & a_{1n} \\ 1/a_{12} & 1 & \cdots & a_{2n} \\ \vdots & \vdots & \ddots & \vdots \\ 1/a_{1n} & 1/a_{2n} & \cdots & 1 \end{bmatrix} \begin{bmatrix} W_1 \\ W_2 \\ \vdots \\ W_n \end{bmatrix} = \lambda_{max} \begin{bmatrix} W_1 \\ W_2 \\ \vdots \\ W_n \end{bmatrix} \tag{8.5}$$

Solution to (8.5) yields the eigen vector matrix $\overline{W}$, which is the normalized weight that can be assigned to each factor i through n.

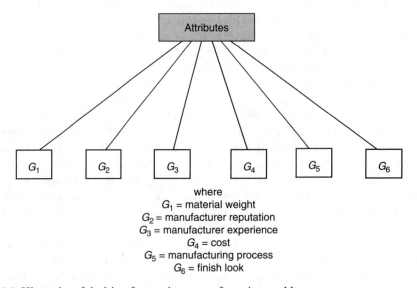

where
G_1 = material weight
G_2 = manufacturer reputation
G_3 = manufacturer experience
G_4 = cost
G_5 = manufacturing process
G_6 = finish look

FIGURE 8.8 Hierarchy of decision factors in a manufacturing problem.

Example 8.6

Consider Example 8.4. Use the AHP process to select the best transportation option.

Solution:

The hierarchy of the decision can be shown as follows:

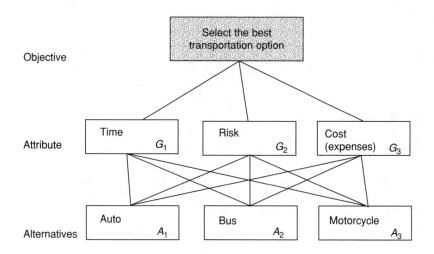

Suppose we judge matrix A (i.e., how attributes relate to each other) as follows, then

$$\overline{A} = \begin{array}{c} \\ G_1 \\ G_2 \\ G_3 \end{array} \begin{array}{c} G_1 \quad G_2 \quad G_3 \\ \begin{bmatrix} 1 & 1/3 & 1/5 \\ 3 & 1 & 1/3 \\ 5 & 3 & 1 \end{bmatrix} \end{array}$$

The eigenvalue problem of (8.4) is

$$\begin{bmatrix} 1 & 1/3 & 1/5 \\ 3 & 1 & 1/3 \\ 5 & 3 & 1 \end{bmatrix} \begin{bmatrix} W_1 \\ W_2 \\ W_3 \end{bmatrix} - \begin{bmatrix} \lambda W_1 \\ \lambda W_2 \\ \lambda W_3 \end{bmatrix} = 0 \quad \text{or} \quad \begin{matrix} W_1 + 1/3 W_2 + 1/5\, W_3 - \lambda W_1 = 0 \\ 3 W_1 + W_2 + 1/3\, W_3 - \lambda W_2 = 0 \\ 5 W_1 + 3\, W_2 + W_3 - \lambda W_3 = 0 \end{matrix}$$

$$\overline{W} = \begin{bmatrix} 0.106 \\ 0.260 \\ 0.634 \end{bmatrix}$$

Now consider each hierarchy separately. For example,

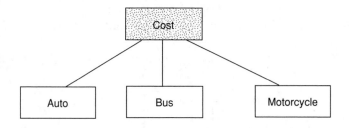

For each hierarchy, set the A matrix and solve the corresponding eigenvalue problem.

$$A = \begin{pmatrix} 1 & 1/3 & 1/2 \\ 3 & 1 & 3 \\ 2 & 1/3 & 1 \end{pmatrix} \begin{pmatrix} W_1 \\ W_2 \\ W_3 \end{pmatrix} = \lambda \begin{pmatrix} W_1 \\ W_2 \\ W_3 \end{pmatrix} \text{ or } \begin{pmatrix} W_1 + 1/3W_2 + 1/2W_3 \\ 2W_1 + W_2 + 3W_3 \\ 2W_1 + 1/3W_2 + W_3 \end{pmatrix} = \begin{pmatrix} \lambda W_1 \\ \lambda W_2 \\ \lambda W_3 \end{pmatrix}$$

Therefore,

$$
\begin{aligned}
(1-\lambda)W_1 + \quad 1/3W_2 + \quad 1/2W_3 &= 0 \\
3W_1 + (1-\lambda)W_2 + \quad 3W_3 &= 0 \\
2W_1 + \quad 1/3W_2 + (1-\lambda)W_3 &= 0
\end{aligned}
$$

The solution to above eigenvalue problem is

$$\lambda_{\max} = 3.05 \quad \text{and} \quad W_1 = 0.16, \quad W_2 = 0.59, \quad W_3 = 0.25.$$

Note that in AHP only $\lambda_{\max}$ is used. The other two lower λ values are not used. $\lambda_{\max}$ is usually close to the size of the A matrix, unless the pairwise estimates are inconsistent.

Similarly for other hierarchies the $\bar{A}$ matrices are:

$$\bar{A}_{G_1} = \begin{array}{c} \\ A_1 \\ A_2 \\ A_3 \end{array} \begin{array}{ccc} A_1 & A_2 & A_3 \\ \left[\begin{array}{ccc} 1 & 3 & 1/5 \\ 1/3 & 1 & 1/7 \\ 7 & 5 & 1 \end{array}\right] \end{array} \quad \bar{A}_{G_2} = \begin{array}{c} \\ A_1 \\ A_2 \\ A_3 \end{array} \begin{array}{ccc} A_1 & A_2 & A_3 \\ \left[\begin{array}{ccc} 1 & 3 & 1/5 \\ 1/3 & 1 & 1/9 \\ 5 & 9 & 1 \end{array}\right] \end{array} \quad \bar{A}_{G_3} = \begin{array}{c} \\ A_1 \\ A_2 \\ A_3 \end{array} \begin{array}{ccc} A_1 & A_2 & A_3 \\ \left[\begin{array}{ccc} 1 & 1/7 & 1/9 \\ 7 & 1 & 1/3 \\ 9 & 3 & 1 \end{array}\right] \end{array}$$

The total result can be written in form of the following multiplication:

$$
\begin{array}{c} \\ A_1 \\ A_2 \\ A_3 \end{array}
\begin{array}{ccc} G_1 & G_2 & G_3 \\ \left[\begin{array}{ccc} 0.188 & 0.188 & 0.055 \\ 0.081 & 0.080 & 0.290 \\ 0.731 & 0.0731 & 0.655 \end{array}\right] \end{array}
\begin{array}{cc} \left[\begin{array}{c} 0.106 \\ 0.260 \\ 0.634 \end{array}\right] & \begin{array}{c} G_1 \\ G_2 \\ G_3 \end{array} \end{array}
=
\begin{array}{c} \begin{array}{ccc} A_1 & A_2 & A_3 \end{array} \\ \left[\begin{array}{ccc} 0.110 & 0.211 & 0.688 \end{array}\right] \end{array}
$$

Clearly motorcycle has the highest weight in this problem.

If necessary, to revise the largest absolute difference $|a_{ij} - (W_i/W_j)|$ preferably replace a_{ij} with W_i/W_j. For example, suppose $G = [0.77, 0.05, 0.17]$, with $\lambda_{\max} = 3.21$, and C.R. $= 0.18$. Further, suppose the largest absolute difference is between a_{12} and the new value as $W_1/W_2 = 15.4$. Replacing a_{12} with 15.4, yields new $\bar{W}$ with $\bar{W} = [0.76, 0.04, 0.02]$, $\lambda_{\max} = 3.023$, and C.R. $= 0.01$, which is a consistent estimate. It is recommended that the analyst avoids using above inconsistency correction method successively.

Deviation from complete consistency may be represented by *consistency index* (CI):

$$\text{C.I.} = \frac{\lambda_{\max} - n}{n - 1}$$

$$\text{C.R.} = \frac{\text{C.I.}}{\text{R.I.}}, \quad \text{R.I.} = \text{Random Index}$$

Saaty [11] offers the value of RI as follows:

n	1	2	3	4	5	6	7	8	9	10
RI	0.0	0.0	0.58	0.9	1.12	1.24	1.32	1.41	1.45	1.49

If CR $\leq$ 0.1, we have an adequate consistency.

8.3.4 DECISION TREE ANALYSIS

Decision trees are good for helping a risk manager choose between several courses of risk control actions. They are highly effective structures within which one can lay out risk control solutions and investigate the possible outcomes of choosing such solutions. The method also helps form a balanced picture of the risks and benefits associated with each possible course of action. A decision tree evaluates ramifications of possible decisions as well as random events. This method is the predecessor to the event tree method discussed in Chapter 3. Decision tree method considers the possibility that some events do not happen as planned, such as planning to implement a risk control solution, including events that lead to situations worst than the *status quo*.

To construct a decision tree, the decision is decomposed to "subdecisions" and uncertain or random "events." A decision is usually followed by all possible uncertain and random events that could follow and affect the outcome of that decision. For example, a decision to install a new barrier for risk reduction could be followed by the events that it may not perform as desired, the barrier may be bypassed due to certain failure mechanisms, or it may cost more than expected and thus not completed. In this case, the probabilities are determined by using the risk assessment.

The decision nodes are in form of the symbol shown in Figure 8.9. The uncertain and random events that may follow a decision node are shown by the chance node as illustrated in Figure 8.10. The probability of each branch coming out of the chance node is also estimated.

Once all decision nodes and chance nodes are systematically modeled and a complete decision tree is developed, the total outcome of each scenario identified in the tree is

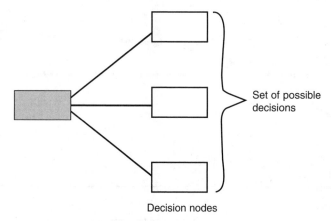

Set of possible decisions

Decision nodes

FIGURE 8.9 Decision nodes in the decision tree.

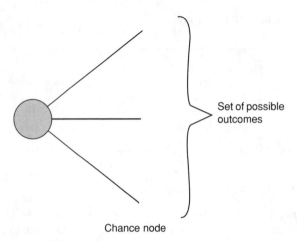

Chance node

FIGURE 8.10 Chance nodes in decision trees.

estimated. The outcome is in form of the losses or gains expected, should the scenario actually happen. The gains and losses may be in any form including monetary or value judgments. For evaluating the tree, using the probability of each chance node, the expected outcome (risk) associated with each decision is calculated. This constitutes the value associated with each decision node. For a decision node, the value associated with highest expected outcome is the choice for representing the expected value of the decision node. Finally, the expected outcome of the main decision node determines whether or not the main decision is acceptable.

To better understand this method, the readers are referred to Hammond et al. [6].

Consider making a decision regarding implementation of a new risk reduction measure. The tree in Figure 8.11 shows the developed decision tree and all subdecisions and events involved. Also, this figure shows the probabilities of the events and possible outcomes of each scenario. Based on this tree, determine whether or not implementation of this risk control measure is justified.

The value associated with each node is computed and shown above each node. For example, the expected value of node 5 is calculated as $V_5 = 0.3 \times 10 + 0.7 \times 30 = 24$ and for node 6, it is calculated as $V_6 = 0.3 \times (-15) + 0.7 \times (-2) = -5.9$.

Assuming the value function shown below applies to the outcomes of all scenarios, repeat the decision tree analysis on the basis of this utility judgmental value function.

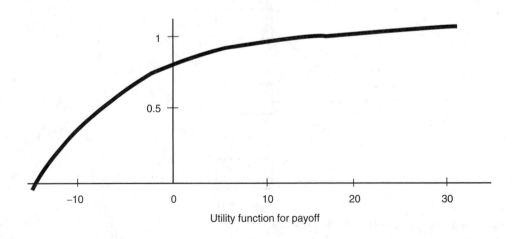

Utility function for payoff

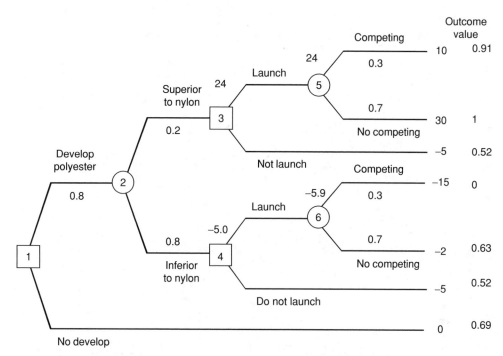

FIGURE 8.11 Developed decision tree and all subdecisions and events involved.

Based on the value function above, the value of each node is summarized.

Node	Expected Value Based on Actual Outcomes	Expected Value Based on the Value Judgment
1	0.8	0.69
2	0.8	0.61
3	24	0.97
4	−5	0.52
5	24	0.97
6	−5.9	0.44

Clearly in both evaluations, the value of node 1 (the main decision) is positive and, therefore, proceeding with the implementation of the proposed risk control solution is warranted.

Example 8.7

For the following decision tree, describe the outcome and the best decision.

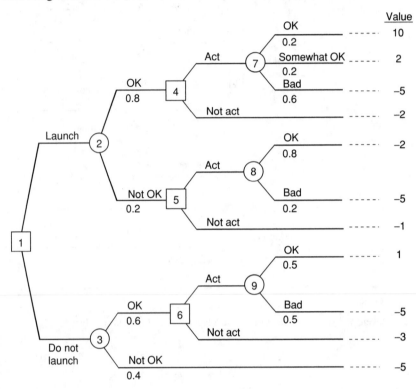

Solution:

The decision nodes (□) are 1, 4, 5, 6 and the chance nodes (○) are 2, 3, 7, 8, 9. For this decision tree, the outcome and the best decision are calculated according to the following:

Multiplying the payoff values by probability for chance nodes 7, 8, and 9:

Node 7: $(0.2 \times 10) + (0.2 \times 2) + (0.6 \times -5) = -0.6$

Node 8: $(0.8 \times -2) + (0.2 \times -5) = -2.6$

Node 9: $(0.5 \times 1) + (0.5 \times -5) = -2.0$

Using the above values and choosing the maximum at the decision nodes 4, 5, and 6:

At *Node 4* (maximum) between −0.6 and −2.0, choose −0.6.

At *Node 5* (maximum) between −2.6 and −1.0, choose −1.0.

At *Node 6* (maximum) between −2.0 and −3.0, choose −2.0.

Then the values at chance nodes 2 and 3 will be:

Node 2: $(-0.6 \times 0.8) + (-1.0 \times 0.2) = -0.68$

Node 3: $(-2.0 \times 0.6) + (0.4 \times -5) = -3.2$

Therefore, the best decision is to "Launch" even though it has a negative payoff, it is still greater than "Do Not Launch" negative payoff.

Decision trees provide an effective method for policy and other decision making problems because they:

- clearly lay out the problem so that all options can be evaluated,
- analyze fully the possible consequences of a decision,
- provide a framework to quantify the values of outcomes and the probabilities of achieving them, and
- help to make the best decisions on the basis of existing information and best guesses.

As with all decision making methods, decision tree analysis should be used with common sense, as decision trees are just one part of the actual risk management and control decision.

Exercises

8.1 A turbine in a generator can cause a great deal of damage if it fails critically. The manager of the plant can choose either to replace the old turbine with a new one, or to leave it in place.

The state of a turbine is categorized as either "good," "acceptable," or "poor." The probability of critical failure per quarter depends on the state of the turbine:

$$P \text{ (failure } | \text{ good)} = 0.0001,$$
$$P \text{ (failure } | \text{ acceptable)} = 0.001,$$
$$P \text{ (failure } | \text{ poor)} = 0.01.$$

The technical department has made a model of the degradation of the turbine. In this model, it is assumed that, if the state is "good" or "acceptable," then at the end of the quarter it stays the same with probability 0.95 or degrades with probability 0.05 (that is, "good" becomes "acceptable," and "acceptable" becomes "poor"). If the state is "poor," then it stays "poor."

(a) Determine the probability that a turbine that is "good," becomes "good," "acceptable," and "poor" in the next three quarters, and does not fail.

The cost of repairing the failure is US$ 1.5 m (including replacement of parts). The cost of a new turbine is US$ 5k.

(b) Consider a decision between two alternatives:

a_1: Install new turbine.

a_2: Continue to use old turbine.

Consider a situation where the old turbine is categorized as "acceptable"; determine the optimal decision using expected monetary loss (risk). What would the decision be if the old turbine was "good" with 10% probability and "acceptable" with 90% probability?

It is possible to carry out two sorts of inspection on the old turbine. A visual inspection costs US$ 200, but does not always give the right diagnosis. The probability of a particular outcome of the inspection given the actual state of the turbine is given in the table below:

Inspection Outcome	Actual State		
	Good	Acceptable	Poor
Good	0.9	0.1	0
Acceptable	0.1	0.8	0.1
Poor	0	0.1	0.9

The second sort of inspection uses an x-ray, and determines the state of the turbine *exactly*. The cost of the x-ray inspection is US$ 800.

(c) Determine the optimal choice between no inspection, a visual inspection, and an x-ray inspection.

8.2 Use the analytic hierarchy process to decide which of the following choices are the best.

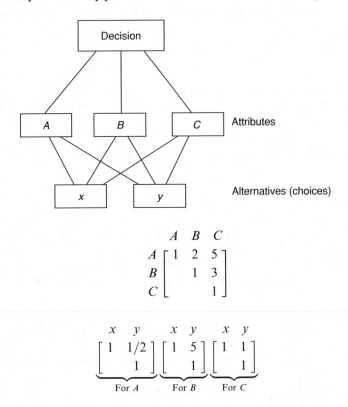

$$
\begin{array}{ccc}
 & A\ \ B\ \ C \\
\begin{array}{c} A \\ B \\ C \end{array}
&
\left[\begin{array}{ccc}
1 & 2 & 5 \\
  & 1 & 3 \\
  &   & 1
\end{array}\right]
\end{array}
$$

$$
\underbrace{\begin{array}{cc} x & y \\ \left[\begin{array}{cc} 1 & 1/2 \\ & 1 \end{array}\right] \end{array}}_{\text{For } A}
\underbrace{\begin{array}{cc} x & y \\ \left[\begin{array}{cc} 1 & 5 \\ & 1 \end{array}\right] \end{array}}_{\text{For } B}
\underbrace{\begin{array}{cc} x & y \\ \left[\begin{array}{cc} 1 & 1 \\ & 1 \end{array}\right] \end{array}}_{\text{For } C}
$$

8.3 Consider a number of hazardous road intersections at a Maryland county where there has been five fatal accidents in 1997, six in 1998, nine in 1999, and seven in 2000. It has been proposed to install traffic monitoring cameras at all of these intersections to reduce the accidents. If the total cost for installment of such equipment in the county is estimated at US$ 1,300,000. Use benefit–cost or risk-effectiveness methods and discuss feasibility of this option if the fatalities is expected to be reduced by 50% after the cameras have been installed.

8.4 Risk of a heart attack is a function of four factors: elevated blood pressure, family history of heart attack, smoking, and high cholesterol levels. Compare heart attack risk of three individuals with the following profiles:

Person 1:	High blood pressure
	No family risk of heart attack
	Nonsmoker
	High cholesterol
Person 2:	High blood pressure
	Family risk of heart attack
	Heavy smoker
	Very low cholesterol
Person 3:	Low blood pressure
	No family risk of heart attack
	Nonsmoker
	Moderately high cholesterol

The following is the weight of each factor (normalized to 1):

$$W = \begin{pmatrix} \text{Blood Pressure} & \text{Family History} & \text{Smoking} & \text{Cholesterol} \\ 0.2 & 0.25 & 0.35 & 0.2 \end{pmatrix}$$

Use the AHP method to perform this analysis.

8.5 Consider a decision making using the AHP. Experts have come up with the following weights for the criteria (C_1, C_2, and C_3):

$$W_{C_1} = 0.58, \quad W_{C_2} = 0.23, \quad W_{C_3} = 0.19$$

(a) Using these data compute the binary comparison matrix A below.

$$A = \begin{bmatrix} a_{11} & a_{12} & a_{13} \\ a_{21} & a_{22} & a_{23} \\ a_{31} & a_{32} & a_{33} \end{bmatrix}$$

8.6 Consider a decision involving optimizing risk and cost using the utility theory. The value is expressed in form of $V = 0.2F_1 + 0.8F_2$, where F_1 is the value of risk and F_2 is the value of cost. Determine an option in terms of exposed risk and cost that may be considered as optimum. The follow functions represent the value of risk and cost.

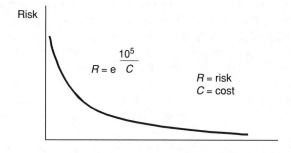

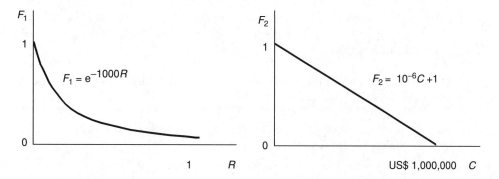

8.7 As a regulator, you need to select between two safety regulations (A and B) aimed at averting risk of exposing some toxic materials to the plant workers. Regulation A reduces the risk by a factor of 100 and Regulation B reduces the risk by a factor 250; however Regulation A costs US$ 100,000 and Regulation B costs US$ 380,000. Which regulation should you adopt?

Use the value theory method to decide which regulation should be adopted: 70% weight is assigned to risk and 30% weight is assigned to cost. Further use the following value plots are proposed for risk and cost.

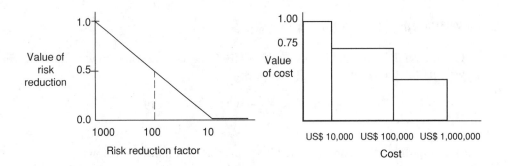

8.8 Using the AHP method, develop your own binary comparison matrices. Consistent with the data provided in Problem 8.7 and using this, determine which regulation should be adopted.

8.9 A 21-year-old student experiences several hours of chest pain. The doctor thinks that she is most likely suffering from a viral syndrome, but he considers the possibility of a pulmonary embolus (PE, a blood clot from her legs lodged in her lungs). The doctor performs a chest x-ray based on which the likelihood of PE is estimated as 50%. In case of PE, the patient should be given a medication to "thin the blood" (anticoagulants) for several months. If the patient has PE, but not given anticoagulants, the chance of remobilization is 50%; if she is properly treated with anticoagulants, the chance is 15%. The major complication of anticoagulants is a hemorrhage. Approximately 5% of patients using these drugs develop hemorrhage.

(a) All other things being equal, should this young woman be treated with anticoagulants? Use the following values: for hemorrhage and remobilization, (-2); for no hemorrhage and remobilization, (-1); for hemorrhage only, (-1), for remobilization only, (-1).

(b) An additional test called pulmonary angiogram can be performed to support the decision of administrating anticoagulants. This test definitely determines whether or not PE is present, but the procedure causes complications in 1% of patients. All scenarios in (a) involving angiogram complication should have a value of (-1) added to them. Should an angiogram be performed?

8.10 The method of structured value analysis is used in risk evaluation. Discuss the problems associated with this method.

8.11 Consider a choice among four alternative safety corrections to a flawed design which may be fatal to an end-user.

	Cost/Single End-Use	Total Risk Reduction Capability
Alternative 1	US$ 100	10%
Alternative 2	US$ 130	15%
Alternative 3	US$ 760	21%
Alternative 4	US$ 1200	34%

(a) Draw a cost-effectiveness plot.

(b) Select the most-cost effective alternative.

8.12 A toxic chemical degrades over time and the annual dosage it produces is

$$d(t) = \exp(-0.05t), (t = \text{years})$$

The annual individual fatality risk probability is $R(d) = d/1000$ and the loss of life cost is $h = \text{US\$ 3 m}$ per person. How much would you be willing to spend to avert this risk, if a protective measure may be added which reduces the fatality risk to a negligible amount (assume a 25-year planned storage for this toxic chemical)?

REFERENCES

1. Glickman, T., and Gough, M. (eds.), *Reading In Risk*, Resources for the Future, Washington, D.C, 1990.
2. Raheja, D., *Assurance Technologies: Principles and Practices*, McGraw-Hill, New York, NY, 1991.
3. Kelman, S., Cost–benefit analysis: an ethical critique, *AEI Journal on Government and Society Regulation*, 5(1), 33, 1981.
4. Butter, G., Calfee, J., and Ippolito, P., Reply to Steven Kelman, *Regulation*, 5(1), 41, 1981.
5. Bernstein, P.L., *Against the Gods: The Remarkable Story of Risk*, John Wiley & Sons, New York, 1996.
6. Hammond, J.S., Keeney R.L., and Raiffa, H., *Smart Choices: A Practical Guide to Making Better Decisions*, Harvard Business School Press, Cambridge, MA, 1999.
7. VanDoren, P., *Cato Handbook for Congress: Policy Recommendations for 108th Congress*, Cato Institute Report, 2003.
8. Tengs, T.O. et al, Five-hundred life-saving interventions and their cost-effectiveness, *Risk Analysis*, 15(3), 369, 1995.
9. Rowe, W.D., *An Anatomy of Risk*, John Wiley & Sons, New York, 1977.
10. Rowe, W.D., *The Application of Structured Value Analysis to Models Using Value Judgment as a Data Source*, The MITRE Corporation, M70-14 McLean, VA, March 1970.
11. Saaty, T.L., *The Analytic Hierarchy Process*, McGraw Hill Company, New York, 1980.

9 Risk Communication

9.1 INTRODUCTION

The basic reason for risk communication is to inform all those with an interest in decision-making about the possible risk scenarios and consequences of various risk management options. Risk communication is not always aimed at decision-making. The general public as a whole, for example, is rarely in the position to make decisions involving public risk, since professionals or politicians normally make such decisions on their behalf, but they will nonetheless be affected by those decisions. Since the public opinion does (at least, in democracies) matter, the actual decision-makers should keep them informed. As such, the public and its attitudes about risks are important aspects of risk analysis.

According to a U.S. National Research Council study on risk communication [1], risk communication is as an interactive process of exchange of information and opinion among individuals, groups, and institutions. It involves multiple messages about the nature of risk, not necessarily about risk, that affects opinions, or reactions to risk messages or to legal and institutional arrangements for risk management. The media also plays an important role in risk communication and in structuring of public opinion. As such, it is also important to understand factors that prompt media coverage.

Over the last few decades, several generic models for communication have been developed. These models are developed on the basis of public and stakeholder views and facilitate two-way exchange of information. Risk communication is not a universal remedy and will not always resolve problems and conflicts. But a poor or absent communication will undoubtedly lead to improper decisions and a failure of risk management.

Two critical parts to risk communication are trust and perception. Establishing trust is the central part of risk communication which is especially challenging for governmental agencies as the public often sees government as not entirely a trustworthy source of information. Understanding perception of risk, which is influenced by a mixture of factors, is also very important in risk communication and is an area of ongoing study by many researchers. The public and experts often differ in their perceptions of risk. Public's value systems also shape people attitudes towards risk, and in combination with the so-called "dread factors," it is possible to determine how individuals will react to and accept risk.

Describing risk assessment and management issues pose a major challenge for risk communicators. Risk managers, regulators, policy–decision makers, and scientists are often reluctant to discuss with the public complex, technical, and scientific information, as they are often misunderstood and misinterpreted. In a regulatory and public policy-related risk management, inputs from both scientific and the public stakeholders ensure a better picture of the information. Risk communication, especially in cases where scientific uncertainty is predominant, should be a key element of final decision.

The fundamental problem in risk communication with the public and lay stakeholders is that they do not typically interpret risk in the same way as experts scientists and engineers. It seems that the public interpretation is usually distorted with complex perceptions which are very difficult to identify. But, in general, hazards with high consequences are ranked by the nonexperts as high risk despite having extremely low probabilities of occurrence (i.e., the low frequency–high consequence events). Two examples serve to show these differing attitudes.

Consider the Three Mile Island nuclear accident of 1979. This accident occupied the news media for a very long time and had ramifications not only in the U.S. nuclear industry, but also in the U.S. and around the world. While no direct loss of life occurred, it led to the modification of regulatory codes and considerable changes in the operation and design of nuclear power plants that cost the industry and ultimately the electricity consumers a lot. It also fundamentally changed the public's attitude towards nuclear power.

As the second example, consider traffic accidents. Approximately, 42,000 people are killed per year in the U.S. in traffic accidents. Most accidents (even some involving fatalities) usually do not make the local news, and certainly not the national news. Unless there is a pattern, they rarely translate to any substantial change in the automobile design and traffic management.

Why should the two cases differ so much? It is because of many factors. For example, the public is used to traffic accidents, but are not used to nuclear accidents. Also, it is due to a feeling of lack of control, far more so in the case of a nuclear accident. Another factor may involve the way the incidents are reported (an important factor here is the role of the media in shaping public opinion). Several other factors will be discussed later in this chapter.

In this chapter, we will first discuss the methods and issues with risk communication and trust building, followed by understanding and methods of quantifying risk perception.

9.2   FORMS OF COMMUNICATION

There are several forms of communications as follows:

Communications from the Public. Communication from the public is frequently aimed at politicians and usually attempts to make use of the news media. Perceived hazards such as the planned construction of a chemical facility in the local neighborhood or the routing of a train carrying hazardous material over local lines typically give rise to public protests, marches, petitions, and organized letters to politicians. In turn, politicians can influence risk management, policy- and regulatory decision-makers. Expressed concerns normally highlight hazards and consequences rather than risk value.

Communications to the Public. Communications to the public are rarely about risk, but typically about various threats or hazards, and about their frequencies. In public domains, there can be some very basic discussion of frequencies and probabilities of major incidents and accidents such as an explosion or rail crash. Discussions about long-term threats such as hazards of smoking or road traffic accidents take a form of mixed qualitative and quantitative information.

Communications by Professionals. Scientists and engineers are the main group of experts who might normally explain the methods and results of risk and hazard identification and analysis. Naturally, they are more open to a more mathematical means of communication than other groups. Methods of communication by this group tend to be more formalized. While science and technology provide reassurances to society, they are increasingly seen as creating risk.

Communication by the News Media. The news media often plays a central role in risk communications. The media plays two fundamentally different roles: as a passive transmitter of information or as an active producer and interpreter of information. This media's information often reflects the concerns of the public, but is not necessarily neutral. The demands on a daily basis on the media mean that they predictably tend to highlight the short-term and report immediate events such as accidents or just-published reports into areas perceived as being of public fear or concern.

9.3   SOME BASIC RULES OF RISK COMMUNICATION

Covello and Allen [2] have formulated seven basic rules of good communications. They are:

- Accept and involve the public as a legitimate partner — a basic tenet of risk communication in a democracy is that people and communities have a right to participate in decisions that affect their lives, their property, and the things they value.
- Plan carefully and evaluate performance — risk communication will be successful only if carefully planned.
- Listen to the public's specific concerns — if you do not listen to people, you cannot expect them to listen to you. Communication is a two-way activity.
- Be honest, frank, and open — in communicating risk information, trust and credibility are the most important elements.
- Coordinate and collaborate with other credible sources — alliances with others can be effective in helping you communicate risk information.
- Meet the needs of the media — the media is a prime transmitter of information on risks; it plays a critical role in setting agendas.
- Speak clearly and with compassion — technical language and jargon are useful as professional shorthand, but they are barriers to successful communication with the public and stakeholders.

9.4   ELEMENTS OF EFFECTIVE RISK COMMUNICATION

Depending on what is to be communicated and to whom, risk communication messages may contain information on the following subjects:

The nature of the risk
- Characteristics and importance of the hazard
- Magnitude and severity of the risk
- Urgency of the situation
- Trends in risk (becoming greater or smaller)
- Probability or frequency of exposure to the hazard
- Distribution of exposure
- Amount of exposure that constitutes a significant outcome
- Nature and size of the population at risk
- Group receiving the greatest risk

The nature of the benefits
- Actual or expected benefits of accepting the risk
- Direct beneficiaries of the risk
- Balance point between risks and benefits
- Magnitude and importance of the benefits
- Total benefit to all affected populations combined

Uncertainties in risk assessment
- Methods used to assess the risk and characterize uncertainties
- Importance of each of the uncertainties
- Reducible (epistemic) uncertainties
- Uncertainties not practical to reduce (aleatory uncertainties)
- Inadequacies, inconsistencies, and inaccuracies in the available data

- Assumptions and their sensitivity to estimated risk
- Effect of changes in the input to risk management decisions

Risk management options
- Risk control actions taken to manage the risk
- Actions individuals may take to reduce personal risk
- Justification for choosing a specific action
- Effectiveness of a specific option
- Benefits of a specific option
- Cost of the option and who pays for it
- Risks that remain or new risks that may occur as a result of the option

9.5   CHARACTERISTICS OF AN EFFECTIVE RISK COMMUNICATION

Several key good practices that characterize a risk communication as successful include:

Know the audience. In composing the messages, the risk communicator should understand characteristics and composition of the audience including their ideas, motivations, and opinions. It is also important to know the groups and preferably individuals involved to understand their concerns and positions. Also, it is important to know and maintain the best channels of communication with the audience.

Involve the experts. Science and engineering experts, must be able to explain the concepts and processes of risk assessment, the results, scientific data, assumptions, and subjective judgments so that decision-makers and other interested parties clearly understand the risk and uncertainties about estimations. Similarly, the risk managers should also be able to clarify how the risk management decisions are made. One of the more significant differences discussed recently is that between the public and what the experts actually do not know; and why experts cannot agree with each other.

Establish expertise in communication. Successful risk communication requires effective transmission of understandable and usable information to the public. Risk assessors, managers, and technical experts sometimes lack the skills needed for effective communication tasks such as responding to the needs of the various audiences (public, industry, media, etc.) and formulating effective messages. It is often advisable to involve people with expertise in risk communication or develop such expertise before interacting with the stakeholders, public, and media. As an example, consider the U.S. National Institutes of Health news release that referred to the research that reported 26% increase in breast cancer risk for patients on particular types of hormone replacement therapy [3]. This statistic was publicized by the media and caused a large degree of public concern and debate. Examination of the original article on which this news release was based, however, tells that while the increase in relative risk was indeed 26%, since the base risk was very low, the increase in the absolute risk was from 30 to 38 in 10,000 person-years [4]. The latter form of communication would have probably been a more appropriate statistic.

Provide credible information. Information from credible sources has a vital impact on shaping and changing public perception of a risk. The credibility associated with a source by a stakeholders and public audience may vary according to the nature of the hazard, culture, and social and economic status. Consistency in messages received from multiple sources reinforces the credibility of the message. Important factors of source credibility include well-known of experts, fairness, impartiality, trust, and timeliness.

Trust and credibility must be cultivated and can be lost through ineffective or inappropriate communication. Recognition of current issues and problems, and timely information enforces credibility. Oversight, distortion, and self-serving statements will also damage credibility.

Share responsibility. Policy-making organizations and regulatory agencies have the duty for risk communication. The public expects these agencies to take part in managing public risks. Even in cases where no action is taken, communication is still necessary to provide reasons as to why taking no action is the best option. The media also plays an essential role in the communication process and shares in these responsibilities. Industry also has a responsibility for risk communication, especially when the risk is a result of their products or processes. All parties involved in the risk communication process (e.g., regulatory and policy agencies, industry, media) have joint responsibilities for the outcome of that communication even though their individual roles may differ.

Make a distinction between science and value judgment. It is essential to make a distinction between "facts" from "values" in considering risk management options. At a practical level, it is useful to report the facts that are known as well as what uncertainties are involved in using those facts in performing risk assessment and in risk management decisions. The risk communicator must explain what facts are known. There are also other assumptions and models that build based on these facts and are built by expert judgment. Uncertainties associated with such models and assumptions must also be clearly stated. Value judgments are those involving the establishment of acceptable levels of risk. Consequently, risk communicators should be able to justify the level of acceptable risk to the public and other stockholders. Note that the acceptable risk is not zero risk, as zero risk is unattainable. Making this clear is an important function of the risk communicator.

Assure transparency. For the public to accept the risk analysis process and its outcomes, the process must be transparent and reproducible. While respecting concerns over protection of confidential information (e.g., proprietary information or data), transparency in risk analysis consists of having the process open and available for scrutiny by the stakeholders.

Put the risk in perspective. One way to put a risk in perspective is to examine it in the context of the benefits associated with the technology or process that poses the risk. Another approach that may be helpful is to compare the risk with other similar, more familiar risks. However, this latter approach can create problems if it appears that the risk comparisons have been intentionally chosen to make the risk at issue seem more acceptable to the public. Comparisons can help put risk in perspective. Benefits should not be used to justify risks, and irrelevant or misleading comparisons can harm trust and credibility. In general, risk comparisons should not be used unless:

- Risk estimates are scientifically and analytically strong.
- Risk estimates are relevant to the specific audience addressed.
- The degree of uncertainty in all risk estimates is similar.
- The substances, products, or activities themselves have similar risk attributes, for example, both or all are voluntary or involuntary in nature.

Covello et al. [4, 5] offer useful guidelines for risk comparison for two or more risks as follows:

Most acceptable: Comparisons of
- risks at two different times
- risk in question with a standard (or acceptability limit)
- different estimates of the same risk

Less desirable: Comparison of
- the risk of doing something versus not doing it
- alternative risk reduction or control solutions to the same problem
- risk with the same risk experienced elsewhere

Even less desirable: Comparison of
- average risk with peak risk at a particular time or location
- risk from one source of an adverse effect with risks from all sources of the same effect

Marginally acceptable: Comparison of
- risk with cost; or the cost/risk ratio with another
- risk with benefit
- health risk with environmental risk
- health risk with safety risk
- safety risk with environmental risk
- risk with other risks from the same source
- risk with other specific causes of the same disease, illness, or injury

Rarely acceptable: Comparison of
- unrelated risks (e.g., chemical plant toxic gas release risk with airplane crash risk, cancer risk due to smoking with fatality risk due to transportation accidents)

Note the factors that affect people's perception of risks (discussed in Section 9.6). A good comparison considers these factors for a more effective risk comparison and risk communication in general.

9.6   RISK PERCEPTION

Risk perception refers to people's intuitive judgment of both aspects of risk: the frequency of occurrence and the severity of the resulting consequences from exposure to hazards. Risk perception is a reflection of opinion. Opinions can be difficult to change, principally as people feel they have adequate knowledge about the risk. When people expect benefits from exposure to a risk, they may be more receptive to that risk. People are more apt to judge the probability or frequency of a hazard higher than actual, if they can readily remember an occurrence of it or something similar. Also, exposures to hazards that lead to severe consequences, even if the frequency of such events is low, attract considerable public and media attention. It is imperative to recognize that all people, regardless of their role in society, use speculative frameworks to make sense of the world around them and form selective judgment in their responses to risk. Slovic [6] reports this discrepancy as illustrated in Table 9.1.

Shark attacks are a good example of events whose probability is greatly overestimated by the general public due to the newsworthiness of such events and the visceral nature of the individual's reaction to the occurrence. According to Burgess [7] of University of Florida's International Shark Attack File, only three people were killed by unprovoked shark attacks in 2002. More than 150 human deaths are attributed to being hit by falling coconuts. However, shark attacks make big news, but not fatalities attributed to falling coconuts. However, people's perception of shark attack risk is universally far higher than that of coconut-impact dangers.

Risk perceptions of the magnitude of risk revealed by the public are influenced by factors other than historical and formal assessment of risk. For example, regardless of the frequency

TABLE 9.1
Factors Influencing Public Risk Perception

Risk Factor

Delayed	–	Immediate
Necessary	–	Luxury
Ordinary	–	Catastrophic
Uncontrollable	–	Controllable
Voluntary	–	Involuntary
Natural	–	Man-made
Occasional	–	Continuous
Old	–	New
Clear	–	Unclear

and consequence of a hazard, factors that actually influence people's perception of risk include:

- Risks perceived to be voluntary are more readily accepted than risks perceived as involuntary (i.e., risks imposed by others).
- Risks perceived to be under an individual's control are more accepted than risks perceived to be controlled by others (i.e., uncontrollable by the individual).
- Risks perceived to be natural are more accepted than risks perceived to be human-made.
- Risks perceived to be common are more accepted than risks perceived to be catastrophic (i.e., risks involving large consequences).
- Risks perceived to be familiar are more accepted than risks perceived to be unfamiliar (i.e., new risks).

According to Covello [5,8], psychological research has identified 47 known factors that influence the perception of risk, issues like control, benefit, whether a risk is voluntarily taken by the individual or group. Another important factor is trust that the individual or public places on the policy-makers, regulators, scientist and engineers, and risk communicators involved. These factors can help explain why risk recipients and stakeholders are concerned about risks that scientists and engineers estimate as trivial. Table 9.1 discusses some of more important risk perception decotious factors.

Clearly, the actual risk would not change, but the risk perception can; and in most domains, perception is reality [1,4]. Understanding risk perception is critical to formulating a message and determining how best to manage and communicate risks.

Several researchers have tried to quantify the degree of distortion associated with each of the risk perception factors. The early work in this area is attributed to Starr [9] who attempted to offer a scientific basis for thresholds of risk imposed by nuclear power plants which were accepted by the public. Others such as the Rasmussen study or WASH-1400 [10] tried to establish general principles of public risk acceptability of nuclear power. This was usually based on familiar mortality statistics and the de minimis risk principle, which argued that if a risk is effectively lower than otherwise accepted risks such as one-in-a-million fatality risk, the risk is effectively zero [1].

In the 1980s, several groups developed models that incorporated the value systems of individuals, groups, and societies into risk communication theory [11,12,13]. This generated

TABLE 9.2
Rankings of Perceived Risks for 30 Activities and Technologies

LWV Ranking	Activity or Technology	Expert's Ranking
1	Nuclear power	20
2	Motor vehicles	1
3	Handguns	4
4	Smoking	2
5	Motorcycles	6
6	Alcoholic beverages	3
7	Private aviation	12
8	Police work	17
9	Pesticides	8
10	Surgery	5
11	Firefighting	18
12	Large construction	13
13	Hunting	23
14	Spray cans	26
15	Mountain climbing	29
16	Bicycles	15
17	Commercial aviation	16
18	Electric power (nonnuclear)	9
19	Swimming	10
20	Contraceptives	11
21	Skiing	30
22	X-rays	7
23	High school and college football	27
24	Railroads	19
25	Food preservatives	14
26	Food coloring	21
27	Power mowers	28
28	Prescription antibiotics	24
29	Home appliances	22
30	Vaccinations	25

Based on a survey of a group of experts and a group of informed members of the League of Women Voters (LWV) in the United States. A ranking of 1 denotes the highest level of perceived risk.

broad agreement that risks are evaluated according to their perceived threat to familiar social relationships and practices, and not by numbers alone.

Public polls indicate that societal perception of risks associated with certain unfamiliar or incorrectly publicized activities is far out of proportion to the actual damage or risk measure. Slovic [13] reports a polling result and the discrepancies in risk perception as shown in Table 9.2. According to Litai and Rasmussen [14], using a psychometric approach and data such as the ones shown in Table 9.3, one can measure the degree of public bias of risks. For example, for motor vehicle and aviation accidents, this bias reveals a risk perception which is less than the actual value by a factor of 10 to 100, but the risk of nuclear power and food coloring is overestimated by a factor of greater than 10,000. Risk conversion and compensating factors must often be applied to determine risk tolerance thresholds accurately. Risk conversion factors (RCFs) account for public bias against risks

TABLE 9.3
Comparing Historical Risk Data of Some Voluntary and Involuntary Activities

Voluntary		Involuntary	
Activity	Risk of Death per Person-Year ($\times 10^{-5}$)	Activity	Risk of Death per Person-Year ($\times 10^{-5}$)
Smoking (20 cigarettes/day)	500	Run over by vehicles	500
Drinking (1 bottle wine/day)	75	Floods	22
Sport	120	Earthquake (cadence)	17
Rock climbing	4	Tornados	22
Car driving	17	Explosion of pressure vessel	2

that are unfamiliar (by a factor of 10), catastrophic (by a factor of 30), involuntary (by a factor of 100), uncontrollable (by a factor of 5 to 10), or have immediate consequences (by a factor of 30). For example, people perceive a voluntary action to be less risky by a factor of 100 than an identical involuntary action. Although the exact values of above RCFs involve uncertainties, they generally show the direction and the degree of bias in people's perception. These factors are valuable information to prepare and communicate the message with public and stakeholders.

As a result of the average values of RCFs measuring the public risk bias, for example, different risk acceptance standards may be employed in the workplace, where risk exposure is voluntary as compared to the general public where the same exposure is involuntary. A good example of this is the U.S. radiation exposure standards that are more than a factor of 10 less for the general public than for the workforce. The general guide to risk standards is that occupational risk should be small compared with natural sources of risk. Some industrial and voluntary risks may be further decreased by strict enforcement or adequate implementation of known risk-avoidance measures (e.g., wearing seat belts, not drinking alcohol, or avoid smoking). Therefore, some of these risks are controllable by the individual (who can choose whether to fly, to work, to drive, or to smoke), while others are not (e.g., chemical dumps, severe floods, and earthquakes).

Exercises

9.1 Discuss problems associated with predicting risk acceptance of a new technology by the society.

9.2 Compare two risks: electromagnetic radiation cancers due to living near electric power lines and automobile accidents. What is the acceptable risk for electromagnetic radiation if the annual risk of automobile fatality per person is 2×10^{-4}?

9.3 Risk of an activity with 1-in-10,000 chance of 10,000 affected is equal to 1-in-10 chance of 10 people affected, yet some may view these risks differently. Describe why. What factors affect the process of their risk acceptability for these two risk situations?

9.4 Considering the RCFs discussed in this chapter:
     Compare air travel accidents with automobile accidents for reaching the same destination.
9.5 (a) It is stated that "people are less interested in risk estimation than in risk
         reduction." Explain what this implies about the importance of risk assessment.
         How does the applicability of the statement depend on the magnitude of the risk?
     (b) Summarize the most important principles of risk communications.
9.6 Radon risk (the radioactive gas that accumulates in residential basement and can
     causes lung cancer) has characteristics that are different from risks related to hazardous waste facilities. In what ways do these differences influence public risk perception?
     What implications do these differences have for risks communications?

REFERENCES

1. U.S. National Research Council, *Improving Risk Communication. Committee on Risk Perception and Communication*, National Academy Press, Washington, DC, 1989, p. 332.
2. Covello, V.T. and Allen, F., *Seven Cardinal Rules of Risk Communication*, U.S. Environmental Protection Agency, Office of Policy Analysis, Washington, DC, 1988.
3. National Institute of Health, National Heart, Lung and Blood Institute, NHLBI stops trial of estrogen plus progestin, NIH New Release, July 9, 2002.
4. Covello, V.T., Slovic, P., and von Winterfeldt, D., Risk communication: a review of the literature, *Risk Abstracts*, 3, 171, 1986.
5. Covello, V.T., McCallum, D.B., and Pavlova, M.T. (Eds.), *Effective Risk Communication: The Role and Responsibility of Government and Nongovernment Organizations*, Plenum Press, New York, 1989.
6. Slovic, P., Informing and educating the public about risk, *Risk Analysis*, 6(4), 403, 1986.
7. Burgess, G., University of Florida International Shark Attack File, http://www.flmnh.ufl.edu/fish/Sharks/Statistics/2002attacksummary.htm, 2002.
8. Covello, V.T., Risk communication: an emerging area of health communication research, in Deetz, S. (Ed.), *Communication Yearbook 15*, Sage Publications, Newbury Park and London, 1992, p. 359.
9. Starr, C., Social benefit versus technical risk, *Science*, 165, 1232–1238, 1969 (U.K. Advisory Council on Science and Technology, 1990, Developments in Biotechnology, HMSO, London, U.K.).
10. U.S. Nuclear Regulatory Commission, Reactor safety study: an assessment of accidents in U.S. commercial nuclear power plants, U.S. Regulatory Commission, WASH-1400, Washington, DC, 1975.
11. Vlek, C.A.J. and Stallen, P.J., Rational and personal aspects of risk, *Acta Psychologica*, 45, 273, 1980.
12. Douglas, M., *Risk Acceptability According to the Social Sciences*, Russel Sage, New York, 1986.
13. Slovic, P., Perception of risk, *Science*, 236, 280, 1987.
14. Litai, D. and Rasmussen, N., The public perception of risk, the analysis of actual versus perceived risks, in Covello, V.T. et al. (Eds.), *The Analysis of Actual Versus Perceived Risks*, Plenum Press, New York, NY, 1983, p. 213.

Appendix A: Basic Reliability Mathematics: Review of Probability and Statistics

A.1 INTRODUCTION

In this appendix, we discuss the elements of mathematical theory that are relevant to the study of reliability of physical objects. We begin with a presentation of basic concepts of probability. Then we briefly consider some fundamental concepts of statistics that are used in reliability data analysis.

A.2 ELEMENTS OF PROBABILITY

Probability is a concept that people use formally and casually everyday. The weather forecasts are probabilistic in nature. People use probability in their casual conversations to show their perception of the likely occurrence or nonoccurrence of particular events. Odds are given for the outcome of sport events, and are used in gambling.

Formal use of probability concepts is widespread in science, for example, astronomy, biology, and engineering. In this appendix, we discuss the formal application of probability theory in the field of reliability engineering.

A.2.1 SETS AND BOOLEAN ALGEBRA

To perform operations associated with probability, it is often necessary to use sets. A set is a collection of items or elements, each with some specific characteristics. A set that includes all items of interest is referred to as a *universal set*, denoted by Ω. A *subset* refers to a collection of items that belong to a universal set. For example, if set Ω represents the collection of all pumps in a power plant, then the collection of electrically driven pumps is a subset E of Ω. Graphically, the relationship between subsets and sets can be illustrated through Venn diagrams. The Venn diagram in Figure A.2 shows the universal set Ω by a rectangle, and subsets E_1 and E_2 by circles. It can also be seen that E_2 is a subset of E_1. The relationship between subsets E_1 and E_2 and the universal set can be symbolized by $E_2 \subset E_1 \subset \Omega$.

The *complement* of a set E, denoted by $\bar{E}$ and called *E not*, is the set of all items (or more specifically events) in the universal set that do not belong to set E. In Figure A.1, the nonshaded area outside of the set E_2 bounded by the rectangle represents $\bar{E}_2$. It is clear that sets E_2 and $\bar{E}_2$ together comprise Ω.

The *union* of two sets, E_1 and E_2, is a set that contains all items that belong to E_1 or E_2. The union is symbolized either by $E_1 \cup E_2$ or $E_1 + E_2$, and is read E_1 or E_2. That is, the set $E_1 \cup E_2$ represents all elements that are in E_1, E_2 or both E_1 and E_2. The shaded area in Figure A.2 shows the union of sets E_1 and E_2.

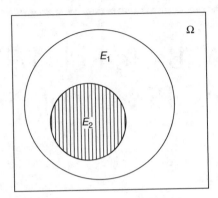

FIGURE A.1 Venn diagram.

Suppose E_1 and E_2 represent positive odd and even numbers between 1 and 10, respectively then

$$E_1 = \{1, 3, 5, 7, 9\}$$
$$E_2 = \{2, 4, 6, 8, 10\}$$

The union of these two sets is

$$E_1 \cup E_2 = \{1, 2, 3, 4, 5, 6, 7, 8, 9, 10\}$$

or, if $E_1 = \{x, y, z\}$ and $E_2 = \{x, t, z\}$, then

$$E_1 \cup E_2 = \{x, y, z, t\}$$

Note that element x is in both sets E_1 and E_2.

The *intersection* of two sets, E_1 and E_2, is the set of items that are common to both E_1 and E_2. This set is symbolized by $E_1 \cap E_2$ or $E_1 \cdot E_2$, and is read E_1 and E_2. In Figure A.3, the shaded area represents the intersection of E_1 and E_2.

Suppose E_1 is a set of manufactured devices that operate for $t > 0$, but fail before 1000 h of operation. If set E_2 represents a set of devices that operate between 500 and 2000 h, then $E_1 \cap E_2$ can be obtained as follows:

$$E_1 = \{t|0 < t < 1000\}$$
$$E_2 = \{t|500 < t < 2000\}$$
$$E_1 \cap E_2 = \{t|500 < t < 1000\}$$

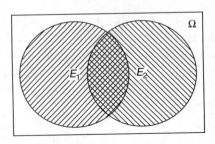

FIGURE A.2 Union of two sets, E_1 and E_2.

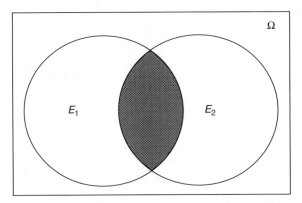

FIGURE A.3 Intersection of two sets E_1 and E_2.

Also, if sets $E_1 = \{x, y, z\}$ and $E_2 = \{x, t, z\}$, then

$$E_1 \cap E_2 = \{x, z\}$$

Note that the first two sets in this example represent "continuous" elements, and the second two sets represent "discrete" elements. This concept will be discussed in more detail further in this appendix.

A *null* or empty set, $\varnothing$, refers to a set that contains no items. One can easily see that the complement of a universal set is a null set, and vice versa. That is,

$$\bar{\Omega} = \varnothing$$
$$\Omega = \bar{\varnothing}$$
$$(A.1)$$

Two sets, E_1 and E_2, are termed *mutually exclusive* or *disjoint* when $E_1 \cap E_2 = \varnothing$. In this case, there are no elements common to E_1 and E_2. Two mutually exclusive sets are illustrated in Figure A.4.

From the discussions so far, as well as from the examination of the Venn diagram, the following conclusions can be drawn:

The intersection of set E and a null set is a null set:

$$E \cap \varnothing = \varnothing$$
$$(A.2)$$

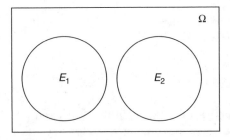

FIGURE A.4 Mutually exclusive sets, E_1 and E_2.

The union of set E and a null set is E:

$$E \cup \varnothing = E \qquad\qquad (A.3)$$

The intersection of set E and the complement of E is a null set:

$$E \cap \bar{E} = \varnothing \qquad\qquad (A.4)$$

The intersection of set E and a universal set is E:

$$E \cap \Omega = E \qquad\qquad (A.5)$$

The union of set E and a universal set is the universal set:

$$E \cup \Omega = \Omega \qquad\qquad (A.6)$$

The complement of the complement of set E is E:

$$\bar{\bar{E}} = E \qquad\qquad (A.7)$$

The union of two identical sets E is E:

$$E \cup E = E \qquad\qquad (A.8)$$

The intersection of two identical sets E is E:

$$E \cap E = E \qquad\qquad (A.9)$$

Boolean algebra provides a means of evaluating sets. The rules are fairly simple. The sets of axioms in Table A.1 provide all the major relations of interest in Boolean algebra including some of the expressions discussed in (A.1)–(A.9).

A.2.2   Basic Laws of Probability

In probability theory, the elements that comprise a set are outcomes of an experiment. Thus, the universal set Ω represents the mutually exclusive listing of all possible outcomes of the experiment and is referred to as the *sample space* of the experiment. In examining the outcomes of rolling a die, the sample space is $S = \{1, 2, 3, 4, 5, 6\}$. This sample space consists of six items (elements) or *sample points*. In probability concepts, a combination of several sample points is called an event. An event is, therefore, a subset of the sample space. For example, the event of "an odd outcome when rolling a die" represents a subset containing sample points 1, 3, and 5.

Associated with any event E of a sample space S is a probability shown by $\Pr(E)$ and obtained from the following equation:

$$\Pr(E) = \frac{m(E)}{m(S)} \qquad\qquad (A.10)$$

where $m(\cdot)$ denotes the number of elements in the set $(\cdot)$ and refers to the size of the set.

The probability of getting an odd number when tossing a die is determined by using $m(odd$ $outcomes) = 3$ and $m(sample\ space) = 6$. In this case, $\Pr(odd\ outcomes) = 3/6 = 0.5$.

TABLE A.1
Laws of Boolean Algebra

$X \cap Y = Y \cap X$ $X \cup Y = Y \cup X$	Commutative law
$X \cap (Y \cap Z) = (Y \cap X) \cap Z$ $X \cup (Y \cup Z) = (Y \cup X) \cup Z$	Associative law
$X \cap (Y \cup Z) = (X \cap Y) \cup (X \cap Z)$	Distributive law
$X \cap X = X$ $X \cup X = X$	Idempotent law
$X \cap (X \cup Y) = X$ $X \cup (X \cap Y) = X$	Absorption law
$X \cap \bar{X} = \emptyset$ $X \cup \bar{X} = \Omega$	Complementation law
$\bar{\bar{X}} = X$ $\overline{(X \cap Y)} = \bar{X} \cup \bar{Y}$ $\overline{(X \cup Y)} = \bar{X} \cap \bar{Y}$	De Morgan's theorem

Note that (A.10) represents a comparison of the relative size of the subset represented by the event E to the sample space S. This is true when all sample points are equally likely to be the outcome. When all sample points are not equally likely to be the outcome, the sample points may be weighted according to their relative frequency of occurrence over many trials or according to expert judgment.

At this point, it is important that the readers appreciate some intuitive differences between three major conceptual interpretations of probability described below.

A.2.2.1 Classical Interpretation of Probability (Equally Likely Concept)

In this interpretation, the probability of an event E can be obtained from (A.10), provided that the sample space contains N equally likely and different outcomes, i.e., $m(S) = N$, n of which have an outcome (event) E, i.e., $m(E) = n$. Thus $\Pr(E) = n/N$. This definition is often inadequate for engineering applications. For example, if failures of a pump to start in a process plant are observed, it is unknown whether all failures are equally likely to occur. Nor is it clear if the whole spectrum of possible events is observed. That case is not similar to rolling a perfect die, with each side having an equal probability of 1/6 at any time in the future.

A.2.2.2 Frequency Interpretation of Probability

In this interpretation, the limitation on the lack of knowledge about the overall sample space is remedied by defining the probability as the limit of n/N as N becomes large. Therefore, $\Pr(E) = \lim_{N \to \infty} (n/N)$. Thus, if we have observed 2000 starts of a pump in which 20 failed, and if we assume that 2000 is a large number, then the probability of the pump failure to start is $20/2000 = 0.01$.

The frequency interpretation is the most widely used classical definition today. However, some argue that because it does not cover cases in which little or no experience (or evidence) is available, nor cases where estimates concerning the observations are intuitive, a broader definition is required. This has led to the third interpretation of probability.

A.2.2.3   Subjective Interpretation of Probability

In this interpretation, $Pr(E)$ is a measure of the degree of belief one holds in a specified event E. To better understand this interpretation, consider the probability of improving a system by making a design change. The designer believes that such a change results in a performance improvement in one out of three missions in which the system is used. It would be difficult to describe this problem through the first two interpretations. That is, the classical interpretation is inadequate since there is no reason to believe that performance is as likely to improve as to not improve. The frequency interpretation is not applicable because no historical data exist to show how often a design change resulted in improving the system. Thus, the subjective interpretation provides a broad definition of the probability concept.

A.2.2.4   Calculus of Probability

The basic rules used to combine and treat the probability of an event are not affected by the interpretations discussed above; we can proceed without adopting any of them. (There is much dispute among probability scholars regarding these interpretations. Readers are referred to Cox [1] for further discussions of this subject.)

In general, the axioms of probability can be defined for a sample space S as follows:

1) $Pr(E) \geq 0$, for every event E such that $E \subset S$,
2) $Pr(E_1 \cup E_2 \cup \cdots \cup E_n) = Pr(E_1) + Pr(E_2) + \cdots + Pr(E_n)$,
   where the events $E_1, E_2, \ldots, E_n$ are such that no two have a sample point in common,
3) $Pr(S) = 1$.

It is important to understand the concept of independent events before attempting to multiply and add probabilities. Two events are independent if the occurrence or nonoccurrence of one does not depend on or change the probability of the occurrence of the other. Mathematically, this can be expressed by

$$Pr(E_1|E_2) = Pr(E_1) \tag{A.11}$$

where $Pr(E_1|E_2)$ reads "the probability of E_1, given that E_2 has occurred." To better illustrate, let us consider the result of a test on 200 manufactured identical parts. It is observed that 23 parts fail to meet the length limitation imposed by the designer, and 18 fail to meet the height limitation. Additionally, 7 parts fail to meet both length and height limitations. Therefore, 152 parts meet both of the specified requirements. Let E_1 represent the event that a part does not meet the specified length, and E_2 represent the event that the part does not meet the specified height. According to (A.10), $Pr(E_1) = (7 + 23)/200 = 0.15$, and $Pr(E_2) = (18 + 7)/200 = 0.125$. Furthermore, among 25 parts $(7 + 18)$ that have at least event E_2, 7 parts also have event E_1. Thus, $Pr(E_1|E_2) = 7/25 = 0.28$. Since $Pr(E_1|E_2) \neq Pr(E_1)$, events E_1 and E_2 are dependent.

We shall now discuss the rules for evaluating the probability of simultaneous occurrence of two or more events, that is, $Pr(E_1 \cap E_2)$. For this purpose, we recognize two facts. First, when E_1 and E_2 are independent, the probability that both E_1 and E_2 occur simultaneously is simply the multiplication of the probabilities that E_1 and E_2 occur individually. That is,

$$Pr(E_1 \cap E_2) = Pr(E_1) \cdot Pr(E_2)$$

Second, when E_1 and E_2 are dependent, the probability that both E_1 and E_2 occur simultaneously is obtained from the following expressions:

$$\Pr(E_1 \cap E_2) = \Pr(E_1) \cdot \Pr(E_2|E_1)$$

We will elaborate further on (A.12) when we discuss Bayes' theorem. It is easy to see that when E_1 and E_2 are independent, and (A.11) is applied, (A.12) reduces to

$$\Pr(E_1 \cap E_2) = \Pr(E_1) \cdot \Pr(E_2) \qquad (A.12)$$

In general, the probability of joint occurrence of n-independent events $E_1, E_2, \ldots, E_n$ is the product of their individual probabilities. That is,

$$\Pr(E_1 \cap E_2 \cap \cdots \cap E_n) = \Pr(E_1) \cdot \Pr(E_2) \cdots \Pr(E_n) = \prod_{i=1}^{n} \Pr(E_i) \qquad (A.13)$$

The probability of joint occurrence of n-dependent events $E_1, E_2, \ldots, E_n$ is obtained from

$$\Pr(E_1 \cap E_2 \cap \cdots \cap E_n) = \Pr(E_1) \cdot \Pr(E_2|E_1) \cdot \Pr(E_3|E_1 \cap E_2) \cdots \Pr(E_n|E_1 \cap E_2 \cap \cdots \cap E_{n-1})$$

$$(A.14)$$

where $\Pr(E_3|E_1 \cap E_2 \cap) \cdots$ denotes the conditional probability of E_3 given the occurrence of both E_1 and E_2 and so on.

The evaluation of the probability of union of two events depends on whether or not these events are mutually exclusive. To illustrate this point, let us consider the 200 electronic parts that we discussed earlier. The union of two events E_1 and E_2 includes those parts that do not meet the length requirement, or the height requirement, or both. That is, a total of $23 + 18 + 7 = 48$. Thus,

$$\Pr(E_1 \cup E_2) = 48/200 = 0.24.$$

In other words, 24% of the parts do not meet one or both of the requirements. We can easily see that $\Pr(E_1 \cup E_2) \neq \Pr(E_1) + \Pr(E_2)$, since $0.24 \neq 0.125 + 0.15$. The reason for this inequality is the fact that the two events E_1 and E_2 are not mutually exclusive. In turn, $\Pr(E_1)$ will include the probability of inclusive events $E_1 \cap E_2$, and $\Pr(E_2)$ will also include events $E_1 \cap E_2$. Thus, joint events are counted twice in the expression $\Pr(E_1) + \Pr(E_2)$. Therefore, $\Pr(E_1 \cap E_2)$ must be subtracted from this expression. This description, which can also be seen in a Venn diagram, leads to the following expression for evaluating the probability of the union of two events that are not mutually exclusive:

$$\Pr(E_1 \cup E_2) = \Pr(E_1) + \Pr(E_2) - \Pr(E_1 \cap E_2) \qquad (A.15)$$

Since $\Pr(E_1 \cap E_2) = 7/200 = 0.035$, then $\Pr(E_1 \cup E_2) = 0.125 + 0.15 - 0.035 = 0.24$, which is what we expect to get. From (A.15) one can easily infer that if E_1 and E_2 are mutually exclusive, then $\Pr(E_1 \cup E_2) = \Pr(E_1) + \Pr(E_2)$. If events E_1 and E_2 are dependent, then by using (A.12), we can write (A.15) in the following form:

$$\Pr(E_1 \cup E_2) = \Pr(E_1) + \Pr(E_2) - \Pr(E_1) \cdot \Pr(E_2|E_1) \qquad (A.16)$$

Equation (A.15) for two events can be logically extended to n events.

$$
\begin{aligned}
\Pr(E_1 \cup E_2 \cup \cdots \cup E_n) =\ & \Pr(E_1) + \Pr(E_2) + \cdots + \Pr(E_n) \\
& - [\Pr(E_1 \cap E_2) + \Pr(E_1 \cap E_3) + \cdots + \Pr(E_{n-1} \cap E_n)] \\
& + [\Pr(E_1 \cap E_2 \cap E_3) + \Pr(E_1 \cap E_2 \cap E_4) + \cdots] \\
& - \cdots - (-1)^{n+1}(E_1 \cap E_2 \cap \cdots \cap E_n)
\end{aligned}
\tag{A.17}
$$

Equation (A.17) consists of 2^{n-1} terms. If events $E_1, E_2, \ldots, E_n$ are mutually exclusive, then

$$
\Pr(E_1 \cup E_2 \cup \cdots \cup E_n) = \Pr(E_1) + \Pr(E_2) + \cdots + \Pr(E_n)
\tag{A.18}
$$

When events $E_1, E_2, \ldots, E_n$ are not mutually exclusive, a useful method known as a *rare event approximation* can be used. In this approximation, (A.18) is used if all $\Pr(E_i)$ are small, e.g., $\Pr(E_i) < (50n)^{-1}$.

For dependent events, (A.17) can also be expanded to the form of (A.16) by using (A.14). If all events are independent, then according to (A.13), (A.15) can be further simplified to

$$
\Pr(E_1 \cup E_2) = \Pr(E_1) + \Pr(E_2) - \Pr(E_1) \cdot \Pr(E_2)
\tag{A.19}
$$

Equation (A.19) can be algebraically reformatted to the easier form of

$$
\Pr(E_1 \cup E_2) = 1 - [1 - \Pr(E_1)] \cdot [1 - \Pr(E_2)]
\tag{A.20}
$$

Equation (A.19) can be expanded in the case of n-independent events to

$$
\Pr(E_1 \cup E_2 \cup \cdots \cup E_n) = 1 - [1 - \Pr(E_1)] \cdot [1 - \Pr(E_2)] \cdots [1 - \Pr(E_n)]
\tag{A.21}
$$

In probability evaluations, it is sometimes necessary to evaluate the probability of the complement of an event, that is, $\Pr(\bar{E})$. To obtain this value, let us begin with (A.10) and recognize that the probability of all events in sample space S is 1. The sample space can also be expressed by event E and its complement $\bar{E}$. That is,

$$
\Pr(S) = 1 = \Pr(E \cup \bar{E})
$$

Since E and $\bar{E}$ are mutually exclusive, $\Pr(E \cup \bar{E}) = \Pr(E) + \Pr(\bar{E})$. Thus, $\Pr(E) + \Pr(\bar{E}) = 1$. By rearrangement, it follows that

$$
\Pr(\bar{E}) = 1 - \Pr(E)
\tag{A.22}
$$

It is important to emphasize the difference between independent events and mutually exclusive events, since these two concepts are sometimes confused. In fact, two events that are mutually exclusive are not independent. Since two mutually exclusive events E_1 and E_2 have no intersection, that is, $E_1 \cap E_2 = \varnothing$, then $\Pr(E_1 \cap E_2) = \Pr(E_1) \cdot \Pr(E_2|E_1) = 0$. This means that $\Pr(E_2|E_1) = 0$, since $\Pr(E_1) = 0$. For two independent events, we expect to have $\Pr(E_2|E_1) = \Pr(E_2)$, which is not zero except for the trivial case of $\Pr(E_2) = 0$. This indicates that two mutually exclusive events are indeed dependent.

A.2.3 BAYES' THEOREM

An important theorem known as Bayes' theorem follows directly from the concept of conditional probability, a form of which is described in (A.12). For example, three forms of (A.12) for events A and E are

$$\Pr(A \cap E) = \Pr(A) \cdot \Pr(E|A)$$

and

$$\Pr(A \cap E) = \Pr(E) \cdot \Pr(A|E)$$

therefore

$$\Pr(A) \cdot \Pr(E|A) = \Pr(E) \cdot \Pr(A|E)$$

By solving for $\Pr(A|E)$, it follows that

$$\Pr(A|E) = \frac{\Pr(A) \cdot \Pr(E|A)}{\Pr(E)} \tag{A.23}$$

This equation is known as Bayes' theorem.

It is easy to prove that if event E depends on some previous events that can occur in one of the n different ways $A_1, A_2, \ldots, A_n$, then (A.23) can be generalized to

$$\Pr(A_j|E) = \frac{\Pr(A_j) \cdot \Pr(E|A_j)}{\sum_{i=1}^{n} \Pr(A_i) \cdot \Pr(E|A_i)} \tag{A.24}$$

The right-hand side of the Bayes' equation consists of two terms: $\Pr(A_j)$, called the *prior probability*, and

$$\frac{\Pr(E|A_j)}{\sum_{i=1}^{n} \Pr(A_i) \cdot \Pr(E|A_i)}$$

the relative likelihood or the factor by which the prior probability is revised based on evidential observations (e.g., limited failure observations). $\Pr(A_j|E)$ is called the posterior probability, that is, given event E, the probability of event A_j can be updated (from prior probability $\Pr(A_j)$). Clearly, when more evidence (in the form of event E) becomes available, $\Pr(A_j|E)$ can be further updated. Bayes' theorem provides a means of changing one's knowledge about an event in light of new evidence related to the event. For further studies about Bayes' theorem, refer to Lindley [5].

A.3 PROBABILITY DISTRIBUTIONS

In this section, we concentrate on basic probability distributions that are used in mathematical theory of reliability and reliability data analysis. A fundamental aspect in describing probability distributions is the concept of a *random variable*. We begin with this concept and then continue with the basics of probability distributions applied in reliability analysis.

A.3.1 RANDOM VARIABLE

Let us consider an experiment with a number of possible outcomes. If the occurrence of each outcome is governed by chance (random outcome), then possible outcomes may be assigned a numerical value.

An uppercase letter (e.g., X, Y) is used to represent a random variable, and a lowercase letter is used to determine the numerical value that the random variable can take. For example, if random variable X represents the number of system breakdowns during a given period of time t (e.g., number of breakdowns per year) in a process plant, then x_i shows the actual number of observed breakdowns.

Random variables can be divided into two classes, namely, *discrete* and *continuous*. A random variable is said to be discrete if its sample space is countable, such as the number of system breakdowns in a given period of time. A random variable is said to be continuous if it can take on a continuum of values. That is, it takes on values from interval(s) as opposed to a specific countable number. Continuous random variables are a result of measured variables as opposed to counted data. For example, the operation of several lightbulbs can be modeled by a random variable T, which takes on a continuous survival time t for each lightbulb. Clearly, time t is not countable.

A.3.2 SOME BASIC DISCRETE DISTRIBUTIONS

Consider a discrete random variable, X. The probability distribution for a discrete random variable is usually denoted by the symbol $\Pr(x_i)$, where x_i is one of the values that random variable X takes on. Let random variable X have the sample space S designating the countable realizations of X, which can be expressed as $S = \{x_1, x_2, \ldots, x_k\}$ where k is a finite or infinite number. The discrete probability distribution for this space is then a function $\Pr(x_i)$, such that

$$\Pr(x_i) \geq 0, \quad i = 1, 2, \ldots, k$$

and

$$\sum_{i=1}^{n} \Pr(x_i)$$

(A.25)

A.3.2.1 Discrete Uniform Distribution

Suppose that all possible k outcomes of an experiment are equally likely. Thus, for the sample space $S = \{x_1, x_2, \ldots, x_k\}$ one can write

$$\Pr(x_i) = p = \frac{1}{k}, \quad i = 1, 2, \ldots, k$$

(A.26)

A traditional model example for this distribution is rolling a die. If random variable X describes the numbered 1 to 6 faces, then the discrete number of outcomes is $k = 6$. Thus, $\Pr(x_i) = 1/6$, $x_i = 1, 2, \ldots, 6$.

A.3.2.2 Binomial Distribution

Consider a random trial having two possible outcomes, for instance, success with probability p, and failure with probability $1-p$. Consider a series of n independent trials with these outcomes. Let random variable X denote the total number of successes. Since the number is a nonnegative integer, the sample space is $S = \{0, 1, 2, \ldots, n\}$. The probability distribution of random variable X is given by the binomial distribution:

$$\Pr(x) = \binom{n}{x} p^x (1-p)^{n-x}, \quad x = 1, 2, \ldots, n \tag{A.27}$$

which gives the probability that a known event or outcome occurs exactly x times out of n trials. In (A.27), x is the number of times that a given outcome has occurred. The parameter p indicates the probability that a given outcome will occur. The symbol $\binom{n}{x}$ denotes the total number of ways that a given outcome can occur without regard to the order of occurrence. By definition,

$$\binom{n}{x} = \frac{n!}{x!(n-x)!} \tag{A.28}$$

where $n! = n(n-1), (n-2), \ldots, 1$, and $0! = 1$.

In the following examples, the binomial probability is treated (in the framework of classical statistical inference approach) as a constant nonrandom quantity.

A.3.2.3 Hypergeometric Distribution

The hypergeometric distribution is the only distribution associated with a finite population, considered in this book. Let us have a finite population of N items among which there are D items of interest, for example, N identical components, among which D components are defective. The probability to find $x(x \leq D)$ objects of interest within a sample (without replacements) of $n(n \leq N)$ items is given by the hypergeometric distribution:

$$\Pr(x; N, D, n) = \frac{\binom{D}{x}\binom{N-D}{n-x}}{\binom{N}{n}} \tag{A.29}$$

where

$$x = 0, 1, 2, \ldots, n; \quad x \leq D; \quad n - x \leq N - D \tag{A.30}$$

The hypergeometric distribution is commonly used in statistical quality control and acceptance–rejection test practice. This distribution approaches the binomial one with parameters $p = D/N$ and n, when the ratio, n/N, becomes small.

A.3.2.4 Poisson Distribution

This model assumes that objects or events of interest are evenly dispersed at random in a time or space domain, with some constant intensity, λ. For example, random variable X can represent the number of failures observed at a process plant per year (time domain), or the number of buses arriving at a given station per hour (time domain), if they arrive randomly and independently in time. It can also represent the number of cracks per unit area of a metal sheet (space domain). It is clear that a random variable X following the Poisson distribution is, in a sense, a number of random events, so that it takes on only integer values. If random variable X follows the Poisson distribution, then

$$\Pr(x) = \frac{\rho^x \exp(-\rho)}{x!}, \quad \rho > 0, \ x = 0, 1, 2, \ldots \qquad (A.31)$$

where ρ is the only parameter of the distribution, which is also its mean. For example, if X is a number of events observed in a nonrandom time interval, t, then $\rho = \lambda t$, where λ is the so-called *rate* (time domain) or *intensity* (space domain) *of occurrence* of Poisson events.

The Poisson distribution can be used as an approximation to the binomial distribution when the parameter, p, of the binomial distribution is small (e.g., when $p \leq 0.1$) and parameter n is large. In this case, the parameter of the Poisson distribution, ρ, is substituted by np in (A.31). Besides, it should be noted that as n increases, the Poisson distribution approaches the normal distribution with mean and variance of ρ. This asymptotical property is used as the normal approximation for the Poisson distribution.

A.3.2.5   Geometric Distribution

Consider a series of binomial trials with probability of success, p. Introduce a random variable, X, equal to the length of a series (number of trials) of successes before the first failure is observed. The distribution of random variable X is given by the geometrical distribution:

$$\Pr(x) = p(1-p)^{x-1}, \quad x = 1, 2, \ldots \qquad (A.32)$$

The term $(1-p)^{x-1}$ is the probability that the failure will not occur in the first $(x-1)$ trials. When multiplied by p, it accounts for the probability of the failure in the xth trial.

The books by Johnson and Kotz [4], Hahn and Shapiro [2], and Nelson [7] are good references for other discrete probability distributions.

A.3.3   SOME BASIC CONTINUOUS DISTRIBUTIONS

In this section, we present certain continuous probability distributions that are fundamental to risk analysis. A continuous random variable X has a probability of zero of assuming one of the exact values of its possible outcomes. For example, if random variable T represents the time interval within which a given emergency action is performed by a pilot, then the probability that a given pilot will perform this emergency action, for example, in *exactly* 2 min is equal to zero. In this situation, it is appropriate to introduce the probability associated with a small range of values that the random variable can take on. For example, one can determine $\Pr(t_1 < T < t_2)$, i.e., the probability that the pilot would perform the emergency action sometime between 1.5 and 2.5 min. To define probability that a random variable assumes a value less than given, one can introduce the so-called cdf, $F(t)$, of a continuous random variable T as

$$F(x) = \Pr(T \leq t) \qquad (A.33)$$

Similarly, the cdf of a discrete random variable X is defined as

$$F(x_i) = \Pr(X \leq x_i) = \sum_{\text{all } x_i \leq x} \Pr(x_i) \qquad (A.34)$$

For a continuous random variable, X, a pdf, $f(t)$, is defined as

$$f(t) = \frac{dF(t)}{dt}$$

It is obvious that the cdf of a random variable t, $F(t)$, can be expressed in terms of its pdf, $f(t)$, as

$$F(t) = \Pr(T \le t) = \int_0^t f(\xi)d\xi \qquad (A.35)$$

$f(t)dt$ is called the *probability element*, which is the probability associated with a small interval dt of a continuous random variable T.

The cdf of any continuous random variable must satisfy the following conditions:

$$F(-\infty) = 0 \qquad (A.36)$$

$$F(\infty) = \int_{-\infty}^{\infty} f(\xi)d\xi = 1 \qquad (A.37)$$

$$0 \le F(t) = \int_{-\infty}^t f(\xi)d\xi \le 1 \qquad (A.38)$$

and

$$F(x_1) = \int_{-\infty}^{x_1} f(\xi)d\xi \le F(x_2) = \int_{-\infty}^{x_2} f(\xi)d\xi, \quad x_1 \le x_2 \qquad (A.39)$$

The cdf is used to determine the probability that a random variable, T, falls in an interval (a, b):

$$\Pr(a < T < b) = \int_{-\infty}^b f(\xi)d\xi - \int_{-\infty}^a f(\xi)d\xi = \int_a^b f(\xi)d\xi \qquad (A.40)$$

or

$$\Pr(a < T < b) = F(b) - F(a)$$

For a discrete random variable X, the cdf is defined in a similar way, so that the analogous equations are

$$\Pr(c < T < d) = \sum_{x_i \le d} p(x_i) - \sum_{x_i \le c} p(x_i) = F(d) - F(c)$$

It is obvious that the pdf $f(t)$ of a continuous random variable T must have the following properties:

$$f(T) \ge 0 \quad \text{for all } t, \quad \int_{-\infty}^{\infty} f(t)dt = 1$$

and

$$\int_{t_1}^{t_2} f(t)dt = \Pr(t_1 < T < t_2)$$

A.3.3.1 Normal Distribution

Perhaps the most well-known and important continuous probability distribution is the normal distribution (sometimes called Gaussian distribution). A normal pdf has the symmetric bell-shaped curve shown in Figure A.5, called the normal curve.

In 1733, De Moivre developed the mathematical representation of the normal pdf, as follows:

$$f(t) = \frac{1}{\sqrt{2\pi}\sigma} \exp\left[-\frac{1}{2}\left(\frac{t-\mu}{\sigma}\right)^2\right], \quad -\infty < t < \infty, \quad -\infty < \mu < \infty \quad (A.41)$$

where μ and σ are the parameters of the distribution, $\sigma > 0$.

From (A.41), it is evident that once μ and σ are specified, the normal curve can be determined.

According to (A.41), the probability that the random variable T takes on a value between abscissas $t = t_1$ and $t = t_2$ is given by

$$\Pr(t_1 < T < t_2) = \frac{1}{\sqrt{2\pi}\sigma} \int_{t_1}^{t_2} \exp\left[-\frac{1}{2}\left(\frac{t-\mu}{\sigma}\right)^2\right] dt \quad (A.42)$$

The integral cannot be evaluated in a closed form, so the numerical integration and tabulation of normal cdf are required. However, it would be impractical to provide a separate table for every conceivable value of μ and σ. One way to get around this difficulty is to use the transformation of the normal pdf to the so-called standard normal pdf, which has a mean of zero ($\mu = 0$) and a standard deviation of 1 ($\sigma = 1$). This can be achieved by means of the random variable transformation Z, such that

$$Z = \frac{T-\mu}{\sigma} \quad (A.43)$$

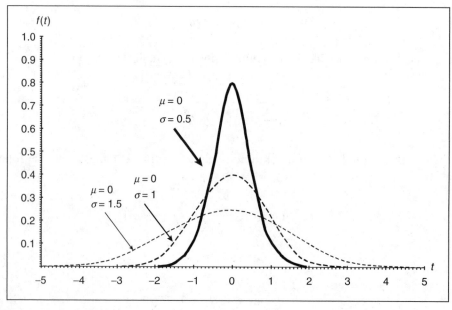

FIGURE A.5 Normal distribution.

That is, whenever random variable T takes on a value t, the corresponding value of random variable Z is given by $z = (t-\mu)/\sigma$. Therefore, if T takes on values $t = t_1$ or $t = t_2$, the random variable Z takes on values $z_1 = (t_1-\mu)/\sigma$, and $z_2 = (t_2-\mu)/\sigma$. Based on this transformation, we can write

$$\Pr(t_1 < T < t_2) = \frac{1}{\sqrt{2\pi\sigma}} \int_{t_1}^{t_2} \exp\left[-\frac{1}{2}\left(\frac{t-\mu}{\sigma}\right)^2\right] dt$$

$$\Pr(t_1 < T < t_2) = \frac{1}{\sqrt{2\pi\sigma}} \int_{z_1}^{z_2} \exp\left(-\frac{1}{2}Z^2\right) dZ \qquad (A.44)$$

$$\Pr(t_1 < T < t_2) = \Pr(z_1 < Z < z_2)$$

where Z, also, has the normal pdf with a mean of zero and a standard deviation of 1. Since the standard normal pdf is characterized by fixed mean and standard deviation, only one table is necessary to provide the areas under the normal pdf curves. Table B.1 presents the area under the standard normal curve corresponding to $\Pr(a < Z < \infty)$.

A.3.3.2 Lognormal Distribution

A positively defined random variable is said to be lognormally distributed if its logarithm is normally distributed. The lognormal distribution has considerable applications in engineering. One major application of this distribution is to represent a random variable that is the result of multiplication of many independent random variables.

If T is a normally distributed random variable, the transformation $Y = \exp(T)$ transforms the normal pdf representing random variable T with mean μ_t and standard deviation σ_t to a lognormal pdf, $f(y)$, which is given by

$$f(y) = \frac{1}{\sqrt{2\pi}\sigma_t y} \exp\left[-\frac{1}{2}\left(\frac{\ln y - \mu_t}{\sigma_t}\right)^2\right], \quad y > 0, \quad -\infty < \mu_t < \infty, \ \sigma_t > 0 \qquad (A.45)$$

Figure A.6 shows the pdfs of the lognormal distribution for different values of μ_t and σ_t.
The area under the lognormal pdf curve $f(y)$ between two points, y_1 and y_2, which is equal to the probability that random variable Y takes a value between y_1 and y_2, can be determined using a procedure similar to that outlined for the normal distribution. Since logarithm is a monotonous function and $\ln y$ is normally distributed, the standard normal random variable with

$$z_1 = \frac{\ln y_1 - \mu_t}{\sigma_t}, \quad z_2 = \frac{\ln y_2 - \mu_t}{\sigma_t} \qquad (A.46)$$

provides the necessary transformation to calculate the probability as follows:

$$\Pr(y_1 < Y < y_2) = \Pr(\ln y_1 < \ln Y < \ln y_2)$$
$$\Pr(y_1 < Y < y_2) = \Pr(\ln y_1 < T < \ln y_2)$$
$$\Pr(y_1 < Y < y_2) = \Pr(z_1 < Z < z_2)$$

If μ_t and σ_t are not known, but μ_y and σ_y are known, the following equations can be used to obtain μ_t and σ_t:

$$\mu_t = \ln \frac{\mu_y}{\sqrt{1 + \dfrac{\sigma_y^2}{\mu_y^2}}} \qquad (A.47)$$

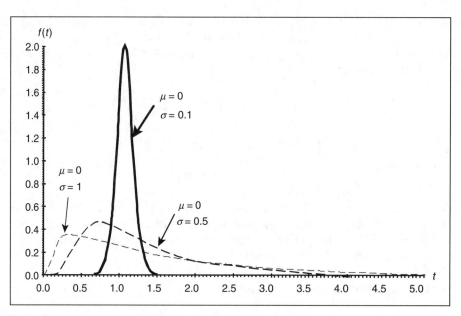

FIGURE A.6 Lognormal distribution.

$$\sigma_t = \sqrt{\ln 1 + \frac{\sigma_y^2}{\mu_y^2}} \qquad (A.48)$$

From (A.47) and (A.48), μ_y and σ_y can also be determined in terms of μ_t and σ_t.

$$\mu_y = \exp\left(\mu_t + \frac{\sigma_t^2}{2}\right) \qquad (A.49)$$

$$\sigma_y = \sqrt{\exp(\sigma_t^2) - 1} \cdot \mu_y \qquad (A.50)$$

A.3.3.3 Exponential Distribution

This distribution was historically the first distribution used as a model of time to failure (TTF) distribution of components and systems, and it is one of the most widely used in reliability problems. The distribution has one-parameter pdf given by

$$f(t) = \begin{cases} \lambda \, \exp(-\lambda t), & \lambda, t > 0 \\ 0, & t \leq 0 \end{cases} \qquad (A.51)$$

Figure A.7 illustrates the exponential pdf. In reliability engineering applications, the parameter λ is referred to as the *failure rate*.

The exponential distribution is closely associated with the Poisson distribution. Consider the following test. A unit is placed on test at $t = 0$. When the unit fails it is instantaneously replaced by an identical new one, which, in turn, is instantaneously replaced on its failure by another identical new unit, etc. The test is terminated at nonrandom time T. It can be shown that if the number of failures during the test is distributed according to the Poisson distribution with the mean of λT, then the time between successive failures (including the time to the

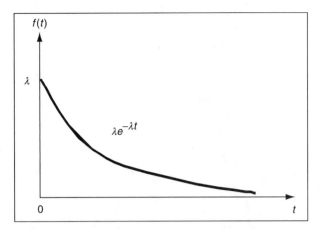

FIGURE A.7 Exponential distribution.

first failure) has the exponential distribution with parameter λ. The test considered is an example of, the so-called, homogenous poisson process (HPP).

A.3.3.4 Weibull Distribution

This distribution is widely used to represent the TTF or life duration of components as well as systems. The continuous random variable T representing the TTF follows a Weibull distribution if its pdf is given by

$$f(t) = \begin{cases} \dfrac{\beta t^{\beta-1}}{\alpha^{\beta}} \exp\left[-\left(\dfrac{t}{\alpha}\right)^{\beta}\right], & t, \alpha, \beta > 0 \\ 0, & \text{otherwise} \end{cases} \qquad (A.52)$$

Figure A.8 shows the Weibull pdfs with various values of parameters of α and β.

A careful inspection of these graphs reveals that the parameter β determines the shape of the distribution pdf. Therefore, β is referred to as the *shape parameter*. The parameter α, on the other hand, controls the scale of the distribution. For this reason, α is referred to as the *scale parameter*. In the case when $\beta = 1$, the Weibull distribution is reduced to the exponential distribution with $\lambda = 1/\alpha$, so the exponential distribution is a particular case of the Weibull distribution. For the values of $\beta > 1$, the distribution becomes bell-shaped with some skew.

A.3.3.5 Gamma Distribution

The gamma distribution can be thought of as a generalization of the exponential distribution. For example, if the time T_i between successive failures of a system has the exponential distribution, then a random variable T such that $T = T_1 + T_2 + \cdots + T_n$, follows the gamma distribution. In the given context, T represents the cumulative time to the nth failure.

A different way to interpret this distribution is to consider a situation in which a system is subjected to shocks occurring according to the Poisson process (with parameter λ). If the system fails after receiving n shocks, then the TTF of such a system follows a gamma distribution.

The pdf of the gamma distribution with parameters β and α is given by

$$f(t) = \frac{1}{\beta^{\alpha}\Gamma(\alpha)} t^{\alpha-1} \exp\left(-\frac{t}{\beta}\right), \quad \alpha, \beta, t \geq 0 \qquad (A.53)$$

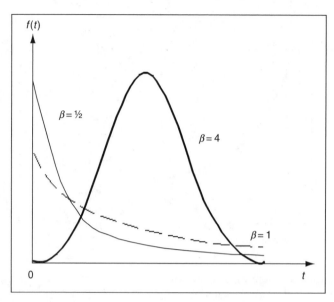

FIGURE A.8 Weibull distribution.

where $\Gamma(\alpha)$ denotes the so-called gamma function defined as

$$\Gamma(\alpha) = \int_0^\infty x^{\alpha-1} e^{-x} dx \qquad (A.54)$$

Note that when α is a positive integer, $\Gamma(\alpha) = (\alpha-1)!$, but in general the parameter α is not necessarily an integer.

The mean and the variance of gamma distribution are

$$E(t) = \alpha\beta$$
$$\text{var}(t) = \alpha\beta^2$$

The parameter α is referred to as the shape parameter and the parameter β is referred to as the scale parameter. It is clear that if $\alpha = 1$, (A.53) is reduced to the exponential distribution. Another important special case of the gamma distribution is the case when $\beta = 2$ and $\alpha = n/2$, where n is a positive integer, referred to as the number of *degrees of freedom*. This one-parameter distribution is known as the chi-square distribution. This distribution is widely used in reliability data analysis.

Figure A.9 shows the gamma distribution pdf curves for some values of α and β.

A.3.3.6 Beta Distribution

The beta distribution is a useful model for random variables that are distributed in a finite interval. The pdf of the standard beta distribution is defined over the interval (0, 1) as

$$f(t; \alpha, \beta) = \begin{cases} \dfrac{\Gamma(\alpha+\beta)}{\Gamma(\alpha)\Gamma(\beta)} t^{\alpha-1}(1-t)^{\beta-1}, & 0 \leq t \leq 1, \qquad \alpha > 0, \beta \geq 0 \\ 0, & \text{otherwise} \end{cases} \qquad (A.55a)$$

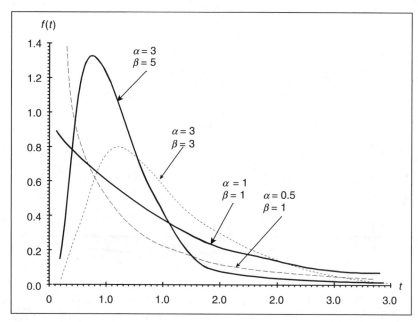

FIGURE A.9 Gamma distribution.

Similar to the gamma distribution, the cdf of the beta distribution cannot be written in closed form. It is expressed in terms of, the so-called incomplete beta function, $I_t(\alpha, \beta)$, i.e.,

$$f(t; \alpha, \beta) = \begin{cases} 0 & t < 0 \\ \dfrac{\Gamma(\alpha + \beta)}{\Gamma(\alpha)\Gamma(\beta)} \displaystyle\int_0^t x^{\alpha-1}(1-x)^{\beta-1}\mathrm{d}x & 0 \le t \le 1, \ \alpha > 0, \beta \ge 0 \\ 1 & t > 1 \end{cases} \qquad (A.56)$$

The mean value and the variance of the beta distribution are

$$E(t) = \frac{\alpha}{\alpha + \beta}$$

$$\mathrm{var}(t) = \frac{\alpha\beta}{(\alpha + \beta)^2(\alpha + \beta + 1)}$$

For the special case of $\alpha = \beta = 1$, the beta distribution reduces to the standard uniform distribution. Practically, the distribution is not used as a TTF distribution. On the other hand, the beta distribution is widely used as an auxiliary distribution in nonparametric classical statistical distribution estimation, as well as a prior distribution in the Bayesian statistical inference. Figure A.10 shows the beta distribution pdf curves for some selected values of α and β.

A.3.4 JOINT AND MARGINAL DISTRIBUTIONS

Thus far, we have discussed distribution functions that are related to one-dimensional sample spaces. There exist, however, situations in which more than one random variable is simultaneously measured and recorded. For example, in a study of human reliability in a control room situation, one can simultaneously estimate (1) the random variable T representing time that various operators spend to fulfill an emergency action, and (2) the random variable E

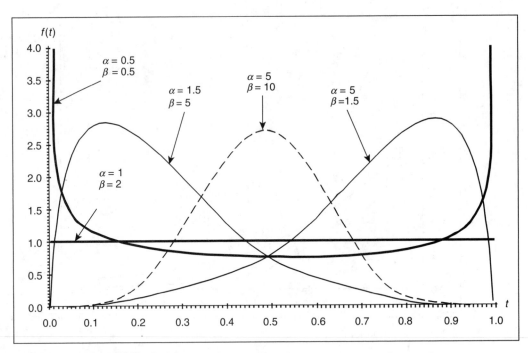

FIGURE A.10 Beta distribution.

representing the level of training that these various operators have had for performing these emergency actions. Since one expects E and T to have some relationships (e.g., more trained operators act faster than less trained ones), a joint distribution of both random variables T and E can be used to express their mutual dispersion.

Let X and Y be two random variables (not necessarily independent). The pdf for their simultaneous occurrence is denoted by $f(x, y)$ and it is called the *joint pdf* of X and Y. If random variables X and Y are discrete, the joint pdf can be denoted by $\Pr(X = x, Y = y)$, or simply $\Pr(x, y)$. Thus, $\Pr(x, y)$ gives the probability that the outcomes x and y occur simultaneously. For example, if random variable X represents the number of circuits of a given type in a process plant, and Y represents the number of failures of the circuit in the most recent year, then $\Pr(7, 1)$ is the probability that a randomly selected process plant has seven circuits and one of them failed once in the most recent year. The function $f(x, y)$ is a joint pdf of continuous random variables X and Y if

1. $f(x, y) \geq 0, -\infty < x, y < \infty$, and
2. $\int_{-\infty}^{\infty} \int_{-\infty}^{\infty} f(x, y) \mathrm{d}x \, \mathrm{d}y = 1$.

Similarly, the function $\Pr(x, y)$ is a joint probability function of the discrete random variables X and Y if

1. $f(x, y) \geq 0$, for all values of x and y, and
2. $\sum_x \sum_y f(x, y) = 1$.

The probability that two or more joint random variables fall within a specified subset of the sample space is given by

$$\Pr(x_1 < X \leq x_2, y_1 < Y \leq y_2) = \int_{x_1}^{x_2} \int_{y_1}^{y_2} f(x, y) \mathrm{d}x \, \mathrm{d}y$$

for continuous random variables, and by

$$\Pr(x_1 < X \leq x_2, y_1 < Y \leq y_2) = \sum_{x_1 < x \leq x_2,\ y_1 < y \leq y_2} \Pr(x, y)$$

for discrete random variables.

The *marginal pdfs* of X and Y are defined respectively as

$$g(x) = \int_{-\infty}^{\infty} f(x, y)\,dy \tag{A.57}$$

and

$$h(y) = \int_{-\infty}^{\infty} f(x, y)\,dx \tag{A.58}$$

for continuous random variables, and by

$$\Pr(x) = \sum_y \Pr(x, y)$$

and

$$\Pr(y) = \sum_x \Pr(x, y) \tag{A.59}$$

for the discrete random variables.

Using (A.12), the conditional probability of an event y, given event x, is

$$\Pr(Y = y | X = x) = \frac{\Pr(X = x \cap Y = y)}{\Pr(X = x)} = \frac{\Pr(x, y)}{\Pr(x)}, \quad \Pr(x) > 0 \tag{A.60}$$

where X and Y are discrete random variables. Similarly, one can extend the same concept to continuous random variables X and Y and write

$$f(x|y) = \frac{f(x, y)}{h(y)}, \quad h(y) > 0$$

or

$$f(x|y) = \frac{f(x, y)}{g(x)}, \quad g(x) > 0 \tag{A.61}$$

where (A.60) and (A.61) are called the *conditional pdfs* of discrete and continuous random variables, respectively. The conditional pdfs have the same properties as any other pdf. Similar to (A.11), if random variable X and random variable Y are independent, then $f(x|y) = f(x)$ for continuous random variables, and $\Pr(x|y) = \Pr(x)$ for discrete random variables. This would lead to the conclusion that for independent random variables X and Y,

$$f(x, y) = g(x) \cdot h(y)$$

if X and Y are continuous, and

$$\Pr(x, y) = \Pr(x) \cdot \Pr(y) \tag{A.62}$$

if X and Y are discrete.

Equation (A.62) can be expanded to a more general case as

$$f(x_1, x_2, \ldots, x_n) = f_1(x_1) \cdot f_2(x_2) \cdots f_n(x_n) \tag{A.63}$$

where $f(x_1, x_2, \ldots, x_n)$ is a joint pdf of random variables $X_1, X_2, \ldots, X_n$, and $f_1(x_1), f_2(x_2), \ldots, f_n(x_n)$ are marginal pdfs of $X_1, X_2, \ldots, X_n$, respectively.

A.4 BASIC CHARACTERISTICS OF RANDOM VARIABLES

In this section, we introduce some other basic characteristics of random variables which are widely used in reliability engineering.

The *expectation* or *expected value* of random variable X is a characteristic applied to continuous as well as discrete random variables. Consider a discrete random variable X that takes on values x_i with corresponding probabilities $\Pr(x_i)$. The expected value of X denoted by $E(X)$ is defined as

$$E(X) = \sum_i x_i \Pr(x_i) \tag{A.64}$$

Analogously, if T is a continuous random variable with a pdf $f(t)$, then the expectation of T is defined as

$$E(T) = \int_{-\infty}^{\infty} t f(t) \mathrm{d}t \tag{A.65}$$

$E(X)$ [or $E(T)$] is a widely used concept in statistics known as the *mean*, or in mechanics known as the center of mass, and sometimes denoted by μ. $E(X)$ is also referred to as the *first moment about the origin*. In general, the kth moment about the origin (ordinary moment) is defined as

$$E(T^k) = \int_{-\infty}^{\infty} t^k f(t) \mathrm{d}t \tag{A.66}$$

for all integer $k \geq 1$.

In general, one can obtain the expected value of any real-value function of a random variable. In the case of a discrete distribution, $\Pr(x_i)$, the expected value of function $g(X)$ is defined as

$$E[g(x)] = \sum_{i=1}^{k} g(x_i) \Pr(x_i) \tag{A.67}$$

Similarly, for a continuous random variable T, the expected value of $g(T)$ is defined as:

$$E[g(T)] = \int_{-\infty}^{\infty} g(t) f(t) \mathrm{d}t \tag{A.68}$$

A measure of dispersion or variation of random variable about its mean is called the variance, and is denoted by $\text{Var}(X)$ or $\sigma^2(X)$. The variance is also referred to as the *second moment about the mean* (which is analogous to the moment of inertia in mechanics), sometimes it is referred to as the *central moment*, and is defined as

$$\text{Var}(X) = \sigma^2(X) = E\left[(X - \mu)^2\right] \tag{A.69}$$

where σ is known as the *standard deviation*, σ^2 is known as the *variance*. In general, the kth moment about the mean is defined (similar to (A.66)) as

$$E\left[(X - \mu)^k\right] \quad \text{for all integer } k > 0 \tag{A.70}$$

Table A.2 represents useful simple algebra associated with expectations. The rules given in Table A.2 can be applied to discrete as well as continuous random variables.

One useful method of determining the moments about the origin of a distribution is use of the Laplace transform. Suppose the Laplace transform of pdf $f(t)$ is $F(S)$, then

$$F(S) = \int_0^\infty f(t) \exp(-St)dt \tag{A.71}$$

and

$$\frac{-\mathrm{d}F(S)}{\mathrm{d}S} = \int_0^\infty tf(t) \exp(-St)dt \tag{A.72}$$

Since for $S = 0$ the right-hand side of (A.72) reduces to the expectation $E(t)$, then

$$E(T) = -\left[\frac{\mathrm{d}F(S)}{\mathrm{d}S}\right]S = 0$$

In general, it is possible to show that

$$E(T^k) = \left[(-1)^k \frac{d^k F(S)}{\mathrm{d}S^k}\right]S = 0 \tag{A.73}$$

Expression (A.73) is useful to determine moments of pdfs whose Laplace transforms are known or can be easily derived.

TABLE A.2
The Algebra of Expectations

1. $E(aX) = aE(X)$, $a = $ constant
2. $E(a) = a$, $a = $ constant
3. $E[g(X) \pm h(X)] = E[g(X)] \pm E[h(X)]$
4. $E[X \pm Y] = E[X] \pm E[Y]$
5. $E[X \cdot Y] = E[X] \cdot E[Y]$, if X and Y are independent

---

The concept of expectation equally applies to joint probability distributions. The expectation of a real-value function h of discrete random variables $X_1, X_2, \ldots, X_n$ is

$$E[h(X_1, X_2, \ldots, X_n)] = \sum_{x_1} \sum_{x_2} \cdots \sum_{x_n} h(x_1, x_2, \ldots, x_n) \Pr(x_1, x_2, \ldots, x_n) \qquad \text{(A.74)}$$

where $\Pr(x_1, x_2, \ldots, x_n)$ is the discrete joint pdf of random variables X_i. When dealing with continuous random variables, the summation terms in (A.74) are replaced with integrals

$$E[h(X_1, X_2, \ldots, X_n)] = \int_{-\infty}^{\infty} \cdots \int_{-\infty}^{\infty} \int_{-\infty}^{\infty} h(x_1, x_2, \ldots, x_n) f(x_1, x_2, \ldots, x_n) \mathrm{d}x_1 \mathrm{d}x_2 \cdots \mathrm{d}x_n$$

where $f(x_1, x_2, \ldots, x_n)$ is the continuous joint pdf of random variables X_i. In the case of a bivariate distribution with two random variables X_1 and X_2, the expectation of the function

$$h(X_1, X_2) = [X_1 - E(X_1)][X_2 - E(X_2)]$$

is called the *covariance* of random variables X_1 and X_2, and is denoted by $\mathrm{Cov}(X_1, X_2)$. Using Table A.2, it is easy to show that

$$\mathrm{Cov}(X_1, X_2) = E(X_1 \cdot X_2) - E(X_1)E(X_2) \qquad \text{(A.75)}$$

A common measure of determining the linear relation between two random variables is a *correlation coefficient*, which carries information about two aspects of the relationship:

1. strength, measured on a scale from 0 to 1; and
2. direction, indicated by the plus or minus sign.

Denoted by $\rho(X_1, X_2)$, the correlation coefficient between random variables X_1 and X_2 is defined as

$$\rho(X_1, X_2) = \frac{\mathrm{Cov}(X_1, X_2)}{\sqrt{\mathrm{Var}(X_1)\mathrm{Var}(X_2)}} \qquad \text{(A.76)}$$

Clearly, if X_1 and X_2 are independent, then from (A.75), $\mathrm{Cov}(X_1, X_2) = 0$, and from (A.76), $\rho(X_1, X_2) = 0$. For a linear function of several random variables, the expectation and variance are given by

$$E\left(\sum_{i=1}^{n} a_i X_i\right) = \sum_{i=1}^{n} a_i E(X_i) \qquad \text{(A.77)}$$

$$\mathrm{Var}\left(\sum_{i=1}^{n} a_i X_i\right) = \sum_{i=1}^{n} a_i \mathrm{Var}(X_i) + 2\sum_{i=1}^{n-1}\sum_{i=1}^{n} a_i a_j \mathrm{Cov}(X_i, X_j) \qquad \text{(A.78)}$$

In cases where random variables are independent, (A.78) becomes simplified to

$$\mathrm{Var}\left(\sum_{i=1}^{n} a_i X_i\right) = \sum_{i=1}^{n} a_i^2 \mathrm{Var}(X_i) \qquad \text{(A.79)}$$

A.5 ESTIMATION AND HYPOTHESIS TESTING

Reliability and performance data obtained from special tests, experiments or practical use of a product provide a basis for performing statistical inference about underlying distribution. Each observed value is considered as a *realization* (or *observation*) of some hypothetical random variable, that is, a value that the random variable, say X, can take on. For example, the number of pump failures following a demand in a large plant can be considered as realization of some random variable.

A set of observations from a distribution is called a *sample*. The number of observations in a sample is called the *sample size*. In the framework of classical statistics, a sample is usually composed of random independently and identically distributed observations. From a practical point of view this assumption means that elements of a given sample are obtained independently and under the same conditions.

To check the applicability of a given distribution (for example, binomial distribution in the pump failure case) and to estimate the parameters of the distribution, one needs to use special statistical procedures known as hypothesis testing and estimation which are very briefly considered below.

A.5.1 POINT ESTIMATION

Point and *interval estimation* are the two basic kinds of estimation procedures considered in statistics. Point estimation provides a single number obtained on the basis of data set (a sample), which represents a parameter of the distribution function or other characteristic of the underlying distribution of interest. As opposed to the interval estimation, the point estimation does not provide any information about its accuracy. Interval estimation is expressed in terms of *confidence intervals*. The confidence interval includes the true value of the parameter with a specified confidence probability.

Suppose, we are interested in estimating a single-parameter distribution $F(X, \theta)$ based on a random sample $x_1, \ldots, x_n$. Let $t(x_1, \ldots, x_n)$ be a single-valued (simple) function of $x_1, x_2, \ldots, x_n$. It is obvious that $t(x_1, x_2, \ldots, x_n)$ is also a random variable, which is referred to as a *statistic*. A point estimate is obtained by using an appropriate statistic and calculating its value based on the sample data. The statistic (as a function) is called the *estimator*, meanwhile its numerical value is called the estimate.

Consider the basic properties of point estimators. An estimator $t(x_1, \ldots, x_n)$ is said to be an unbiased estimator for θ if its expectation coincides with the value of the parameter of interest θ, i.e., $E[t(x_1, x_2, \ldots, x_n)] = \theta$ for any value of θ. Thus, the bias is the difference between the expected value of an estimate and the true parameter value itself. It is obvious that the smaller the bias, the better the estimator is.

Another desirable property of an estimator $t(x_1, x_2, \ldots, x_n)$ is the property of consistency. An estimator t is said to be consistent if, for every $\varepsilon > 0$,

$$\lim_{n \to \infty} \Pr[|t(x_1, x_1, \ldots, x_n) - \theta|\varepsilon] = 1 \tag{A.80}$$

This property implies that as the sample size n increases, the estimator $t(x_1, x_2, \ldots, x_n)$ gets closer to the true value of θ. In some situations several unbiased estimators can be found. A possible procedure for selecting the best one among the unbiased estimators can be based on choosing one having the least variance. An unbiased estimator t of θ, having minimum variance among all unbiased estimators of θ, is called *efficient*.

Another estimation property is sufficiency. An estimator $t(x_1, x_2, \ldots, x_n)$ is said to be a sufficient statistics for the parameter if it contains all the information about that is in the sample $x_1, x_2, \ldots, x_n$. In other words the sample $x_1, x_2, \ldots, x_n$ can be replaced by $t(x_1, x_2, \ldots, x_n)$ without loss of any information about the parameter of interest.

Several methods of estimation are considered in mathematical statistics. In the following section, two of the most common methods, i.e., method of moments and method of maximum likelihood, are briefly discussed.

A.5.1.1 Method of Moments

In the previous section the mean and the variance of a continuous random variable X were defined as the expected value of X and expected value of $(X-\mu)^2$, respectively. Quite naturally, one can define the *sample mean* and *sample variance* as the respective expected values of a sample of size n from the distribution of X, namely, $x_1, x_2, \ldots, x_n$, as follows:

$$\bar{x} = \frac{1}{n} \sum_{i=1}^{n} x_i \qquad (A.81)$$

and

$$S^2 = \frac{1}{n} \sum_{i=1}^{n} (x_i - \bar{x})^2 \qquad (A.82)$$

so that $\bar{x}$ and S^2 can be used as the point estimates of the distribution mean, μ, and variance, σ^2. It should be mentioned that estimator of variance (A.82) is biased, since $\bar{x}$ is estimated from the same sample. However, it can be shown that this bias can be removed by multiplying it by $n/(n-1)$:

$$S^2 = \frac{1}{n-1} \sum_{i=1}^{n} (x_i - \bar{x})^2 \qquad (A.83)$$

Generalizing the examples considered, it can be said that the method of moments is an estimation procedure based on empirically estimated (or *sample*) moments of the random variable. According to this procedure, the sample moments are equated to the corresponding distribution moments. The solutions of the equations obtained provide the estimators of the distribution parameters.

A.5.1.2 Maximum Likelihood Method

This method is one of the most widely used methods of estimation. Consider a continuous random variable, X, with pdf $f(X, \theta)$, where θ is a parameter. Let us have a sample $x_1, x_2, \ldots, x_n$ of size n from the distribution of random variable X. Under the maximum likelihood approach, the estimate of θ is found as the value of θ, which delivers the highest (or most likely) probability density of observing the particular set $x_1, x_2, \ldots, x_n$. The likelihood of obtaining this particular set of sample values is proportional to the joint pdf $f(x, \theta)$ calculated at the sample points $x_1, x_2, \ldots, x_n$. The likelihood function for a continuous distribution is introduced as

$$L(x_1, x_1, \ldots, x_n; \theta_0) = f(x_1, \theta_0)f(x_2, \theta_0) \cdots f(x_n, \theta_0) \qquad (A.84)$$

Generally speaking, the definition of the likelihood function is based on the probability (for a discrete random variable) or the pdf (for continuous random variable) of the joint occurrence of n events, $X = x_1, x_2, \ldots, X = x_n$. The maximum likelihood estimate, $\hat{\theta}_0$, is chosen as one that maximizes the likelihood function, $L(x_1, x_2, \ldots, x_n; \theta_0)$, with respect to θ_0.

The standard way to find a maximum of a parameter is to calculate the first derivative with respect to this parameter and equate it to zero. This yields the equation

$$\frac{\partial L(x_1, x_1, \ldots, x_n; \theta_0)}{\partial \theta_0} = 0 \tag{A.85}$$

from which the maximum likelihood estimate $\hat{\theta}_0$ can be obtained.

Due to the multiplicative form of the likelihood function, it turns out, in many cases, to be more convenient to maximize the logarithm of the likelihood function instead, i.e., to solve the following equation:

$$\frac{\partial \log L(x_1, x_1, \ldots, x_n; \theta_0)}{\partial \theta_0} = 0 \tag{A.86}$$

Because the logarithm is monotonous transformation, the estimate of θ_0 obtained from this equation is the same as that obtained from (A.85). For some cases (A.85) or (A.86) can be solved analytically, for other cases they have to be solved numerically.

Under some general conditions, the maximum likelihood estimates are consistent, asymptotically efficient, and asymptotically normal.

A.5.2 Interval Estimation and Hypothesis Testing

A two-sided confidence interval for an unknown distribution parameter θ of continuous random variable X, based on a sample $x_1, x_2, \ldots, x_n$ of size n from the distribution of X is introduced in the following way. Consider two statistics $\theta_f(x_1, x_2, \ldots, x_n)$ and $\theta_u(x_1, x_2, \ldots, x_n)$ chosen in such a way that the probability that parameter 0 lies in an interval $[\theta_l, \theta_u]$ is

$$\Pr[\theta_f(x_1, x_1, \ldots, x_n) < \theta_0 < \theta_u(x_1, x_1, \ldots, x_n)] = 1 - \alpha \tag{A.87}$$

The *random* interval $[l, u]$ is called a $100(1-\alpha)\%$ *confidence interval* for the parameter θ_0. The endpoints l and u are referred to as the $100(1-\alpha)\%$ upper and lower confidence limits of θ_0; $(1-\alpha)$ is called the *confidence coefficient* or *confidence level*. The most commonly used values for α are 0.10, 0.05, and 0.01. In the case when $\theta_0 > \theta_l$ with the probability of 1, θ_u is called the *one-sided upper confidence limit* for θ_0. In the case when $\theta_0 < \theta_u$ with probability of 1, θ_l is the *one-sided lower confidence limit* for θ_0. A $100(1-\alpha)\%$ confidence interval for an unknown parameter θ_0 is interpreted as follows: if a series of repetitive experiments (tests) yields random samples from the same distribution and the same confidence interval is calculated for each sample, then $100(1-\alpha)\%$ of the constructed intervals will, *in the long run*, contain the true value of θ_0.

Consider a typical example illustrating the basic idea of confidence limits construction. Consider a procedure for constructing confidence intervals for the mean of a normal distribution with known variance. Let $x_1, x_2, \ldots, x_n$, be a random sample from the normal distribution, $N(\mu, \sigma^2)$, in which μ is unknown, and σ^2 is assumed to be known. It can be shown that the sample mean $\overline{X}$ (as a statistic) has the normal distribution $N(\mu, \sigma^2/n)$. Thus, $(\overline{X}-\mu)/(n)^{1/2}/\sigma$ has the standard normal distribution. Using this distribution one can write

$$\Pr\left(-z_{1-\alpha/2} \leq \frac{\overline{X} - \mu}{\sigma/\sqrt{n}} \leq z_{1-\alpha/2}\right) = 1 - \alpha \tag{A.88}$$

where $z_{1-\alpha/2}$ is the $100(1-\alpha/2)$th percentile of the standard normal distribution, which can be obtained from Table B.1. After simple algebraic transformations, the inequalities inside the parentheses of (A.88) can be rewritten as

$$\Pr\left(\overline{X} - z_{1-\alpha/2}\frac{\sigma}{\sqrt{n}} \leq \mu \leq \overline{X} + z_{1-\alpha/2}\frac{\sigma}{\sqrt{n}}\right) = 1 - \alpha \qquad (A.89)$$

Equation (A.89) provides the symmetric $(1-\alpha)$ confidence interval of interest. Generally, a two-sided confidence interval is wider for a higher confidence level $(1-\alpha)$. As the sample size n increases, the confidence interval becomes shorter for the same confidence coefficient $(1-\alpha)$.

In the case when σ^2 is unknown, and it is estimated using (A.83), the respective confidence interval is given by

$$\Pr\left(\overline{x} - t_{\alpha/2}\frac{S}{\sqrt{n}} < \mu < \overline{x} + t_{\alpha/2}\frac{S}{\sqrt{n}}\right) = 1 - \alpha \qquad (A.90)$$

where $t_{\alpha/2}$ is the percentile of t-student distribution with $(n-1)$ degrees of freedom. Values of t_α for different numbers of degrees of freedom are given in Table B.2. Confidence intervals for σ^2 for a normal distribution can be obtained as

$$\frac{(n-1)S^2}{\chi^2_{1-\alpha/2}(n-1)} < \sigma^2 < \frac{(n-1)S^2}{\chi^2_{\alpha/2}(n-1)} \qquad (A.91)$$

where $\chi^2_{1-\alpha/2}(n-1)$ is the percentile of χ^2 distribution with $(n-1)$ degrees of freedom which are given in Table B.3. For more discussion on this topic see reference [3].

A.5.2.1 Hypothesis Testing

Interval estimation and hypothesis testing may be viewed as mutually inverse procedures. Let us consider a random variable X with a known pdf $f(x; \theta)$. Using a random sample from this distribution one can obtain a point estimate $\hat{\theta}$ of the parameter θ. Let θ have a hypothesized value of $\theta = \theta_0$. Under these quite realistic conditions, the following question can be raised: Is the $\hat{\theta}$ estimate compatible with the hypothesized value θ_0? In terms of *statistical hypothesis* testing the statement $\theta = \theta_0$ is called the *null hypothesis*, which is denoted by H_0. For the case considered it is written as

$$H_0 : \theta = \theta_0$$

The null hypothesis is always tested against an *alternative hypothesis*, denoted by H_1, which for the case considered might be the statement $\theta \neq \theta$, which is written as

$$H_1 : \theta \neq \theta_0$$

The null and alternative hypotheses are also classified as *simple* (or exact when they specify exact parameter values) and *composite* (or *inexact* when they specify an interval of parameter values). In the considered example, H_0 is simple and H_1 is composite. An example of a simple alternative hypothesis might be $H_1 : \theta = \theta^*$.

For testing statistical hypotheses *test statistics* are used. In many situations, the test statistic is the point estimator of the unknown distribution. In this case, as in the case of the interval estimation, one has to obtain the distribution of the test statistic used.

Recall the example considered above. Let $x_1, x_2, \ldots, x_n$ be a random sample from the normal distribution, $N(\mu, \sigma^2)$, in which μ is an unknown parameter, and σ^2 is assumed to be known. One has to test the simple null hypothesis

$$H_0{:}\mu = \mu*$$

against the composite alternative

$$H_0{:}\mu \neq \mu*$$

As the test statistic, use the same (A.81) sample mean, $\bar{x}$, which has the normal distribution $N(\mu, \sigma^2/n)$. Having the value of the test statistic $\bar{x}$, one can construct the confidence interval using (A.89) and see whether the value of $\mu*$ falls inside the interval. This is the test of the null hypothesis. If the confidence interval includes $\mu*$, the null hypothesis is not rejected at *significance level* α.

In terms of hypothesis testing, the confidence interval considered is called the *acceptance region*, the upper and the lower limits of the acceptance region are called the *critical values*, while the significance level α is referred to as a probability of *type I error*. In making a decision about whether or not to reject the null hypothesis, it is possible to commit the following errors: (i) reject H_0 when it is true (type I error), (ii) do not reject H_0 when it is false (type II error).

The probability of the *type II error* is designated by β. These situations are traditionally represented by the following table:

Decision	State of Nature (True Situation)	
	H_0 is true	H_0 is false
Reject H_0	Type I error	No error
Do not reject H_0	No error	Type II error

It is clear that increasing the acceptance region, decreases α, and simultaneously results in increasing β. The traditional approach to this problem is to keep the probability of type I error at a low level (0.01, 0.05, or 0.10) and to minimize the probability a *type II error* as much as possible. The probability of not making of *type II error* is referred to as the *power of the test*. Examples of a special class of hypothesis testing are considered in Section A.2.7.

A.6 FREQUENCY TABLES AND HISTOGRAMS

When studying distributions, it becomes convenient to start with some preliminary procedures useful for data editing and detecting outliers by constructing *empirical distributions* and *histograms*. Such preliminary data analysis procedures might be useful themselves (the data speak for themselves), as well as they may be used for other elaborate analyses (the goodness-of-fit testing, for instance).

The measure of interest is often the probability associated with each interval of TTF data. This can be obtained using (A.10), i.e., by dividing each interval frequency by the total number of devices observed or tested. Sometimes, it is important in reliability and risk estimation to indicate how well a set of observed data fits a known distribution, i.e., to determine whether a hypothesis that the data originate from a known distribution is true. For this purpose, it is necessary to calculate the expected frequencies of failures from the known distribution, and compare them with the observed frequencies. Several methods exist to determine the adequacy of such a fit. We discuss these methods further in the section below.

A.7   GOODNESS-OF-FIT TESTS

Consider the problem of determining whether a sample belongs to a hypothesized theoretical distribution. For this purpose, we need to perform a test that estimates the adequacy of a fit by determining the difference between the frequency of occurrence of a random variable characterized by an observed sample, and the expected frequencies obtained from the hypothesized distribution. For this purpose, the so-called goodness-of-fit tests are used.

Below we briefly consider two procedures often used as goodness-of-fit tests: the chi-square and Kolmogorov goodness-of-fit tests.

A.7.1   CHI-SQUARE TEST

As the name implies, this test is based on a statistic that has an approximate chi-square distribution. To perform this test, an observed sample taken from the population representing a random variable X must be split into k ($k \geq 5$) nonoverlapping intervals (the lower limit for the first interval can be $-\infty$, as well as the upper limit for the last interval can be $+\infty$). The assumed (hypothesized) distribution model is then used to determine the probabilities p_i that the random variable X would fall into each interval i ($i = 1, 2, \ldots, k$). This process was described to some extent in Section A.2.6. By multiplying p_i by the sample size n, we get the expected frequency for each interval. Denote the expected frequency as e_i. It is obvious that $e_i = np_i$. If the observed frequency for each interval i of the sample is denoted by o_i, then the magnitude of differences between e_i and o_i can characterize the adequacy of the fit. The chi-square test uses the statistic χ^2 which is defined as

$$W = \chi^2 = \sum_{i=1}^{k} \frac{(o_i - e_i)^2}{e_i} \tag{A.92}$$

The χ^2 statistic approximately follows the chi-square distribution (mentioned in Section A.2.3). If the observed frequencies o_i differ considerably from the expected frequencies e_i, then W will be large and the fit is considered to be poor. A good fit would not obviously lead to rejecting the hypothesized distribution, whereas a poor fit leads to the rejection. It is important to note that one can only fail to support the hypothesis, so the person rejects it rather than positively affirm its truth. Therefore, the hypothesis is either *rejected* or *not rejected* as opposed to accepted or not accepted. The test can be summarized as follows:

Step 1.   Choose a hypothesized distribution for the given sample.
Step 2.   Select a specified significance level of the test denoted by α.
Step 3.   Define the rejection region $R \geq \chi^2_{1-\alpha}(k - m - 1)$, where $\chi^2_{1-\alpha}(k - m - 1)$ is the $(1 - \alpha)100$ percentile of the chi-square distribution with $k - m - 1$ degrees of freedom (the percentiles are given in Table B.3), k is the number of intervals, and m is the number of parameters estimated from the sample. If the parameters of the distribution were estimated without using the given sample, then $m = 0$.
Step 4.   Calculate the value of the chi-square statistic, W (A.92).
Step 5.   If $W > R$, reject the hypothesized distribution; otherwise do not reject the distribution.

It is important at this point to specify the role of α in the chi-square test. Suppose the calculated value of W in (A.92) exceeds the 95th percentile, $\chi^2_{0.95}(\cdot)$ given in Table B.3. This indicates that chances are lower than 1 in 20 that the observed data are from the hypothesized distribution. In this case, the model should be rejected (by not rejecting the model, one makes the type II error discussed above). On the other hand, if the calculated value of W is smaller than $\chi^2_{0.95}(\cdot)$, chances are greater than 1 in 20 that the observed data match the hypothesized distribution model. In this case, the model should not be rejected (by rejecting the model, one makes the type I error discussed above).

One instructive step in chi-square testing is to compare the observed data with the expected frequencies to note which classes (intervals) contributed most to the value of W. This sometimes could help to indicate the nature of deviations.

A.7.2   KOLMOGOROV TEST

In the framework of this test, the individual sample components are treated without clustering them into intervals. Similar to the chi-square test, a hypothesized cdf is compared with its estimate known as *empirical* (or *sample*) *cdf*.

A sample cdf is defined for an ordered sample $t_{(1)} < t_{(2)} < t_{(3)} < \cdots < t_{(n)}$ as

$$S_n(t) = \begin{cases} 0 & -\infty < t < t_{(1)} \\ \frac{i}{n} & t_{(i)} \le t < t_{(i+1)} \quad i = 1, 2, \ldots, n-1 \\ 1 & t_{(n)} \le t < \infty \end{cases} \tag{A.93}$$

Statistic $K-S$ used in the Kolmogorov test to measure the maximum difference between $S_n(t)$ and a hypothesized cdf, $F(t)$, is introduced as

$$K - S = \max_i \left[|F(t_i) - S_n(t_i)|, |F(t_i) - S_n(t_{i-1})| \right] \tag{A.94}$$

Similar to the chi-square test, the following steps compose the test:

Step 1. Choose a hypothesized cumulative distribution $F(T)$ for the given sample.
Step 2. Select a specified significance level of the test, α.
Step 3. Define the rejection region $R > D_n(\alpha)$, where $D_n(\alpha)$ can be obtained from Table B.4.
Step 4. If $K - S > D_n(\alpha)$, reject the hypothesized distribution and conclude that $F(t)$ does not fit the data; otherwise, do not reject the hypothesis.

A.8   REGRESSION ANALYSIS

In Section A.2.7, we mainly dealt with one or two random variables. However, reliability and risk assessment problems often require relationships among several random variables or between random and nonrandom variables. For example, TTF of electrical generator can depend on its age, environmental temperature, and power capacity. In this case, we can consider the TTF as a random variable Y, which is a function of the variables x_1 (age), x_2 (temperature), and x_3 (power capacity).

In regression analysis, one refers to Y as the *dependent variable* and to $x_1, x_2, \ldots, x_k$ as the *independent variables, explanatory variables,* or *factors*. Generally speaking, independent variables $x_1, x_2, \ldots, x_k$ might be random or nonrandom variables whose values are known or chosen by the experimenter (in the case of the, so-called, *design of experiments* (DoE)). The conditional expectation of Y for any given values of $x_1, x_2, \ldots, x_k$, $E(Y|x_1, x_2, \ldots, x_k)$ is known as the *regression* of Y on $x_1, x_2, \ldots, x_k$. In other words, regression analysis estimates the average value for the dependent variable corresponding to each value of the independent variable.

In the case when the regression of Y is a linear function with respect to the independent variables $x_1, x_2, \ldots, x_k$, it can be written in the form

$$E(Y|x_1, x_2, \ldots, x_k) = \beta_0 + \beta_1 x_1 + \beta_2 x_2 + \cdots + \beta_k x_k \tag{A.95}$$

The coefficients $\beta_0, \beta_1, \beta_2, \ldots, \beta_k$ are called *regression coefficients* or *parameters*. When the expectation of Y is nonrandom, the relationship (A.95) is a deterministic one. The corresponding regression model for the random variable Y can be written in the following form:

$$Y = \beta_0 + \beta_1 x_1 + \beta_2 x_2 + \cdots + \beta_k x_k + \varepsilon \tag{A.96}$$

where ε is the *random error*, assumed to be independent (for all combinations of x considered) random variable distributed with mean $E(\varepsilon) = 0$ and finite variance σ^2. If it is normally distributed, one deals with the *normal* regression.

A.8.1 SIMPLE LINEAR REGRESSION

Consider the regression model for the simple deterministic relationship

$$Y = \beta_0 + \beta_1 x_1 \tag{A.97}$$

Let us have n pairs of observations $(x_1, y_1), (x_2, y_2), \ldots, (x_n, y_n)$. Also, assume that for any given value x, the dependent variable Y is related to the value of x by

$$Y = \beta_0 + \beta_1 x + \varepsilon \tag{A.98}$$

where ε is normally distributed with mean β_0 and variance σ^2. The random variable Y has, for a given x, normal distribution with mean $\beta_0 + \beta_1 x$ and variance σ^2. Also suppose that for any given values $x_1, x_2, \ldots, x_n$, random variables $Y_1, Y_2, \ldots, Y_n$ are independent. For the above n pairs of observations, the joint pdf of $y_1, y_2, \ldots, y_n$ is given by

$$f_n\left(y|x, \beta_0, \beta_1, \sigma^2\right) = \frac{1}{\sqrt{(2\pi)^n}\sigma^n} \exp\left[-\frac{1}{2\sigma^2}\sum_{i=1}^{n}(y_i - \beta_0 - \beta_1 x_i)^2\right] \tag{A.99}$$

Function (A.99) is the likelihood function (discussed in Section A.2.5) for the parameters β_0 and β_1. Maximizing this function with respect to β_0 and β_1 reduces the problem to minimizing the sum of squares

$$S(\beta_0, \beta_1) = \sum_{i=1}^{n}(y_i - \beta_0 - \beta_1 x_i)^2$$

with respect to β_0 and β_1.

Thus, the maximum likelihood estimation of the parameters β_0 and β_1 is the estimation by the *method of least squares*. The values of β_0 and β_1 minimizing $S(\beta_0, \beta_1)$ are those for which the derivatives

$$\frac{\partial S(\beta_0, \beta_1)}{\partial \beta_0} = 0, \quad \frac{\partial S(\beta_0, \beta_1)}{\partial \beta_1} = 0 \tag{A.100}$$

The solution of the above equations yields the least squares estimates of the parameters β_0 and β_1 (denoted $\hat{\beta}_0$ and $\hat{\beta}_1$) as

$$\hat{\beta}_0 = \bar{y} - \hat{\beta}_1 \bar{x}, \quad \hat{\beta}_1 = \frac{\sum\limits_{i=1}^{n}(x_i - \bar{x})y_i}{\sum\limits_{i=1}^{n}(x_i - \bar{x})^2} \tag{A.101}$$

where

$$\bar{y} = \frac{1}{n}\sum_{i=1}^{n} y_i, \quad \bar{x} = \frac{1}{n}\sum_{i=1}^{n} x_i$$

Note that the estimates are linear functions of the observations y_i, they are also unbiased and have the minimum variance among all unbiased estimates.

The estimate of the dependent variable variance σ^2 can be found as

$$S^2 = \frac{\sum\limits_{i=1}^{n}(y_i - \hat{y}_i)^2}{n - 2} \tag{A.102}$$

where

$$\hat{Y}_i = \hat{\beta}_0 + \hat{\beta}_1 x_i \tag{A.103}$$

are predicted by the regression model values for the dependent variable, $(n - 2)$ is the number of degrees of freedom (2 is the number of the estimated parameters of the model). The estimate of variance of Y (A.102) is also called the *residual variance* and it is used as a measure of accuracy of model fitting as well. The positive square root of S^2 in (A.102) is called the *standard error of the estimate* of Y and the numerator in (A.102) is called the *residual sum of squares*. For more detailed discussion on reliability applications of regression analysis see Lawless [8].

REFERENCES

1. Cox, R.T., Probability, frequency and reasonable expectation, *American Journal of Physics*, 14:1, 1946.
2. Hahn, G.J. and Shapiro, S.S., *Statistical Models in Engineering*, John Wiley & Sons, New York, 1967.
3. Hill, H.E. and Prane, J.W., *Applied Techniques in Statistics for Selected Industries: Coatings, Paints and Pigments*, John Wiley & Sons, New York, 1984.
4. Johnson, N.L. and Kotz, S., *Distribution in Statistics*, 2 Volumes, John Wiley & Sons, New York, 1970.
5. Lindley, D.V., *Introduction to Probability and Statistics from a Bayesian Viewpoint*, 2 Volumes, Cambridge Press, Cambridge, 1965.
6. Mann, N., Schafer, R.E., and Singpurwalla, N.D., *Methods for Statistical Analysis of Reliability and Life Data*, John Wiley & Sons, New York, 1974.
7. Nelson, W., *Applied Life Data Analysis*, John Wiley & Sons, New York, 1982.
8. Lawless, J.F., *Statistical Models and Methods for Life Time Data*, John Wiley & Sons, New York, 1982.

Appendix B

TABLE B.1
Standard Normal Cumulative Distribution Function

z	$\Phi(z)$	z	$\Phi(z)$	z	$\Phi(z)$	z	$\Phi(z)$
−4.00	0.00003	−3.50	0.00023	−3.00	0.00135	−2.50	0.00621
−3.99	0.00003	−3.49	0.00024	−2.99	0.00139	−2.49	0.00639
−3.98	0.00003	−3.48	0.00025	−2.98	0.00144	−2.48	0.00657
−3.97	0.00004	−3.47	0.00026	−2.97	0.00149	−2.47	0.00676
−3.96	0.00004	−3.46	0.00027	−2.96	0.00154	−2.46	0.00695
−3.95	0.00004	−3.45	0.00028	−2.95	0.00159	−2.45	0.00714
−3.94	0.00004	−3.44	0.00029	−2.94	0.00164	−2.44	0.00734
−3.93	0.00004	−3.43	0.00030	−2.93	0.00169	−2.43	0.00755
−3.92	0.00004	−3.42	0.00031	−2.92	0.00175	−2.42	0.00776
−3.91	0.00005	−3.41	0.00032	−2.91	0.00181	−2.41	0.00798
−3.90	0.00005	−3.40	0.00034	−2.90	0.00187	−2.40	0.00820
−3.89	0.00005	−3.39	0.00035	−2.89	0.00193	−2.39	0.00842
−3.88	0.00005	−3.38	0.00036	−2.88	0.00199	−2.38	0.00866
−3.87	0.00005	−3.37	0.00038	−2.87	0.00205	−2.37	0.00889
−3.86	0.00006	−3.36	0.00039	−2.86	0.00212	−2.36	0.00914
−3.85	0.00006	−3.35	0.00040	−2.85	0.00219	−2.35	0.00939
−3.84	0.00006	−3.34	0.00042	−2.84	0.00226	−2.34	0.00964
−3.83	0.00006	−3.33	0.00043	−2.83	0.00233	−2.33	0.00990
−3.82	0.00007	−3.32	0.00045	−2.82	0.00240	−2.32	0.01017
−3.81	0.00007	−3.31	0.00047	−2.81	0.00248	−2.31	0.01044
−3.80	0.00007	−3.30	0.00048	−2.80	0.00256	−2.30	0.01072
−3.79	0.00008	−3.29	0.00050	−2.79	0.00264	−2.29	0.01101
−3.78	0.00008	−3.28	0.00052	−2.78	0.00272	−2.28	0.01130
−3.77	0.00008	−3.27	0.00054	−2.77	0.00280	−2.27	0.01160
−3.76	0.00008	−3.26	0.00056	−2.76	0.00289	−2.26	0.01191
−3.75	0.00009	−3.25	0.00058	−2.75	0.00298	−2.25	0.01222
−3.74	0.00009	−3.24	0.00060	−2.74	0.00307	−2.24	0.01255
−3.73	0.00010	−3.23	0.00062	−2.73	0.00317	−2.23	0.01287
−3.72	0.00010	−3.22	0.00064	−2.72	0.00326	−2.22	0.01321
−3.71	0.00010	−3.21	0.00066	−2.71	0.00336	−2.21	0.01355
−3.70	0.00011	−3.20	0.00069	−2.70	0.00347	−2.20	0.01390
−3.69	0.00011	−3.19	0.00071	−2.69	0.00357	−2.19	0.01426
−3.68	0.00012	−3.18	0.00074	−2.68	0.00368	−2.18	0.01463
−3.67	0.00012	−3.17	0.00076	−2.67	0.00379	−2.17	0.01500
−3.66	0.00013	−3.16	0.00079	−2.66	0.00391	−2.16	0.01539
−3.65	0.00013	−3.15	0.00082	−2.65	0.00402	−2.15	0.01578
−3.64	0.00014	−3.14	0.00084	−2.64	0.00415	−2.14	0.01618
−3.63	0.00014	−3.13	0.00087	−2.63	0.00427	−2.13	0.01659
−3.62	0.00015	−3.12	0.00090	−2.62	0.00440	−2.12	0.01700
−3.61	0.00015	−3.11	0.00094	−2.61	0.00453	−2.11	0.01743
−3.60	0.00016	−3.10	0.00097	−2.60	0.00466	−2.10	0.01786
−3.59	0.00017	−3.09	0.00100	−2.59	0.00480	−2.09	0.01831
−3.58	0.00017	−3.08	0.00104	−2.58	0.00494	−2.08	0.01876

TABLE B.1
Standard Normal Cumulative Distribution Function — *continued*

z	Φ(z)	z	Φ(z)	z	Φ(z)	z	Φ(z)
−3.57	0.00018	−3.07	0.00107	−2.57	0.00508	−2.07	0.01923
−3.56	0.00019	−3.06	0.00111	−2.56	0.00523	−2.06	0.01970
−3.55	0.00019	−3.05	0.00114	−2.55	0.00539	−2.05	0.02018
−3.54	0.00020	−3.04	0.00118	−2.54	0.00554	−2.04	0.02068
−3.53	0.00021	−3.03	0.00122	−2.53	0.00570	−2.03	0.02118
−3.52	0.00022	−3.02	0.00126	−2.52	0.00587	−2.02	0.02169
−3.51	0.00022	−3.01	0.00131	−2.51	0.00604	−2.01	0.02222
−2.00	0.02275	−1.50	0.06681	−1.00	0.15866	−0.50	0.30854
−1.99	0.02330	−1.49	0.06811	−0.99	0.16109	−0.49	0.31207
−1.98	0.02385	−1.48	0.06944	−0.98	0.16354	−0.48	0.31561
−1.97	0.02442	−1.47	0.07078	−0.97	0.16602	−0.47	0.31918
−1.96	0.02500	−1.46	0.07215	−0.96	0.16853	−0.46	0.32276
−1.95	0.02559	−1.45	0.07353	−0.95	0.17106	−0.45	0.32636
−1.94	0.02619	−1.44	0.07493	−0.94	0.17361	−0.44	0.32997
−1.93	0.02680	−1.43	0.07636	−0.93	0.17619	−0.43	0.33360
−1.92	0.02743	−1.42	0.07780	−0.92	0.17879	−0.42	0.33724
−1.91	0.02807	−1.41	0.07927	−0.91	0.18141	−0.41	0.34090
−1.90	0.02872	−1.40	0.08076	−0.90	0.18406	−0.40	0.34458
−1.89	0.02938	−1.39	0.08226	−0.89	0.18673	−0.39	0.34827
−1.88	0.03005	−1.38	0.08379	−0.88	0.18943	−0.38	0.35197
−1.87	0.03074	−1.37	0.08534	−0.87	0.19215	−0.37	0.35569
−1.86	0.03144	−1.36	0.08691	−0.86	0.19489	−0.36	0.35942
−1.85	0.03216	−1.35	0.08851	−0.85	0.19766	−0.35	0.36317
−1.84	0.03288	−1.34	0.09012	−0.84	0.20045	−0.34	0.36693
−1.83	0.03362	−1.33	0.09176	−0.83	0.20327	−0.33	0.37070
−1.82	0.03438	−1.32	0.09342	−0.82	0.20611	−0.32	0.37448
−1.81	0.03515	−1.31	0.09510	−0.81	0.20897	−0.31	0.37828
−1.80	0.03593	−1.30	0.09680	−0.80	0.21186	−0.30	0.38209
−1.79	0.03673	−1.29	0.09853	−0.79	0.21476	−0.29	0.38591
−1.78	0.03754	−1.28	0.10027	−0.78	0.21770	−0.28	0.38974
−1.77	0.03836	−1.27	0.10204	−0.77	0.22065	−0.27	0.39358
−1.76	0.03920	−1.26	0.10383	−0.76	0.22363	−0.26	0.39743
−1.75	0.04006	−1.25	0.10565	−0.75	0.22663	−0.25	0.40129
−1.74	0.04093	−1.24	0.10749	−0.74	0.22965	−0.24	0.40517
−1.73	0.04182	−1.23	0.10935	−0.73	0.23270	−0.23	0.40905
−1.72	0.04272	−1.22	0.11123	−0.72	0.23576	−0.22	0.41294
−1.71	0.04363	−1.21	0.11314	−0.71	0.23885	−0.21	0.41683
−1.70	0.04457	−1.20	0.11507	−0.70	0.24196	−0.20	0.42074
−1.69	0.04551	−1.19	0.11702	−0.69	0.24510	−0.19	0.42465
−1.68	0.04648	−1.18	0.11900	−0.68	0.24825	−0.18	0.42858
−1.67	0.04746	−1.17	0.12100	−0.67	0.25143	−0.17	0.43251
−1.66	0.04846	−1.16	0.12302	−0.66	0.25463	−0.16	0.43644
−1.65	0.04947	−1.15	0.12507	−0.65	0.25785	−0.15	0.44038
−1.64	0.05050	−1.14	0.12714	−0.64	0.26109	−0.14	0.44433
−1.63	0.05155	−1.13	0.12924	−0.63	0.26435	−0.13	0.44828
−1.62	0.05262	−1.12	0.13136	−0.62	0.26763	−0.12	0.45224
−1.61	0.05370	−1.11	0.13350	−0.61	0.27093	−0.11	0.45620
−1.60	0.05480	−1.10	0.13567	−0.60	0.27425	−0.10	0.46017
−1.59	0.05592	−1.09	0.13786	−0.59	0.27760	−0.09	0.46414
−1.58	0.05705	−1.08	0.14007	−0.58	0.28096	−0.08	0.46812
−1.57	0.05821	−1.07	0.14231	−0.57	0.28834	−0.07	0.47210
−1.56	0.05938	−1.06	0.14457	−0.56	0.28774	−0.06	0.47608
−1.55	0.06057	−1.05	0.14686	−0.55	0.29116	−0.05	0.48006

continued

TABLE B.1
Standard Normal Cumulative Distribution Function — *continued*

z	Φ(z)	z	Φ(z)	z	Φ(z)	z	Φ(z)
−1.54	0.06178	−1.04	0.14917	−0.54	0.29460	−0.04	0.48405
−1.53	0.06301	−1.03	0.15150	−0.53	0.29806	−0.03	0.48803
−1.52	0.06426	−1.02	0.15386	−0.52	0.30153	−0.02	0.49202
−1.51	0.06552	−1.01	0.15625	−0.51	0.30503	−0.01	0.49601
+0.00	0.50000	+0.50	0.69146	+1.00	0.84134	+1.50	0.93319
+0.01	0.50399	+0.51	0.69497	+1.01	0.84375	+1.51	0.93448
+0.02	0.50798	+0.52	0.69847	+1.02	0.84614	+1.52	0.93574
+0.03	0.51197	+0.53	0.70194	+1.03	0.84850	+1.53	0.93766
+0.04	0.51595	+0.54	0.70540	+1.04	0.85083	+1.54	0.93822
+0.05	0.51994	+0.55	0.70884	+1.05	0.85314	+1.55	0.93943
+0.06	0.52392	+0.56	0.71226	+1.06	0.85543	+1.56	0.94062
+0.07	0.52790	+0.57	0.71566	+1.07	0.85769	+1.57	0.94179
+0.08	0.53188	+0.58	0.71904	+1.08	0.85993	+1.58	0.94295
+0.09	0.53586	+0.59	0.72240	+1.09	0.86214	+1.59	0.94408
+0.10	0.53983	+0.60	0.72575	+1.10	0.86433	+1.60	0.94520
+0.11	0.54380	+0.61	0.72907	+1.11	0.86650	+1.61	0.94630
+0.12	0.54776	+0.62	0.73237	+1.12	0.86864	+1.62	0.94738
+0.13	0.55172	+0.63	0.73565	+1.13	0.87076	+1.63	0.94845
+0.14	0.55567	+0.64	0.73891	+1.14	0.87286	+1.64	0.94950
+0.15	0.55962	+0.65	0.74215	+1.15	0.87493	+1.65	0.95053
+0.16	0.56356	+0.66	0.74537	+1.16	0.87698	+1.66	0.95154
+0.17	0.56749	+0.67	0.74857	+1.17	0.87900	+1.67	0.95254
+0.18	0.57142	+0.68	0.75175	+1.18	0.88100	+1.68	0.95352
+0.19	0.57535	+0.69	0.75490	+1.19	0.88298	+1.69	0.95449
+0.20	0.57926	+0.70	0.75804	+1.20	0.88493	+1.70	0.95543
+0.21	0.58317	+0.71	0.76115	+1.21	0.88686	+1.71	0.95637
+0.22	0.58706	+0.72	0.76424	+1.22	0.88877	+1.72	0.95728
+0.23	0.59095	+0.73	0.76730	+1.23	0.89065	+1.73	0.95818
+0.24	0.59483	+0.74	0.77035	+1.24	0.89251	+1.74	0.95907
+0.25	0.59871	+0.75	0.77337	+1.25	0.89435	+1.75	0.95994
+0.26	0.60257	+0.76	0.77637	+1.26	0.89617	+1.76	0.96080
+0.27	0.60642	+0.77	0.77935	+1.27	0.89796	+1.77	0.96164
+0.28	0.61206	+0.78	0.78230	+1.28	0.89973	+1.78	0.96246
+0.29	0.61409	+0.79	0.78524	+1.29	0.90147	+1.79	0.96327
+0.30	0.61791	+0.80	0.78814	+1.30	0.90320	+1.80	0.96407
+0.31	0.62172	+0.81	0.79103	+1.31	0.90490	+1.81	0.96485
+0.32	0.62552	+0.82	0.79389	+1.32	0.90658	+1.82	0.96562
+0.33	0.62930	+0.83	0.79673	+1.33	0.90824	+1.83	0.96638
+0.34	0.63307	+0.84	0.79955	+1.34	0.90988	+1.84	0.96712
+0.35	0.63683	+0.85	0.80234	+1.35	0.91149	+1.85	0.96784
+0.36	0.64958	+0.86	0.80511	+1.36	0.91309	+1.86	0.96856
+0.37	0.64431	+0.87	0.80785	+1.37	0.91466	+1.87	0.96926
+0.38	0.64803	+0.88	0.81057	+1.38	0.91621	+1.88	0.96995
+0.39	0.65173	+0.89	0.81327	+1.39	0.91774	+1.89	0.97062
+0.40	0.65542	+0.90	0.81594	+1.40	0.91924	+1.90	0.97128
+0.41	0.65910	+0.91	0.81859	+1.41	0.92073	+1.91	0.97193
+0.42	0.66276	+0.92	0.82121	+1.42	0.92220	+1.92	0.97257
+0.43	0.66640	+0.93	0.82381	+1.43	0.92364	+1.93	0.97320
+0.44	0.67003	+0.94	0.82639	+1.44	0.92507	+1.94	0.97381
+0.45	0.67364	+0.95	0.82894	+1.45	0.92647	+1.95	0.97441
+0.46	0.67724	+0.96	0.83147	+1.46	0.92785	+1.96	0.97500
+0.47	0.68082	+0.97	0.83398	+1.47	0.92922	+1.97	0.97558
+0.48	0.68439	+0.98	0.83646	+1.48	0.93056	+1.98	0.97615

TABLE B.1
Standard Normal Cumulative Distribution Function — *continued*

z	Φ(z)	z	Φ(z)	z	Φ(z)	z	Φ(z)
+0.49	0.68793	+0.99	0.83891	+1.49	0.93189	+1.99	0.97670
+2.00	0.97725	+2.50	0.99379	+3.00	0.99865	+3.50	0.99977
+2.01	0.97778	+2.51	0.99396	+3.01	0.99869	+3.51	0.99978
+2.02	0.97831	+2.52	0.99413	+3.02	0.99874	+3.52	0.99978
+2.03	0.97882	+2.53	0.99430	+3.03	0.99878	+3.53	0.99979
+2.04	0.97932	+2.54	0.99446	+3.04	0.99882	+3.54	0.99980
+2.05	0.97982	+2.55	0.99461	+3.05	0.99886	+3.55	0.99981
+2.06	0.98030	+2.56	0.99477	+3.06	0.99889	+3.56	0.99981
+2.07	0.98077	+2.57	0.99492	+3.07	0.99893	+3.57	0.99982
+2.08	0.98124	+2.58	0.99506	+3.08	0.99897	+3.58	0.99983
+2.09	0.98169	+2.59	0.99520	+3.09	0.99900	+3.59	0.99983
+2.10	0.98214	+2.60	0.99534	+3.10	0.99903	+3.60	0.99984
+2.11	0.98257	+2.61	0.99547	+3.11	0.99906	+3.61	0.99985
+2.12	0.98300	+2.62	0.99560	+3.12	0.99910	+3.62	0.99985
+2.13	0.98341	+2.63	0.99573	+3.13	0.99913	+3.63	0.99986
+2.14	0.98382	+2.64	0.99585	+3.14	0.99916	+3.64	0.99986
+2.15	0.98422	+2.65	0.99698	+3.15	0.99918	+3.65	0.99987
+2.16	0.98461	+2.66	0.99609	+3.16	0.99921	+3.66	0.99987
+2.17	0.98500	+2.67	0.99621	+3.17	0.99924	+3.67	0.99988
+2.18	0.98537	+2.68	0.99632	+3.18	0.99926	+3.68	0.99988
+2.19	0.98574	+2.69	0.99643	+3.19	0.99929	+3.69	0.99989
+2.20	0.98610	+2.70	0.99653	+3.20	0.99931	+3.70	0.99989
+2.21	0.98645	+2.71	0.99664	+3.21	0.99934	+3.71	0.99990
+2.22	0.98679	+2.72	0.99674	+3.22	0.99936	+3.72	0.99990
+2.23	0.98713	+2.73	0.99683	+3.23	0.99938	+3.73	0.99990
+2.24	0.98745	+2.74	0.99693	+3.24	0.99940	+3.74	0.99991
+2.25	0.98778	+2.75	0.99702	+3.25	0.99942	+3.75	0.99991
+2.26	0.98809	+2.76	0.99711	+3.26	0.99944	+3.76	0.99992
+2.27	0.98840	+2.77	0.99720	+3.27	0.99946	+3.77	0.99992
+2.28	0.98870	+2.78	0.99728	+3.28	0.99948	+3.78	0.99992
+2.29	0.98999	+2.79	0.99736	+3.29	0.99950	+3.79	0.99992
+2.30	0.98928	+2.80	0.99744	+3.30	0.99952	+3.80	0.99993
+2.31	0.98956	+2.81	0.99752	+3.31	0.99953	+3.81	0.99993
+2.32	0.98983	+2.82	0.99760	+3.32	0.99955	+3.82	0.99993
+2.33	0.99010	+2.83	0.99767	+3.33	0.99957	+3.83	0.99994
+2.34	0.99036	+2.84	0.99774	+3.34	0.99958	+3.84	0.99994
+2.35	0.99061	+2.85	0.99781	+3.35	0.99960	+3.85	0.99994
+2.36	0.99086	+2.86	0.99788	+3.36	0.99961	+3.86	0.99994
+2.37	0.99111	+2.87	0.99795	+3.37	0.99962	+3.87	0.99995
+2.38	0.99134	+2.88	0.99801	+3.38	0.99964	+3.88	0.99995
+2.39	0.99158	+2.89	0.99807	+3.39	0.99965	+3.89	0.99995
+2.40	0.99180	+2.90	0.99813	+3.40	0.99966	+3.90	0.99995
+2.41	0.99202	+2.91	0.99819	+3.41	0.99968	+3.91	0.99995
+2.42	0.99224	+2.92	0.99825	+3.42	0.99969	+3.92	0.99996
+2.43	0.99245	+2.93	0.99831	+3.43	0.99970	+3.93	0.99996
+2.44	0.99266	+2.94	0.99836	+3.44	0.99971	+3.94	0.99996
+2.45	0.99286	+2.95	0.99841	+3.45	0.99972	+3.95	0.99996
+2.46	0.99305	+2.96	0.99846	+3.46	0.99973	+3.96	0.99996
+2.47	0.99324	+2.97	0.99851	+3.47	0.99974	+3.97	0.99996
+2.48	0.99343	+2.98	0.99856	+3.48	0.99975	+3.98	0.99997
+2.49	0.99361	+2.99	0.99861	+3.49	0.99976	+3.99	0.99997

TABLE B.2
Critical Values of Student's t Distribution

One-Sided Limit (Read Down)

ν	$t_{0.80}$	$t_{0.90}$	$t_{0.95}$	$t_{0.975}$	$t_{0.99}$	$t_{0.995}$	$t_{0.999}$	$t_{0.9995}$
1	1.3764	3.0777	6.3138	12.7062	31.8205	63.6567	318.3088	636.6192
2	1.0607	1.8856	2.9200	4.3027	6.9646	9.9248	22.3271	31.5991
3	0.9785	1.6377	2.3534	3.1824	4.5407	5.8409	10.2145	12.9240
4	0.9410	1.5332	2.1318	2.7764	3.7469	4.6041	7.1732	8.6103
5	0.9195	1.4759	2.0150	2.5706	3.3649	4.0321	5.8934	6.8688
6	0.9057	1.4398	1.9432	2.4469	3.1427	3.7074	5.2076	5.9588
7	0.8960	1.4149	1.8946	2.3646	2.9980	3.4995	4.7853	5.4079
8	0.8889	1.3968	1.8595	2.3060	2.8965	3.3554	4.5008	5.0413
9	0.8834	1.3830	1.8331	2.2622	2.8214	3.2498	4.2968	4.7809
10	0.8791	1.3722	1.8125	2.2281	2.7638	3.1693	4.1437	4.5869
11	0.8755	1.3634	1.7959	2.2010	2.7181	3.1058	4.0247	4.4370
12	0.8726	1.3562	1.7823	2.1788	2.6810	3.0545	3.9296	4.3178
13	0.8702	1.3502	1.7709	2.1604	2.6503	3.0123	3.8520	4.2208
14	0.8681	1.3450	1.7613	2.1448	2.6245	2.9768	3.7874	4.1405
15	0.8662	1.3406	1.7531	2.1314	2.6025	2.9467	3.7328	4.0728
16	0.8647	1.3368	1.7459	2.1199	2.5835	2.9208	3.6862	4.0150
17	0.8633	1.3334	1.7396	2.1098	2.5669	2.8982	3.6458	3.9651
18	0.8620	1.3304	1.7341	2.1009	2.5524	2.8784	3.6105	3.9216
19	0.8610	1.3277	1.7291	2.0930	2.5395	2.8609	3.5794	3.8834
20	0.8600	1.3253	1.7247	2.0860	2.5280	2.8453	3.5518	3.8495
21	0.8591	1.3232	1.7207	2.0796	2.5176	2.8314	3.5272	3.8193
22	0.8583	1.3212	1.7171	2.0739	2.5083	2.8188	3.5050	3.7921
23	0.8575	1.3195	1.7139	2.0687	2.4999	2.8073	3.4850	3.7676
24	0.8569	1.3178	1.7109	2.0639	2.4922	2.7969	3.4668	3.7454
25	0.8562	1.3163	1.7081	2.0595	2.4851	2.7874	3.4502	3.7251
26	0.8557	1.3150	1.7056	2.0555	2.4786	2.7787	3.4350	3.7066
27	0.8551	1.3137	1.7033	2.0518	2.4727	2.7707	3.4210	3.6896
28	0.8546	1.3125	1.7011	2.0484	2.4671	2.7633	3.4082	3.6739
29	0.8542	1.3114	1.6991	2.0452	2.4620	2.7564	3.3962	3.6594
30	0.8538	1.3104	1.6973	2.0423	2.4573	2.7500	3.3852	3.6460
31	0.8534	1.3095	1.6955	2.0395	2.4528	2.7440	3.3749	3.6335
32	0.8530	1.3086	1.6939	2.0369	2.4487	2.7385	3.3653	3.6218
33	0.8526	1.3077	1.6924	2.0345	2.4448	2.7333	3.3563	3.6109
34	0.8523	1.3070	1.6909	2.0322	2.4411	2.7284	3.3479	3.6007
35	0.8520	1.3062	1.6896	2.0301	2.4377	2.7238	3.3400	3.5911
36	0.8517	1.3055	1.6883	2.0281	2.4345	2.7195	3.3326	3.5821
37	0.8514	1.3049	1.6871	2.0262	2.4314	2.7154	3.3256	3.5737
38	0.8512	1.3042	1.6860	2.0244	2.4286	2.7116	3.3190	3.5657
38	0.8512	1.3042	1.6860	2.0244	2.4286	2.7116	3.3190	3.5657
40	0.8507	1.3031	1.6839	2.0211	2.4233	2.7045	3.3069	3.5510
	$t_{0.60}$	$t_{0.80}$	$t_{0.90}$	$t_{0.95}$	$t_{0.98}$	$t_{0.99}$	$t_{0.998}$	$t_{0.999}$

Two-Sided Limit (Read Up)

TABLE B.2
Critical Values of Student's t Distribution — *continued*

One-Sided Limit (Read Down)

ν	$t_{0.80}$	$t_{0.90}$	$t_{0.95}$	$t_{0.975}$	$t_{0.99}$	$t_{0.995}$	$t_{0.999}$	$t_{0.9995}$
41	0.8505	1.3025	1.6829	2.0195	2.4208	2.7012	3.3013	3.5442
42	0.8503	1.3020	1.6820	2.0181	2.4185	2.6981	3.2960	3.5377
43	0.8501	1.3016	1.6811	2.0167	2.4163	2.6951	3.2909	3.5316
44	0.8499	1.3011	1.6802	2.0154	2.4141	2.6923	3.2861	3.5258
45	0.8497	1.3006	1.6794	2.0141	2.4121	2.6896	3.2815	3.5203
46	0.8495	1.3002	1.6787	2.0129	2.4102	2.6870	3.2771	3.5150
47	0.8493	1.2998	1.6779	2.0117	2.4083	2.6846	3.2729	3.5099
48	0.8492	1.2994	1.6772	2.0106	2.4066	2.6822	3.2689	3.5051
49	0.8490	1.2991	1.6766	2.0096	2.4049	2.6800	3.2651	3.5004
50	0.8489	1.2987	1.6759	2.0086	2.4033	2.6778	3.2614	3.4960
55	0.8482	1.2971	1.6730	2.0040	2.3961	2.6682	3.2451	3.4764
60	0.8477	1.2958	1.6706	2.0003	2.3901	2.6603	3.2317	3.4602
65	0.8472	1.2947	1.6686	1.9971	2.3851	2.6536	3.2204	3.4466
70	0.8468	1.2938	1.6669	1.9944	2.3808	2.6479	3.2108	3.4350
75	0.8464	1.2929	1.6654	1.9921	2.3771	2.6430	3.2025	3.4250
80	0.8461	1.2922	1.6641	1.9901	2.3739	2.6387	3.1953	3.4163
85	0.8459	1.2916	1.6630	1.9883	2.3710	2.6349	3.1889	3.4087
90	0.8456	1.2910	1.6620	1.9867	2.3685	2.6316	3.1833	3.4019
95	0.8454	1.2905	1.6611	1.9853	2.3662	2.6286	3.1782	3.3959
100	0.8452	1.2901	1.6602	1.9840	2.3642	2.6259	3.1737	3.3905
110	0.8449	1.2893	1.6588	1.9818	2.3607	2.6213	3.1660	3.3812
120	0.8446	1.2886	1.6577	1.9799	2.3578	2.6174	3.1595	3.3735
130	0.8444	1.2881	1.6567	1.9784	2.3554	2.6142	3.1541	3.3669
140	0.8442	1.2876	1.6558	1.9771	2.3533	2.6114	3.1495	3.3614
150	0.8440	1.2872	1.6551	1.9759	2.3515	2.6090	3.1455	3.3566
∞	0.8416	1.2816	1.6449	1.9600	2.3263	2.5758	3.0902	3.2905
	$t_{0.60}$	$t_{0.80}$	$t_{0.90}$	$t_{0.95}$	$t_{0.98}$	$t_{0.99}$	$t_{0.998}$	$t_{0.999}$

Two-Sided Limit (Read Up)

TABLE B.3
Critical Values of Chi-Square (χ_α^2) Distribution Function

					α			
ν	$\chi_{0.0005}^2$	$\chi_{0.001}^2$	$\chi_{0.005}^2$	$\chi_{0.01}^2$	$\chi_{0.025}^2$	$\chi_{0.05}^2$	$\chi_{0.1}^2$	$\chi_{0.25}^2$
1	0.0^639	0.0^516	0.0^4393	0.0^3157	0.0^3982	0.00393	0.0158	0.1015
2	0.0010	0.0020	0.0100	0.0201	0.0506	0.1026	0.2107	0.5754
3	0.0153	0.0243	0.0717	0.1148	0.2158	0.3518	0.5844	1.2125
4	0.0639	0.0908	0.2070	0.2971	0.4844	0.7107	1.0636	1.9226
5	0.1581	0.2102	0.4117	0.5543	0.8312	1.1455	1.6103	2.6746
6	0.2994	0.3811	0.6757	0.8721	1.2373	1.6354	2.2041	3.4546
7	0.4849	0.5985	0.9893	1.2390	1.6899	2.1673	2.8331	4.2549
8	0.7104	0.8571	1.3444	1.6465	2.1797	2.7326	3.4895	5.0706
9	0.9717	1.1519	1.7349	2.0879	2.7004	3.3251	4.1682	5.8988
10	1.2650	1.4787	2.1559	2.5582	3.2470	3.9403	4.8652	6.7372
11	1.5868	1.8339	2.6032	3.0535	3.8157	4.5748	5.5778	7.5841
12	1.9344	2.2142	3.0738	3.5706	4.4038	5.2260	6.3038	8.4384
13	2.3051	2.6172	3.5650	4.1069	5.0088	5.8919	7.0415	9.2991
14	2.6967	3.0407	4.0747	4.6604	5.6287	6.5706	7.7895	10.1653
15	3.1075	3.4827	4.6009	5.2293	6.2621	7.2609	8.5468	11.0365
16	3.5358	3.9416	5.1422	5.8122	6.9077	7.9616	9.3122	11.9122
17	3.9802	4.4161	5.6972	6.4078	7.5642	8.6718	10.0852	12.7919
18	4.4394	4.9048	6.2648	7.0149	8.2307	9.3905	10.8649	13.6753
19	4.9123	5.4068	6.8440	7.6327	8.9065	10.1170	11.6509	14.5620
20	5.3981	5.9210	7.4338	8.2604	9.5908	10.8508	12.4426	15.4518
21	5.8957	6.4467	8.0337	8.8972	10.2829	11.5913	13.2396	16.3444
22	6.4045	6.9830	8.6427	9.5425	10.9823	12.3380	14.0415	17.2396
23	6.9237	7.5292	9.2604	10.1957	11.6886	13.0905	14.8480	18.1373
24	7.4527	8.0849	9.8862	10.8564	12.4012	13.8484	15.6587	19.0373
25	7.9910	8.6493	10.5197	11.5240	13.1197	14.6114	16.4734	19.9393
26	8.5379	9.2221	11.1602	12.1981	13.8439	15.3792	17.2919	20.8434
27	9.0932	9.8028	11.8076	12.8785	14.5734	16.1514	18.1139	21.7494
28	9.6563	10.3909	12.4613	13.5647	15.3079	16.9279	18.9392	22.6572
29	10.2268	10.9861	13.1211	14.2565	16.0471	17.7084	19.7677	23.5666
30	10.8044	11.5880	13.7867	14.9535	16.7908	18.4927	20.5992	24.4776
35	13.7875	14.6878	17.1918	18.5089	20.5694	22.4650	24.7967	29.0540
40	16.9062	17.9164	20.7065	22.1643	24.4330	26.5093	29.0505	33.6603
44	19.4825	20.5763	23.5837	25.1480	27.5746	29.7875	32.4871	37.3631
50	23.4610	24.6739	27.9907	29.7067	32.3574	34.7643	37.6886	42.9421
60	30.3405	31.7383	35.5345	37.4849	40.4817	43.1880	46.4589	52.2938
70	37.4674	39.0364	43.2752	45.4417	48.7576	51.7393	55.3289	61.6983
80	44.7910	46.5199	51.1719	53.5401	57.1532	60.3915	64.2778	71.1445
90	52.2758	54.1552	59.1963	61.7541	65.6466	69.1260	73.2911	80.6247
100	59.8957	61.9179	67.3276	70.0649	74.2219	77.9295	82.3581	90.1332
120	75.4665	77.7551	83.8516	86.9233	91.5726	95.7046	100.6236	109.2197
$>\nu$	$\frac{1}{2}[A-3.29]^2$	$\frac{1}{2}[A-3.09]^2$	$\frac{1}{2}[A-2.58]^2$	$\frac{1}{2}[A-2.33]^2$	$\frac{1}{2}[A-1.96]^2$	$\frac{1}{2}[A-1.64]^2$	$\frac{1}{2}[A-1.28]^2$	$\frac{1}{2}[A-0.67]^2$
	$A=(2\nu-1)^{1/2}$	$A=(2\nu-1)^{1/2}$	$A=(2\nu-1)^{1/2}$	$A=(2\nu-1)^{1/2}$	$A=(2\nu-1)^{1/2}$	$A=(2\nu-1)^{1/2}$	$A=(2\nu-1)^{1/2}$	$A=(2\nu-1)^{1/2}$

Note: $0.0^639 = 0.00000039$, ν is the number of degrees of freedom.

TABLE B.3
Critical Values of Chi-Square (χ_α^2) Distribution Function — *continued*

ν	$\chi_{0.75}^2$	$\chi_{0.90}^2$	$\chi_{0.95}^2$	$\chi_{0.025}^2$	$\chi_{0.99}^2$	$\chi_{0.995}^2$	$\chi_{0.999}^2$	$\chi_{0.9995}^2$
1	1.3233	2.7055	3.8415	5.0239	6.6349	7.8794	10.8276	12.1157
2	2.7726	4.6052	5.9915	7.3778	9.2103	10.5966	13.8155	15.2018
3	4.1083	6.2514	7.8147	9.3484	11.3449	12.8382	16.2662	17.7300
4	5.3853	7.7794	9.4877	11.1433	13.2767	14.8603	18.4668	19.9974
5	6.6257	9.2364	11.0705	12.8325	15.0863	16.7496	20.5150	22.1053
6	7.8408	10.6446	12.5916	14.4494	16.8119	18.5476	22.4577	24.1028
7	9.0371	12.0170	14.0671	16.0128	18.4753	20.2777	24.3219	26.0178
8	10.2189	13.3616	15.5073	17.5345	20.0902	21.9550	26.1245	27.8680
9	11.3888	14.6837	16.9190	19.0228	21.6660	23.5894	27.8772	29.6658
10	12.5489	15.9872	18.3070	20.4832	23.2093	25.1882	29.5883	31.4198
11	13.7007	17.2750	19.6751	21.9200	24.7250	26.7568	31.2641	33.1366
12	14.8454	18.5493	21.0261	23.3367	26.2170	28.2995	32.9095	34.8213
13	15.9839	19.8119	22.3620	24.7356	27.6882	29.8195	34.5282	36.4778
14	17.1169	21.0641	23.6848	26.1189	29.1412	31.3193	36.1233	38.1094
15	18.2451	22.3071	24.9958	27.4884	30.5779	32.8013	37.6973	39.7188
16	19.3689	23.5418	26.2962	28.8454	31.9999	34.2672	39.2524	41.3081
17	20.4887	24.7690	27.5871	30.1910	33.4087	35.7185	40.7902	42.8792
18	21.6049	25.9894	28.8693	31.5264	34.8053	37.1565	42.3124	44.4338
19	22.7178	27.2036	30.1435	32.8523	36.1909	38.5823	43.8202	45.9731
20	23.8277	28.4120	31.4104	34.1696	37.5662	39.9968	45.3147	47.4985
21	24.9348	29.6151	32.6706	35.4789	38.9322	41.4011	46.7970	49.0108
22	26.0393	30.8133	33.9244	36.7807	40.2894	42.7957	48.2679	50.5111
23	27.1413	32.0069	35.1725	38.0756	41.6384	44.1813	49.7282	52.0002
24	28.2412	33.1962	36.4150	39.3641	42.9798	45.5585	51.1786	53.4788
25	29.3389	34.3816	37.6525	40.6465	44.3141	46.9279	52.6197	54.9475
26	30.4346	35.5632	38.8851	41.9232	45.6417	48.2899	54.0520	56.4069
27	31.5284	36.7412	40.1133	43.1945	46.9629	49.6449	55.4760	57.8576
28	32.6205	37.9159	41.3371	44.4608	48.2782	50.9934	56.8923	59.3000
29	33.7109	39.0875	42.5570	45.7223	49.5879	52.3356	58.3012	60.7346
30	34.7997	40.2560	43.7730	46.9792	50.8922	53.6720	59.7031	62.1619
35	40.2228	46.0588	49.8018	53.2033	57.3421	60.2748	66.6188	69.1986
40	45.6160	51.8051	55.7585	59.3417	63.6907	66.7660	73.4020	76.0946
45	50.9849	57.5053	61.6562	65.4102	69.9568	73.1661	80.0767	82.8757
50	56.3336	63.1671	67.5048	71.4202	76.1539	79.4900	86.6608	89.5605
60	66.9815	74.3970	79.0819	83.2977	88.3794	91.9517	99.6072	102.6948
70	77.5767	85.5270	90.5312	95.0232	100.4252	104.2149	112.3169	115.5776
80	88.1303	96.5782	101.8795	106.6286	112.3288	116.3211	124.8392	128.2613
90	98.6499	107.5650	113.1453	118.1359	124.1163	128.2989	137.2084	140.7823
100	109.1412	118.4980	124.3421	129.5612	135.8067	140.1695	149.4493	153.1670
120	130.0546	140.2326	146.5674	152.2114	158.9502	163.6482	173.6174	177.6029
$>\nu$	$\frac{1}{2}[A+0.67]^2$ $A = (2\nu-1)^{1/2}$	$\frac{1}{2}[A+1.28]^2$ $A = (2\nu-1)^{1/2}$	$\frac{1}{2}[A+1.64]^2$ $A = (2\nu-1)^{1/2}$	$\frac{1}{2}[A+1.96]^2$ $A = (2\nu-1)^{1/2}$	$\frac{1}{2}[A+2.33]^2$ $A = (2\nu-1)^{1/2}$	$\frac{1}{2}[A+2.58]^2$ $A = (2\nu-1)^{1/2}$	$\frac{1}{2}[A+3.09]^2$ $A = (2\nu-1)^{1/2}$	$\frac{1}{2}[A+3.29]^2$ $A = (2\nu-1)^{1/2}$

TABLE B.4

Critical Values of the Kolmogorov–Smirnov Statistic $D_n = \sup_x \left[\, |F_n(x) - F_0(x)| \,\right]$ D_n is the Least Upper Bound of All Pointwise Differences $|F_n(x) - F_0(x)|$

	$\alpha = 1 - \Pr(D_n \leq d)$				
n	0.20	0.15	0.10	0.05	0.01
1	0.900	0.925	0.950	0.975	0.995
2	0.684	0.725	0.776	0.842	0.929
3	0.565	0.597	0.636	0.708	0.829
4	0.493	0.525	0.565	0.624	0.734
5	0.447	0.474	0.510	0.563	0.669
6	0.410	0.436	0.468	0.519	0.617
7	0.381	0.405	0.436	0.483	0.576
8	0.358	0.381	0.410	0.454	0.542
9	0.339	0.360	0.387	0.430	0.513
10	0.323	0.342	0.369	0.409	0.489
11	0.308	0.326	0.352	0.391	0.468
12	0.296	0.313	0.338	0.375	0.449
13	0.285	0.302	0.325	0.361	0.432
14	0.275	0.292	0.314	0.349	0.418
15	0.266	0.283	0.304	0.338	0.404
16	0.258	0.274	0.295	0.327	0.392
17	0.250	0.266	0.286	0.318	0.381
18	0.244	0.259	0.279	0.309	0.371
19	0.237	0.252	0.271	0.301	0.361
20	0.232	0.246	0.265	0.294	0.352
21	0.226	0.249	0.259	0.287	0.344
22	0.221	0.243	0.253	0.281	0.337
23	0.216	0.238	0.247	0.275	0.330
24	0.212	0.233	0.242	0.269	0.323
25	0.208	0.228	0.238	0.264	0.317
26	0.204	0.224	0.233	0.259	0.311
27	0.200	0.219	0.229	0.254	0.305
28	0.197	0.215	0.225	0.250	0.300
29	0.193	0.212	0.221	0.246	0.295
30	0.190	0.208	0.218	0.242	0.290
31	0.187	0.205	0.214	0.238	0.285
32	0.184	0.201	0.211	0.234	0.281
33	0.182	0.198	0.208	0.231	0.277
34	0.179	0.195	0.205	0.227	0.273
35	0.177	0.193	0.201	0.225	0.269
36	0.175	0.190	0.199	0.219	0.265
37	0.173	0.187	0.196	0.219	0.261
38	0.170	0.185	0.193	0.216	0.258
39	0.168	0.182	0.191	0.213	0.255
40	0.166	0.180	0.188	0.210	0.251
>40	$1.05/(n)^{1/2}$	$1.14/(n)^{1/2}$	$1.19/(n)^{1/2}$	$1.33/(n)^{1/2}$	$1.59/(n)^{1/2}$

n is the number of trials.

TABLE B.5
F-Cumulative Distribution Function, Upper 1 Percentage Points

$F_{0.01, \nu_1, \nu_2}$ ν_2	ν_1								
	1	2	3	4	5	6	7	8	9
1	4052.18	4999.50	5403.35	5624.58	5763.65	5858.99	5928.36	5981.07	6022.47
2	98.5025	99.0000	99.1662	99.2494	99.2993	99.3326	99.3564	99.3742	99.3881
3	34.1162	30.8165	29.4567	28.7099	28.2371	27.9107	27.6717	27.4892	27.3452
4	21.1977	18.0000	16.6944	15.9770	15.5219	15.2069	14.9758	14.7989	14.6591
5	16.2582	13.2739	12.0600	11.3919	10.9670	10.6723	10.4555	10.2893	10.1578
6	13.7450	10.9248	9.7795	9.1483	8.7459	8.4661	8.2600	8.1017	7.9761
7	12.2464	9.5466	8.4513	7.8466	7.4604	7.1914	6.9928	6.8400	6.7188
8	11.2586	8.6491	7.5910	7.0061	6.6318	6.3707	6.1776	6.0289	5.9106
9	10.5614	8.0215	6.9919	6.4221	6.0569	5.8018	5.6129	5.4671	5.3511
10	10.0443	7.5594	6.5523	5.9943	5.6363	5.3858	5.2001	5.0567	4.9424
11	9.6460	7.2057	6.2167	5.6683	5.3160	5.0692	4.8861	4.7445	4.6315
12	9.3302	6.9266	5.9525	5.4120	5.0643	4.8206	4.6395	4.4994	4.3875
13	9.0738	6.7010	5.7394	5.2053	4.8616	4.6204	4.4410	4.3021	4.1911
14	8.8616	6.5149	5.5639	5.0354	4.6950	4.4558	4.2779	4.1399	4.0297
15	8.6831	6.3589	5.4170	4.8932	4.5556	4.3183	4.1415	4.0045	3.8948
16	8.5310	6.2262	5.2922	4.7726	4.4374	4.2016	4.0259	3.8896	3.7804
17	8.3997	6.1121	5.1850	4.6690	4.3359	4.1015	3.9267	3.7910	3.6822
18	8.2854	6.0129	5.0919	4.5790	4.2479	4.0146	3.8406	3.7054	3.5971
19	8.1849	5.9259	5.0103	4.5003	4.1708	3.9386	3.7653	3.6305	3.5225
20	8.0960	5.8489	4.9382	4.4307	4.1027	3.8714	3.6987	3.5644	3.4567
21	8.0166	5.7804	4.8740	4.3688	4.0421	3.8117	3.6396	3.5056	3.3981
22	7.9454	5.7190	4.8166	4.3134	3.9880	3.7583	3.5867	3.4530	3.3458
23	7.8811	5.6637	4.7649	4.2636	3.9392	3.7102	3.5390	3.4057	3.2986
24	7.8229	5.6136	4.7181	4.2184	3.8951	3.6667	3.4959	3.3629	3.2560
25	7.7698	5.5680	4.6755	4.1774	3.8550	3.6272	3.4568	3.3239	3.2172
26	7.7213	5.5263	4.6366	4.1400	3.8183	3.5911	3.4210	3.2884	3.1818
27	7.6767	5.4881	4.6009	4.1056	3.7848	3.5580	3.3882	3.2558	3.1494
28	7.6356	5.4529	4.5681	4.0740	3.7539	3.5276	3.3581	3.2259	3.1195
29	7.5977	5.4204	4.5378	4.0449	3.7254	3.4995	3.3303	3.1982	3.0920
30	7.5625	5.3903	4.5097	4.0179	3.6990	3.4735	3.3045	3.1726	3.0665
35	7.4191	5.2679	4.3957	3.9082	3.5919	3.3679	3.2000	3.0687	2.9630
40	7.3141	5.1785	4.3126	3.8283	3.5138	3.2910	3.1238	2.9930	2.8876
50	7.1706	5.0566	4.1993	3.7195	3.4077	3.1864	3.0202	2.8900	2.7850
60	7.0771	4.9774	4.1259	3.6490	3.3389	3.1187	2.9530	2.8233	2.7185
80	6.9627	4.8807	4.0363	3.5631	3.2550	3.0361	2.8713	2.7420	2.6374
100	6.8953	4.8239	3.9837	3.5127	3.2059	2.9877	2.8233	2.6943	2.5898

continued

TABLE B.5
F*-Cumulative Distribution Function, Upper 1 Percentage Points — *continued

$F_{0.01, \nu_1, \nu_2}$ ν_2	ν_1								
	10	11	12	15	20	30	40	50	100
1	6055.85	6083.32	6106.32	6157.28	6208.73	6260.65	6286.78	6302.52	6334.11
2	99.3992	99.4083	99.4159	99.4325	99.4492	99.4658	99.4742	99.4792	99.4892
3	27.2287	27.1326	27.0518	26.8722	26.6898	26.5045	26.4108	26.3542	26.2402
4	14.5459	14.4523	14.3736	14.1982	14.0196	13.8377	13.7454	13.6896	13.5770
5	10.0510	9.9626	9.8883	9.7222	9.5526	9.3793	9.2912	9.2378	9.1299
6	7.8741	7.7896	7.7183	7.5590	7.3958	7.2285	7.1432	7.0915	6.9867
7	6.6201	6.5382	6.4691	6.3143	6.1554	5.9920	5.9084	5.8577	5.7547
8	5.8143	5.7343	5.6667	5.5151	5.3591	5.1981	5.1156	5.0654	4.9633
9	5.2565	5.1779	5.1114	4.9621	4.8080	4.6486	4.5666	4.5167	4.4150
10	4.8491	4.7715	4.7059	4.5581	4.4054	4.2469	4.1653	4.1155	4.0137
11	4.5393	4.4624	4.3974	4.2509	4.0990	3.9411	3.8596	3.8097	3.7077
12	4.2961	4.2198	4.1553	4.0096	3.8584	3.7008	3.6192	3.5692	3.4668
13	4.1003	4.0245	3.9603	3.8154	3.6646	3.5070	3.4253	3.3752	3.2723
14	3.9394	3.8640	3.8001	3.6557	3.5052	3.3476	3.2656	3.2153	3.1118
15	3.8049	3.7299	3.6662	3.5222	3.3719	3.2141	3.1319	3.0814	2.9772
16	3.6909	3.6162	3.5527	3.4089	3.2587	3.1007	3.0182	2.9675	2.8627
17	3.5931	3.5185	3.4552	3.3117	3.1615	3.0032	2.9205	2.8694	2.7639
18	3.5082	3.4338	3.3706	3.2273	3.0771	2.9185	2.8354	2.7841	2.6779
19	3.4338	3.3596	3.2965	3.1533	3.0031	2.8442	2.7608	2.7093	2.6023
20	3.3682	3.2941	3.2311	3.0880	2.9377	2.7785	2.6947	2.6430	2.5353
21	3.3098	3.2359	3.1730	3.0300	2.8796	2.7200	2.6359	2.5838	2.4755
22	3.2576	3.1837	3.1209	2.9779	2.8274	2.6675	2.5831	2.5308	2.4217
23	3.2106	3.1368	3.0740	2.9311	2.7805	2.6202	2.5355	2.4829	2.3732
24	3.1681	3.0944	3.0316	2.8887	2.7380	2.5773	2.4923	2.4395	2.3291
25	3.1294	3.0558	2.9931	2.8502	2.6993	2.5383	2.4530	2.3999	2.2888
26	3.0941	3.0205	2.9578	2.8150	2.6640	2.5026	2.4170	2.3637	2.2519
27	3.0618	2.9882	2.9256	2.7827	2.6316	2.4699	2.3840	2.3304	2.2180
28	3.0320	2.9585	2.8959	2.7530	2.6017	2.4397	2.3535	2.2997	2.1867
29	3.0045	2.9311	2.8685	2.7256	2.5742	2.4118	2.3253	2.2714	2.1577
30	2.9791	2.9057	2.8431	2.7002	2.5487	2.3860	2.2992	2.2450	2.1307
35	2.8758	2.8026	2.7400	2.5970	2.4448	2.2806	2.1926	2.1374	2.0202
40	2.8005	2.7274	2.6648	2.5216	2.3689	2.2034	2.1142	2.0581	1.9383
50	2.6981	2.6250	2.5625	2.4190	2.2652	2.0976	2.0066	1.9490	1.8248
60	2.6318	2.5587	2.4961	2.3523	2.1978	2.0285	1.9360	1.8772	1.7493
80	2.5508	2.4777	2.4151	2.2709	2.1153	1.9435	1.8489	1.7883	1.6548
100	2.5033	2.4302	2.3676	2.2230	2.0666	1.8933	1.7972	1.7353	1.5977

TABLE B.6
F-Cumulative Distribution Function, Upper 5 Percentage Points

$F_{0.05, \nu_1, \nu_2}$ ν_2	ν_1								
	1	2	3	4	5	6	7	8	9
1	161.4476	199.5000	215.7073	224.5832	230.1619	233.9860	236.7684	238.8827	240.5433
2	18.5128	19.0000	19.1643	19.2468	19.2964	19.3295	19.3532	19.3710	19.3848
3	10.1280	9.5521	9.2766	9.1172	9.0135	8.9406	8.8867	8.8452	8.8123
4	7.7086	6.9443	6.5914	6.3882	6.2561	6.1631	6.0942	6.0410	5.9988
5	6.6079	5.7861	5.4095	5.1922	5.0503	4.9503	4.8759	4.8183	4.7725
6	5.9874	5.1433	4.7571	4.5337	4.3874	4.2839	4.2067	4.1468	4.0990
7	5.5914	4.7374	4.3468	4.1203	3.9715	3.8660	3.7870	3.7257	3.6767
8	5.3177	4.4590	4.0662	3.8379	3.6875	3.5806	3.5005	3.4381	3.3881
9	5.1174	4.2565	3.8625	3.6331	3.4817	3.3738	3.2927	3.2296	3.1789
10	4.9646	4.1028	3.7083	3.4780	3.3258	3.2172	3.1355	3.0717	3.0204
11	4.8443	3.9823	3.5874	3.3567	3.2039	3.0946	3.0123	2.9480	2.8962
12	4.7472	3.8853	3.4903	3.2592	3.1059	2.9961	2.9134	2.8486	2.7964
13	4.6672	3.8056	3.4105	3.1791	3.0254	2.9153	2.8321	2.7669	2.7144
14	4.6001	3.7389	3.3439	3.1122	2.9582	2.8477	2.7642	2.6987	2.6458
15	4.5431	3.6823	3.2874	3.0556	2.9013	2.7905	2.7066	2.6408	2.5876
16	4.4940	3.6337	3.2389	3.0069	2.8524	2.7413	2.6572	2.5911	2.5377
17	4.4513	3.5915	3.1968	2.9647	2.8100	2.6987	2.6143	2.5480	2.4943
18	4.4139	3.5546	3.1599	2.9277	2.7729	2.6613	2.5767	2.5102	2.4563
19	4.3807	3.5219	3.1274	2.8951	2.7401	2.6283	2.5435	2.4768	2.4227
20	4.3512	3.4928	3.0984	2.8661	2.7109	2.5990	2.5140	2.4471	2.3928
21	4.3248	3.4668	3.0725	2.8401	2.6848	2.5727	2.4876	2.4205	2.3660
22	4.3009	3.4434	3.0491	2.8167	2.6613	2.5491	2.4638	2.3965	2.3419
23	4.2793	3.4221	3.0280	2.7955	2.6400	2.5277	2.4422	2.3748	2.3201
24	4.2597	3.4028	3.0088	2.7763	2.6207	2.5082	2.4226	2.3551	2.3002
25	4.2417	3.3852	2.9912	2.7587	2.6030	2.4904	2.4047	2.3371	2.2821
26	4.2252	3.3690	2.9752	2.7426	2.5868	2.4741	2.3883	2.3205	2.2655
27	4.2100	3.3541	2.9604	2.7278	2.5719	2.4591	2.3732	2.3053	2.2501
28	4.1960	3.3404	2.9467	2.7141	2.5581	2.4453	2.3593	2.2913	2.2360
29	4.1830	3.3277	2.9340	2.7014	2.5454	2.4324	2.3463	2.2783	2.2229
30	4.1709	3.3158	2.9223	2.6896	2.5336	2.4205	2.3343	2.2662	2.2107
35	4.1213	3.2674	2.8742	2.6415	2.4851	2.3718	2.2852	2.2167	2.1608
40	4.0847	3.2317	2.8387	2.6060	2.4495	2.3359	2.2490	2.1802	2.1240
50	4.0343	3.1826	2.7900	2.5572	2.4004	2.2864	2.1992	2.1299	2.0734
60	4.0012	3.1504	2.7581	2.5252	2.3683	2.2541	2.1665	2.0970	2.0401
80	3.9604	3.1108	2.7188	2.4859	2.3287	2.2142	2.1263	2.0564	1.9991
100	3.9361	3.0873	2.6955	2.4626	2.3053	2.1906	2.1025	2.0323	1.9748

continued

TABLE B.6
F-Cumulative Distribution Function, Upper 5 Percentage Points — *continued*

$F_{0.05,\nu_1,\nu_2}$ ν_2	ν_1								
	10	11	12	15	20	30	40	50	100
1	241.8817	242.9835	243.9060	245.9499	248.0131	250.0951	251.1432	251.7742	253.0411
2	19.3959	19.4050	19.4125	19.4291	19.4458	19.4624	19.4707	19.4757	19.4857
3	8.7855	8.7633	8.7446	8.7029	8.6602	8.6166	8.5944	8.5810	8.5539
4	5.9644	5.9358	5.9117	5.8578	5.8025	5.7459	5.7170	5.6995	5.6641
5	4.7351	4.7040	4.6777	4.6188	4.5581	4.4957	4.4638	4.4444	4.4051
6	4.0600	4.0274	3.9999	3.9381	3.8742	3.8082	3.7743	3.7537	3.7117
7	3.6365	3.6030	3.5747	3.5107	3.4445	3.3758	3.3404	3.3189	3.2749
8	3.3472	3.3130	3.2839	3.2184	3.1503	3.0794	3.0428	3.0204	2.9747
9	3.1373	3.1025	3.0729	3.0061	2.9365	2.8637	2.8259	2.8028	2.7556
10	2.9782	2.9430	2.9130	2.8450	2.7740	2.6996	2.6609	2.6371	2.5884
11	2.8536	2.8179	2.7876	2.7186	2.6464	2.5705	2.5309	2.5066	2.4566
12	2.7534	2.7173	2.6866	2.6169	2.5436	2.4663	2.4259	2.4010	2.3498
13	2.6710	2.6347	2.6037	2.5331	2.4589	2.3803	2.3392	2.3138	2.2614
14	2.6022	2.5655	2.5342	2.4630	2.3879	2.3082	2.2664	2.2405	2.1870
15	2.5437	2.5068	2.4753	2.4034	2.3275	2.2468	2.2043	2.1780	2.1234
16	2.4935	2.4564	2.4247	2.3522	2.2756	2.1938	2.1507	2.1240	2.0685
17	2.4499	2.4126	2.3807	2.3077	2.2304	2.1477	2.1040	2.0769	2.0204
18	2.4117	2.3742	2.3421	2.2686	2.1906	2.1071	2.0629	2.0354	1.9780
19	2.3779	2.3402	2.3080	2.2341	2.1555	2.0712	2.0264	1.9986	1.9403
20	2.3479	2.3100	2.2776	2.2033	2.1242	2.0391	1.9938	1.9656	1.9066
21	2.3210	2.2829	2.2504	2.1757	2.0960	2.0102	1.9645	1.9360	1.8761
22	2.2967	2.2585	2.2258	2.1508	2.0707	1.9842	1.9380	1.9092	1.8486
23	2.2747	2.2364	2.2036	2.1282	2.0476	1.9605	1.9139	1.8848	1.8234
24	2.2547	2.2163	2.1834	2.1077	2.0267	1.9390	1.8920	1.8625	1.8005
25	2.2365	2.1979	2.1649	2.0889	2.0075	1.9192	1.8718	1.8421	1.7794
26	2.2197	2.1811	2.1479	2.0716	1.9898	1.9010	1.8533	1.8233	1.7599
27	2.2043	2.1655	2.1323	2.0558	1.9736	1.8842	1.8361	1.8059	1.7419
28	2.1900	2.1512	2.1179	2.0411	1.9586	1.8687	1.8203	1.7898	1.7251
29	2.1768	2.1379	2.1045	2.0275	1.9446	1.8543	1.8055	1.7748	1.7096
30	2.1646	2.1256	2.0921	2.0148	1.9317	1.8409	1.7918	1.7609	1.6950
35	2.1143	2.0750	2.0411	1.9629	1.8784	1.7856	1.7351	1.7032	1.6347
40	2.0772	2.0376	2.0035	1.9245	1.8389	1.7444	1.6928	1.6600	1.5892
50	2.0261	1.9861	1.9515	1.8714	1.7841	1.6872	1.6337	1.5995	1.5249
60	1.9926	1.9522	1.9174	1.8364	1.7480	1.6491	1.5943	1.5590	1.4814
80	1.9512	1.9105	1.8753	1.7932	1.7032	1.6017	1.5449	1.5081	1.4259
100	1.9267	1.8857	1.8503	1.7675	1.6764	1.5733	1.5151	1.4772	1.3917

TABLE B.7
F-Cumulative Distribution Function, Upper 10 Percentage Points

$F_{0.1, \nu_1, \nu_2}$ ν_2	ν_1								
	1	2	3	4	5	6	7	8	9
1	39.8635	49.5000	53.5932	55.8330	57.2401	58.2044	58.9060	59.4390	59.8576
2	8.5263	9.0000	9.1618	9.2434	9.2926	9.3255	9.3491	9.3668	9.3805
3	5.5383	5.4624	5.3908	5.3426	5.3092	5.2847	5.2662	5.2517	5.2400
4	4.5448	4.3246	4.1909	4.1072	4.0506	4.0097	3.9790	3.9549	3.9357
5	4.0604	3.7797	3.6195	3.5202	3.4530	3.4045	3.3679	3.3393	3.3163
6	3.7759	3.4633	3.2888	3.1808	3.1075	3.0546	3.0145	2.9830	2.9577
7	3.5894	3.2574	3.0741	2.9605	2.8833	2.8274	2.7849	2.7516	2.7247
8	3.4579	3.1131	2.9238	2.8064	2.7264	2.6683	2.6241	2.5893	2.5612
9	3.3603	3.0065	2.8129	2.6927	2.6106	2.5509	2.5053	2.4694	2.4403
10	3.2850	2.9245	2.7277	2.6053	2.5216	2.4606	2.4140	2.3772	2.3473
11	3.2252	2.8595	2.6602	2.5362	2.4512	2.3891	2.3416	2.3040	2.2735
12	3.1765	2.8068	2.6055	2.4801	2.3940	2.3310	2.2828	2.2446	2.2135
13	3.1362	2.7632	2.5603	2.4337	2.3467	2.2830	2.2341	2.1953	2.1638
14	3.1022	2.7265	2.5222	2.3947	2.3069	2.2426	2.1931	2.1539	2.1220
15	3.0732	2.6952	2.4898	2.3614	2.2730	2.2081	2.1582	2.1185	2.0862
16	3.0481	2.6682	2.4618	2.3327	2.2438	2.1783	2.1280	2.0880	2.0553
17	3.0262	2.6446	2.4374	2.3077	2.2183	2.1524	2.1017	2.0613	2.0284
18	3.0070	2.6239	2.4160	2.2858	2.1958	2.1296	2.0785	2.0379	2.0047
19	2.9899	2.6056	2.3970	2.2663	2.1760	2.1094	2.0580	2.0171	1.9836
20	2.9747	2.5893	2.3801	2.2489	2.1582	2.0913	2.0397	1.9985	1.9649
21	2.9610	2.5746	2.3649	2.2333	2.1423	2.0751	2.0233	1.9819	1.9480
22	2.9486	2.5613	2.3512	2.2193	2.1279	2.0605	2.0084	1.9668	1.9327
23	2.9374	2.5493	2.3387	2.2065	2.1149	2.0472	1.9949	1.9531	1.9189
24	2.9271	2.5383	2.3274	2.1949	2.1030	2.0351	1.9826	1.9407	1.9063
25	2.9177	2.5283	2.3170	2.1842	2.0922	2.0241	1.9714	1.9292	1.8947
26	2.9091	2.5191	2.3075	2.1745	2.0822	2.0139	1.9610	1.9188	1.8841
27	2.9012	2.5106	2.2987	2.1655	2.0730	2.0045	1.9515	1.9091	1.8743
28	2.8938	2.5028	2.2906	2.1571	2.0645	1.9959	1.9427	1.9001	1.8652
29	2.8870	2.4955	2.2831	2.1494	2.0566	1.9878	1.9345	1.8918	1.8568
30	2.8807	2.4887	2.2761	2.1422	2.0492	1.9803	1.9269	1.8841	1.8490
35	2.8547	2.4609	2.2474	2.1128	2.0191	1.9496	1.8957	1.8524	1.8168
40	2.8354	2.4404	2.2261	2.0909	1.9968	1.9269	1.8725	1.8289	1.7929
50	2.8087	2.4120	2.1967	2.0608	1.9660	1.8954	1.8405	1.7963	1.7598
60	2.7911	2.3933	2.1774	2.0410	1.9457	1.8747	1.8194	1.7748	1.7380
80	2.7693	2.3701	2.1535	2.0165	1.9206	1.8491	1.7933	1.7483	1.7110
100	2.7564	2.3564	2.1394	2.0019	1.9057	1.8339	1.7778	1.7324	1.6949

continued

TABLE B.7
F*-Cumulative Distribution Function, Upper 10 Percentage Points — *continued

$F_{0.1, \nu_1, \nu_2}$ ν_2	ν_1								
	10	11	12	15	20	30	40	50	100
1	60.1950	60.4727	60.7052	61.2203	61.7403	62.2650	62.5291	62.6881	63.0073
2	9.3916	9.4006	9.4081	9.4247	9.4413	9.4579	9.4662	9.4712	9.4812
3	5.2304	5.2224	5.2156	5.2003	5.1845	5.1681	5.1597	5.1546	5.1443
4	3.9199	3.9067	3.8955	3.8704	3.8443	3.8174	3.8036	3.7952	3.7782
5	3.2974	3.2816	3.2682	3.2380	3.2067	3.1741	3.1573	3.1471	3.1263
6	2.9369	2.9195	2.9047	2.8712	2.8363	2.8000	2.7812	2.7697	2.7463
7	2.7025	2.6839	2.6681	2.6322	2.5947	2.5555	2.5351	2.5226	2.4971
8	2.5380	2.5186	2.5020	2.4642	2.4246	2.3830	2.3614	2.3481	2.3208
9	2.4163	2.3961	2.3789	2.3396	2.2983	2.2547	2.2320	2.2180	2.1892
10	2.3226	2.3018	2.2841	2.2435	2.2007	2.1554	2.1317	2.1171	2.0869
11	2.2482	2.2269	2.2087	2.1671	2.1230	2.0762	2.0516	2.0364	2.0050
12	2.1878	2.1660	2.1474	2.1049	2.0597	2.0115	1.9861	1.9704	1.9379
13	2.1376	2.1155	2.0966	2.0532	2.0070	1.9576	1.9315	1.9153	1.8817
14	2.0954	2.0729	2.0537	2.0095	1.9625	1.9119	1.8852	1.8686	1.8340
15	2.0593	2.0366	2.0171	1.9722	1.9243	1.8728	1.8454	1.8284	1.7929
16	2.0281	2.0051	1.9854	1.9399	1.8913	1.8388	1.8108	1.7934	1.7570
17	2.0009	1.9777	1.9577	1.9117	1.8624	1.8090	1.7805	1.7628	1.7255
18	1.9770	1.9535	1.9333	1.8868	1.8368	1.7827	1.7537	1.7356	1.6976
19	1.9557	1.9321	1.9117	1.8647	1.8142	1.7592	1.7298	1.7114	1.6726
20	1.9367	1.9129	1.8924	1.8449	1.7938	1.7382	1.7083	1.6896	1.6501
21	1.9197	1.8956	1.8750	1.8271	1.7756	1.7193	1.6890	1.6700	1.6298
22	1.9043	1.8801	1.8593	1.8111	1.7590	1.7021	1.6714	1.6521	1.6113
23	1.8903	1.8659	1.8450	1.7964	1.7439	1.6864	1.6554	1.6358	1.5944
24	1.8775	1.8530	1.8319	1.7831	1.7302	1.6721	1.6407	1.6209	1.5788
25	1.8658	1.8412	1.8200	1.7708	1.7175	1.6589	1.6272	1.6072	1.5645
26	1.8550	1.8303	1.8090	1.7596	1.7059	1.6468	1.6147	1.5945	1.5513
27	1.8451	1.8203	1.7989	1.7492	1.6951	1.6356	1.6032	1.5827	1.5390
28	1.8359	1.8110	1.7895	1.7395	1.6852	1.6252	1.5925	1.5718	1.5276
29	1.8274	1.8024	1.7808	1.7306	1.6759	1.6155	1.5825	1.5617	1.5169
30	1.8195	1.7944	1.7727	1.7223	1.6673	1.6065	1.5732	1.5522	1.5069
35	1.7869	1.7614	1.7394	1.6880	1.6317	1.5691	1.5346	1.5127	1.4653
40	1.7627	1.7369	1.7146	1.6624	1.6052	1.5411	1.5056	1.4830	1.4336
50	1.7291	1.7029	1.6802	1.6269	1.5681	1.5018	1.4648	1.4409	1.3885
60	1.7070	1.6805	1.6574	1.6034	1.5435	1.4755	1.4373	1.4126	1.3576
80	1.6796	1.6526	1.6292	1.5741	1.5128	1.4426	1.4027	1.3767	1.3180
100	1.6632	1.6360	1.6124	1.5566	1.4943	1.4227	1.3817	1.3548	1.2934

Index